A transcontinental career: Essays in honour of Wim van Binsbergen

Wim van Binsbergen in his study, 2017

A TRANS-CONTINENTAL CAREER

Essays in honour of
Wim van Binsbergen

edited by Pius M. Mosima

PIP-TraCS – Papers in Intercultural Philosophy and Transcontinental Comparative Studies – No. 24

Papers in Intercultural Philosophy and Transcontinental Comparative Studies is a publication initiative of:

Quest: An African Journal of Philosophy / Revue Africaine de Philosophie

volumes in this series enjoy free Internet access at: http://www.quest-journal.net/PIP/index.htm . also see that webpage for information on the series as a whole, its publishers, directions for authors, and ordering of copies especially if a bookseller's or institutional discount is sought

Papers in Intercultural Philosophy and Transcontinental Comparative Studies, and *Quest: An African Journal of Philosophy / Revue Africaine de Philosophie*, are published by Shikanda Press, Hoofddorp, the Netherlands

version: 04-10 2018

ISBN / EAN: 978-90-78382-35-5

NUR-code: 694 *(History of culture and mentality / Cultuur- en mentaliteitsgeschiedenis)*

cover illustrations:

(front cover) Wim van Binsbergen / Tatashikanda (left) with his adoptive cousin Dr Stanford Mayowe (right), kneeling and clapping the royal salute in the setting sun, so as to gain entry to the royal capital of King Mwene Kahare (the former's adoptive father / the latter's uncle) of the Mashasha Nkoya people, Western Zambia, 1992. Note the pointed poles supporting the reed fence – a jealously guarded royal privilege in this part of the world, and one in which transcontinental Sunda influences resonate (photo Dennis Shiyowe);

(back cover) Pius Mosima (right, standing) and Wim van Binsbergen (left, standing) in debate during a seminar they delivered jointly at Nijmegen University, the Netherlands, 2016, on the contentious pioneer of African Philosophy, Placide Tempels (photo: Nijmeegs Instituut voor Missiewetenschappen, with thanks).

Table of contents

Introduction:
A transcontinental career – Essays in honour of Wim van Binsbergen

by Pius M. Mosima

This book brings together essays conceived as a Festschrift or *liber amicorum* in honour of Wim van Binsbergen on the occasion of his 70th birthday to be celebrated on the 25th of February 2017. These essays are devoted to the significance of his works in the domain of intercultural knowledge production. Its contributors range from some of his former students of the Intercontinental network of PhD candidates around the chair of the Foundations of Intercultural Philosophy Erasmus University Rotterdam; in conjunction with the African Studies Centre, Leiden (RETICULUM); some of his former colleagues at the Erasmus University Rotterdam; the African Studies Centre Leiden (ASCL); at *Quest: An African Journal of Philosophy / Revue Africaine de Philosophie*; and those with whom he has been in close contact, and has more or less closely collaborated, over the past two decades. This collection of essays relate with and reflect upon some of the significant contributions in the various fields of scholarship in which he has been very active in the course of his very rich, inspiring, complex and innovative career. Wim van Binsbergen's contributions to African Studies in general and Intercultural Philosophy, in particular, cannot be underestimated. For close to half a century, he has published a lot on Africa and helped to answer the deeply emotional and political questions as to the validity, global relevance and global applicability of African knowledge systems.[1] If we need a brief characterization of the person and work of Wim van Binsbergen, perhaps the blurb of his most recent book *Vicarious Reflections* (2015) may inspire us:

> 'An indispensable, exciting and lavishly illustrated sequel to the author's Intercultural Encounters: African and anthropological lessons towards a philosophy of interculturality(2003),this book Vicarious Reflections leans on dozens of short empirical essays from comparative ethnography, comparative mythology, and long-range linguistics; on many field-work photos and distribution maps; and a bibliography of over 2000 titles. It brings together discussions of virtuality, globalisation, religious anthropology, spirituality, hegemony (illustrated from the study of evil, divination, the Truth and Reconciliation Commission of South Africa, and Islamic terrorism), Afrocentricity, African Christian intellectuals, African knowledge systems, and wisdom. It restores empirical methods (especially anthropological field-work) and social-science theory to the heart of intercultural knowledge production. It offers incisive analyses of the work of Mudimbe, Sandra Harding, Derrida, Guattari, Hebga, Kearney, Devisch, Geschiere, Schoffeleers, Van der Geest – and Aristotle. Van Binsbergen's vicarious, counter-hegemonic approach challenges the usual North-Atlantic thinking down upon Africa. His is a passionate plea to restore an empirical, empathic and dialogical orientation to the heart of intercontinental studies. In transcultural encounter, nothing has proved so pernicious as the shift, away from time-honoured anthropology (sophisticated theory, method, prolonged field-work, humble linguistic and cultural learning, seeking criticism from both locals and peers), and towards facile and complacent reliance on introspection, North Atlantic common-sense categories, linguae francae, furtive data collection, and the Internet. Ironically, such a shift has often been justified in the name of post-modernism, yet the qualified celebration of major postmodern philosophers is the backbone of this book'.

Willibrordus Michiel Johannes van Binsbergen, fondly known as Wim, was trained at Amsterdam University and the Free University, Amsterdam; the Netherlands (Cand. 1968, Drs 1971, Dr 1979, the latter two *cum laude*), as a spe-

---

[1] Wim van Binsbergen's bibliography runs into dozens of books, many dozens of peer-reviewed articles, and hundreds of lesser publications – and even at age seventy continues to expand at an alarming rate. Although a *Festschrift* is the typical place to present an overview of the target's publications, we have opted against this: an extensive though not entirely up-to-date list is available at the Internet (http://www.quest-journal.net/shikanda/publications/index.htm), while similar listings, usually with URLs of free digital versions of the texts in question, may be found in van Binsbergen's recent books, especially *Vicarious Reflections* (2015; see http://www.quest-journal.net/shikanda/topicalities/vicarious/vicariou.htm); *Religion as a Social Construct* (2017; see: http://www.quest-journal.net/shikanda/topicalities/rel%2obk%2ofor%2oweb/webpage%2orelbk.htm); and *Confronting the Sacred: Durkheim vindicated through philosophical analysis, ethnography, archaeology, long-range linguistics, and comparative mythology*, Hoofddorp: Shikanda; see: http://www.quest-journal.net/shikanda/topicalities/naar%2owebsite%208-2018/Table of contents.htm . At the end of the present volume a full list may be found of van Binsbergen's books / independent publications, with clickable links to their freely accessible Internet versions.

cialist in the social-scientific and historical study of popular Islam, folk religion in general, and the anthropology of the present-day Mediterranean region. Yet, due to institutional politics which were far beyond his control, in 1971 his main field of interest became sub-Saharan Africa. Through successful field-work in Tunisia, Zambia and Guinea-Bissau, and through a number of prestigious international publications including the innovatively Marxist and theoretical book *Religious Change in Zambia* (1981), as well as appointments including Simon Professor at Manchester, UK, professorships in Berlin, Amsterdam, and Durban; and others as acting Professor of African Anthropology, Leiden University (1975-1977), and as Head of Political and Historical studies , African Studies Centre (ASC), Leiden (1980-1990), scientific co-director (1980-1989), member of the Management Team (1996-2002), and leader of the Theme Group on Globalisation (1996-2002); had by the mid-1980s firmly established himself in the study of African religious anthropology-which culminated in his serving as President of the Netherlands Association for African Studies (1990-1993), and earned him membership of the Africa Committee of WOTRO (Netherlands Foundation for Tropical Research, a division of the Netherlands Research Foundation NWO). However, beginning with his new field-work (in booming Francistown, urban Botswana, from 1988 on) into Southern African urban culture as an interface between regional cultural traditions and globalisation, major changes took place in his research and writing. In Botswana, under the spell of mounting existential, epistemological and political doubts about the distancing, reductionist, hegemonic and debunking sceptical attitude which was the main stock-in-trade of religious anthropology at the time, van Binsbergen was forced to become a *sangoma* (diviner- healer-priest in the Southern African tradition). This had a profound impact on his personal life and transcontinental career. In becoming a *sangoma* van Binsbergen largely forfeited the methodological basis for conventional research into African religion thereby putting his intellectual production on a new footing in two complementary ways.

First; as a *sangoma*, he became the certified exponent of a local African belief and therapy system, which he exercised not only in Botswana but also in the North Atlantic and other continents, combining this role with his professorial chair in anthropology. His knowledge of *sangoma* divination, which was initially highly problematic and counter paradigmatic given his training as a social anthropologist in North Atlantic scholarship, permitted him extend its intercontinental and long-range historical strands into other continents the African-

9

ist oral-historical, and ethnographic research into ecstatic healing cults, divination systems, leopard-skin symbolism, kingship, *etc.* This new orientation gave him insights into new fields of knowledge and skills which were not readily available in the mainstream scientific comparative literature at the time. When he expands his horizons by entering the world of others, he sees where the local and global realities intersect, and enables him to make bold statements about the human condition. This new line of research which is still continuing include: cosmology, comparative mythology, art history, the ability to read objects of material culture and myths in detail, archaeology, genetics, comparative and historical linguistics, Egyptology, Assyriology, History of Ideas, Biblical Studies- and some of the ancient languages (Greek, Latin, Hebrew, and Arabic) informing these fields of study. These new fields of study forced him to develop from scratch a transcontinental perspective on cultural and cosmological continuity through space and time. When he becomes a *sangoma*, van Binsbergen further crosses borders and gets to know about other divinatory systems in the world such as The New Age intellectual movement, *I Ching*, runic divination, tarot, Arabic geomantic divination, the Zulu bones oracle, Native American varieties, *Ifa* divination and astrology. The variety of so many divinatory systems and orders of knowledge at his disposal is likely to enrich his practice of *sangomahood* and make him a more interculturally relevant *sangoma* than his African co-practitioners. Moreover, van Binsbergen not only uses oracular tablets but has been able to translate *sangoma* divination to a modern setting by making use of information and communicating technologies. In various ways, *sangomahood* and its divinatory practices bring van Binsbergen to the frontiers of interculturality, to newer multicultural forms of knowledge resources, different cultural orientations and their modes of social organiztion. The tension and struggle in dealing with a multiplicity of intercultural contexts, different orders of knowledge and diverse forms of communication proves inspiring. It shows how local African knowledge practices could be of global relevance and could be mediated and articulated interculturally.

Second:this interesting mix of human situations enmeshed in hybridity, cosmopolitanism and difference opened him to more receptive and dialogic, less hegemonic and less implicitly racist form of intercultural encounter. The intersubjective understanding of the lived experiences in different cultural contexts gives him a new urge to frame a dialectical inquiry between the local and the global; the particular and the universal. In this way he is able to elucidate the specificities of African knowledge systems, how they work in the local context and the possibility to apply them globally. This provoked him, in the course of the 1990s, to gradually develop more articulated statements on the unacceptably subordinating nature of North-South knowledge formation with contributions on Afrocentrism and Martin Gardiner's *Black Athena* thesis. This preoccupation made it possible for him to trade, in 1998, his Amsterdam (Free University) chair in the social anthropology of ethnicity, for one in the Foundations of Intercultural Philosophy, within the Philosophical Faculty of Erasmus University Rotterdam. This shift in intellectual research and production from cultural anthropology to intercultural philosophy and comparative mythology has brought in additional comparative and historical approaches to his career. It has given him the possibility to trace the history and the varieties of human thought beyond the Western philosophical tradition, and into regions and periods where most philosophers would hardly perceive any philosophy to speak of. Van Binsbergen's attempts at intercultural philosophy have been informed by transcontinental empirically-grounded research on *the Postulate of the Fundamental Unity of Humankind* (at least, anatomically Modern Humans- the subspecies that emerged in East Africa c. 200 ka BP (kiloannums, = millennia, Before Present) and to which all humans now living belong).

Even before this book project, Wim van Binsbergen's contributions to African philosophy and to the global history of philosophy had been earlier recognized and celebrated. On his 65th birthday, *Quest: An African Journal of Philosophy / Revue Africaine de Philosophie* where he has been Editor since 2002, devoted a combined annual volume XXIII-XXIV (2009-2010) to a long-standing book project of the Editor (van Binsbergen 2012), in recognition of his contributions to that journal and to the sake of African and global philosophy. Moreover, on the 1st of February 2012 Wim van Binsbergen formally marked his retirement from the African Studies Centre, Leiden, with a valedictory international conference under the theme: *Rethinking Africa's transcontinental continuities in pre- and protohistory* (Leiden 12-13 April 2012). All the relevant papers and manuscript of the 2012 Leiden conference have been submitted to the African Studies Centre editorial board for one of its series with Brill Publishing House. Even though he might have formally retired from the Erasmus University Rotterdam and the African Studies Centre, Leiden; he is not tired. At 70, he remains active in the academia- he is a life Honorary Fellow at the African Studies Centre, Leiden; the Netherlands; the Editor of *Quest*; and a qualified and practising *sangoma* diviner-healer. This book provides an opportunity for some of his former PhD students, now working across the globe, and colleagues who could not make it at the valedictory conference in Leiden, to do this collective publication in his honour and encourage him to kick strong with newer and challenging perspectives. Having carefully followed van Binsbergen's career and having read through the contributions in this Festschrift, I concur without hesitation that they capture in significant measure his dream of a shared, fundamental unity of humankind. They reflect this very successful and transcontinental career that saw its own moments of insecurities, doubts, conflicts and ruptures.

The various contributions in this book reflect Wim van Binsbergen's outstanding career and transcontinental scholarly output.

In his article, Boele van Hensbroek re-evaluates van Binsbergen's argument presented during the inaugural ceremony when taking up the chair of the Foundations of Intercultural Philosophy Erasmus University Rotterdam, 21 January, 1999.[2] In "Van Binsbergen's 'Cultures do not exist' re-visited", Boele van Hensbroek thinks it is time to see if the argument has been received or ignored in writing about Africa and African Philosophy. He thinks that van Binsbergen has made important input in his critique of the definition of culture as it helps us rid ourselves of misleading illusions of culture clashes when speaking about cultural phenomena. He also interrogates van Binsbergen's suggestion that there are hidden patterns of thought, ritual and belief that run through African cultural complexes; connect them with Ancient Egypt and even far beyond.

Among the contributors to this book, Sjaak van der Geest certainly belongs to the cohort with the longest standing: his association with Wim van Binsbergen goes back to the mid-1970s, when both belonged to the class of PhD students supervised by the nestor of Amsterdam anthropology, André Köbben. Their debate in *Human Organization*, 1979 (*cf.* van Binsbergen 2015: 169 *f.*), foreshadowed – but with a twist – major themes in van Binsbergen's intercultural philosophy. Van der Geest came to occupy a chair in medical anthropology, and in that capacity he sollicited, in 1990 in the context of a local conference on medical anthropology, van Binsbergen's first argument on divination. In his contribution, "The Belly Open: Fieldwork, Defecation and Literature with a Capital

---

[2]The text was published in Dutch as a small book: van Binsbergen 1999a. The English text was published in *Quest: An African Journal of Philosophy*, 13: 37-114 as van Binsbergen 2002d; and later incorporated into van Binsbergen 2003: 459-522.

L", van der Geest discusses some of the problems encountered in the construction of intercultural knowledge through anthropological fieldwork. In this case, his *point de depart* is the opening scene of van Binsbergen's 1988 novel *Een Buik Openen* (Opening up a Belly), based on his first anthropological fieldwork experiences in the highlands of North-western (Khumiriyya region) Tunisia in 1968. He reflects on the hidden worries, shame and discomfort around defecation among anthropological fieldworkers in other cultures. He questions why this part of life remains largely untold in the many fieldwork accounts that are now being published, and what role literary work can play in the articulation of personal emotion in anthropological fieldwork.

The title of Stephanus Djunatan's article "Becoming Chinese Indonesians: An affirmative approach to social identity" points to an experiential-epistemic approach of having a meaningful social identity both for internal and external groups in a non-violent way. This alternative way out is clearly argued by Wim van Binsbergen (2003: 317-332) when he elaborates the 'sensus communis' and sensus particularis' in his critique to the Kantian argument of 'sensus communis'. The interplay between becoming 'communis' and 'particularis' implies a principle of thinking which intensifies both methods of identification: they are becoming communis and particularis, which suggests this alternative approach known as the principle of affirmation. Djunatan asserts that the identification of Chinese Indonesians as social identity should be deliberated through this approach. The complex identification of their identity will continue to be problematic unless they take into account the strategy to keep the interplay between being the inherent part of the national identity and being the unique ethnic groups within the context of intercultural interaction in the Archipelago.

Emily Lyle's article titled "Facets of the Egyptian Ennead in Relation to a Posited Indo-European and Chinese Ten-God System" is also rooted in van Binsbergen's life-long fascination with the Presocratics (*cf.* van Binsbergen 2012) and with cosmology in general. She explores the theory of the ten-god system by linking the Chinese, Indo-European and Egyptian cosmologies. She posits the Chinese case as that of the ten Heavenly Stems, eight of which, represented at the periphery, are equated with the eight trigrams, while the remaining two are centrally located. In Indo-European terms, the posited system deals with four old cosmic gods (the Indian Asuras) and four young social gods (Devas), plus a king of the living (Indra) and a king of the dead (Yama). Since the gods can be regarded as the head terms of an extensive series of correlations, they are related to directions of space, segments of time, colours, elements, *etc.* Taken together the Indo-European and Chinese patterns may serve to suggest a possible pre-form of the Egyptian nine fold Ennead which, with the living king, Horus, has a tenfold structure.

One of the puzzling results of Wim van Binsbergen's (*e.g.* 2010) grappling with mythology in South Central Africa, among the Nkoya people with whom he did research ever since 1972, was the counter-paradigmatic impression that here resonated familiar and time-honoured themes from Ancient Egypt, Ancient Mesopotamia, and from South Asia in more recent times. The cultural partitioning of the world into distinct continents, even though less than two millennia old, has been reified in modern scholarship, constitutes one of the conditions for the discipline of 'African Studies', and has formed a practically insurmountable barrier for transcontinental analysis. Hence van Binsbergen's great admiration for the work of the German African historian Dierk Lange, who in an endless series of penetrating and convincing arguments has established (as part of the empirical substantiation of the Southbound, trans-Saharan vector in van Binsbergen's Pelasgian Hypothesis) the link between West Asia in the last millennium BCE, and sub-Saharan Africa. In his contribution "The traditions of Gulfeil: Projection of Israelite-Assyrian history on the local conditions", Dierk Lange traces the history of the small Kokoto town of Gulfeil. It shows the impact of migrations and occupations, which Lange delineates in four episodes, on the town at various periods of its history – ultimately going back to Assyria and Israel in *Old Testament* times.

Wim van Binsbergen's work on comparative mythology, especially from 2003 on, brought him in close contact with a network of Asianists focussing on the Department of Sanskrit and Indian Studies, Harvard University, Cambridge MA, USA – and on the International Association for Comparative Mythology, of which he was to be one of the founding members and directors. Among his colleagues in this connection we may count not only Emily Lyle (see above) from Scotland, but also Kazuo Matsumara, from Japan. The latter's contribution "Sacred Deer in Japan", focuses on the phenomenon of the wide-spread worship of deer. The investigation starts from the reason why the deer in Nara Park in contemporary Japan are given the status of a national treasure and the search for that reason leads to the traditional belief about deer in ancient times. The author expands the themes of the religious significance of deer across the Eurasian continent, touching upon the cases of the Chinese, Mongolians, and

Scythians, eventually reaching Ireland, and also compares it with other animals images such as (wolves, dragons, phoenixes, tigers *etc.*) and birds (woodpeckers, eagles).

Thera Rasing's article is titled "Urban female initiation rites in globalised Christian Zambia: an example of virtuality?" She discusses some ideas on globalisation, ethnicity and virtuality, as they have been analysed and used by van Binsbergen. The concept of virtuality has been used by van Binsbergen to explain female initiation rites in urban Zambia. She focuses primarily on female initiation rites in urban Zambia, and examines if and how van Binsbergen's ideas on globalisation, ethnicity and virtuality can be used to understand female initiation rites in contemporary urban Zambia. She highlights the merits and potentials of these concepts, as well as expresses some pitfalls on them. This permits her to explore other concepts to analyse initiation rites in current urban Zambia. She opines that in as much as virtuality can be used to analyse these rites, this is too narrow a concept. These rites should also be seen as the construction of gender and adulthood, and the creation of a female domain, which is neither urban nor rural.

Fred Woudhuizen's contribution falls in line with van Binsbergen's passion for long-range approaches to Bronze-Age Mediterranean proto-history and to comparative mythology. In his "Recurrent Indo-European Ethnonyms", Woudhuizen takes up a theme already prominent in the monumental study of Mediterranean Bronze Age ethnicity (van Binsbergen & Woudhuizen 2011, especially pp. 42 f.) and discusses the recurrent nature ethnonyms, now with specific emphasis on Indo-European as a linguistic phylum. He asserts that basically the same Indo-European ethnonym can be attested in widely differing regions of the Eurasian continent and in widely differing times. Consequently, he argues that awareness of this fact in a number of cases helps to improve our understanding of the linguistic nature of the ethnonym in question.

In "Protohistory, Presocratics and Philosophy", Sanya Osha brings to the fore van Binsbergen's research over Africa's contributions to the study of the global history of human thought and philosophy with a close reading of *Before the Presocratics* (van Binsbergen 2012). He admires van Binsbergen's assessment of regional and global epistemic traditions and configurations before the advent of Ancient Greek thought. In this way he notes van Binsbergen's eclectic assemblage and interrogation of worlds that relate to Afrocentricity with different specialties such as protohistory, archaeology, comparative mythology, comparative linguistics and genetics. This broader scope permits him to caution that rather than viewing different regional epistemic formations as singular and distinct, it is more appropriate to understand them as being part of a global and historical continuum of knowledge traditions that are perpetually subject to migration and transformation- in short, all the elements of transplantation and dispersal. In this light, the strict separation between regional and ethnic knowledge becomes misguided and often preposterous.

In "Les défies de la participation africaine au monde africaniste occidental", Julie Ndaya Tshiteku sketches some of the major dilemmas inherent in conducting anthropological fieldwork. She argues that the theories and methods, used in anthropology and especially during fieldwork in Africa, are conceived in a totally different world from that of the people studied. She is also obliged to use these theories and methods in a bid to give a 'scientific status' to her work. However, she thinks that methods like participant observation and interviews are intrusive, violent and an imposition on the daily lives of the people studied. She proposes the conversational method in which there is equality between the researcher and the research subjects. She thinks that this conversational method could help the African / Africanist researchers to be accepted by the intercontinental scientific establishment and at the same time remaining faithful to her / his own people that are the subjects of the research.

In "Kobia's Clash: *Ubuntu* - and International Management within the World Council of Churches", Frans Dokman relates van Binsbergen's thinking about *ubuntu*, with a domain which is usually less connected to his name – management. He explores the relation between African *ubuntu* management, and Western dominated international management. He presents a case study of the introduction of *Ubuntu* management, by the Kenyan general secretary Dr Samuel Kobia at the World Council of Churches (WCC). Although sharing a Christian identity, there is a disharmony between Southern and Western colleagues of this international organization about religion and management in general and *Ubuntu* management in particular.

Bongasu Tanla Kishani, with the poem titled "Our Lodge of Being-Language" honours his long time friend, Wim. He interrogates Being-*Nothingness*-

Language; all in one, in its metaphysical extension, far beyond the material world or every contingent nature. As such it gives meaning to, and necessarily connotes God as the most enigmatic, fully incomprehensible, immensurable, just and uncaused creating and sustaining source of the whole cosmos. God eternally, rather than tautologically, as the Being, shares and shares *Being* alike positively as *Dasein* from our human perspectives.

The title of Louise Muller's article is "The Graeco-Egyptian origins of Western myths and philosophy, and a note on the magnificence of the creative mind". She addresses the wider question of the Graeco-Egyptian origins of Western myths and philosophy. In an attempt to contribute to the understanding of the birth of Western philosophy (the logos) she asserts that rational thinking and philosophy is not a Greek invention. She opines that even though the ancient Egyptians were predominantly right brain hemisphere oriented, they combined convergent and divergent thinking and valued both myths and logic. She stresses that the ancient Greeks, including the Presocratics and Plato, were not so different from their Egyptian predecessors, whose rich and advanced culture they admired and from which they borrowed elements in the fields of both mythology and proto-scientific thinking. Consequently, she dismisses the stage theories of conscious-ness that stress that logical thinking developed itself at the cost of mythological thinking and that the ancient Egyptians were primitive minded people incapable of producing rational thoughts. She also demonstrates that the multiple soul theory of the ancient Egyptians was shared by the pre-Socratics and that this theory hindered neither of them in thinking logically.

In my own contribution, I indulge in a discussion of the very long, complex and sometimes difficult itinerary with Wim van Binsbergen from anthropology through intercultural philosophy to comparative mythology. I focus on two of his books: *Intercultural Encounters: African and Anthropological Lessons to-wards a Philosophy of Interculturality* (2003) and *Vicarious Reflections: African explorations in empirically- grounded Intercultural Philosophy* (2015). I argue that van Binsbergen has been very passionate and consistent for the past two decades in his output on the epistemological challenges of interculturality; notably in his explicit rendering of, 1) not only of the pitfalls of North Atlantic, potentially hegemonic (and occasionally racist) knowledge formation about past and present African social and cultural realities; but also, 2) articulating on the deeply emotional and political question as to the validity, global relevance and global applicability of African knowledge systems, and finally, 3) contemplating on the possibility, beyond specific Northern or Southern concerns of getting to intercultural forms of knowledge, beyond local cultural boundaries, in a bid to celebrate our common humanity.

Van Binsbergen has indeed had a transcontinental career. In his professional life as an anthropologist and intercultural philosopher, he has been conversant, at the cultural and linguistic level, with at least four regional complexes in Africa and Asia, in addition to the European complex into which he was born. From an Africanist perspective, van Binsbergen has successfully demonstrated the tangle of reference and appropriation linking African knowledge systems, the representation of such knowledges by non-Africans, the adoption of North Atlantic knowledge systems by Africans, and the ways in which all such representations can be more or less faithful to the original, can claim greater or lesser integrity, authenticity, and truth, and can become dominated by, or liberated from, the power games that have informed global North-South interactions for the past half millennium. He has also been very critical of North Atlantic knowledge construction and condescending appropriation on other continents and has forcibly made attempts to cross these borders to arrive as newer forms of knowledge with transcultural relevance. To do this, he shows the importance of intersubjectivity, inter-existence in exploring social relationships, which he calls intercultural encounters, the uninhabitable 'inter' which is where interculturality now roams and permits him to cross cultural boundaries in a bid to argue, and with right reason, for the fundamental unity of humankind. These encounters permit him to learn and let the African worlds be thought in him in a bid to articulate and represent them vis-a-vis both non-Africans and Africans. This attempt enables him to establish some transcontinental continuities, some of which appear in his works as wisdom, myth, therapy and perhaps divination. In this way he has shown the validity, global relevance and global applicability of African knowledge systems.

The broad and diverse character of the book can be seen from the contributors who come from a variety of backgrounds, different continents and write from various perspectives; a reflection of van Binsbergen's way of researching and writing. It is our hope that this scholar who has had a complex and often difficult career will continue to inspire us for decades to come.

Finally, it is my pleasure to thank all the contributors who, even at short notice, have been able to show the richness and continued relevance of van Binsbergen's works: my colleagues of the Intercontinental network of PhD candidates around the chair of Intercultural philosophy Erasmus University Rotterdam; in conjunction with the African Studies Centre, Leiden (RETICULUM); Christopher Ngewoh, Prosper Achingale, George Tansinda and Fabian Lankar for their contribution in the editing process.

References cited

van Binsbergen, Wim M.J., 1981, *Religious Change in Zambia: Exploratory studies*, London / Boston: Kegan Paul International; also as Google Book, at: https://books.google.nl/books?id=slN963EyZrQC&printsec=frontcover&dq=inauthor:%22 Wim+M.+J.+van+Binsbergen%22&hl=en&sa=X&ved=0ahUKEwjIu8iszNXQAhWmKMAKH XQuD4MQ6AEIHTAA#v=onepage&q&f=false

van Binsbergen, Wim M.J., 1988, *Een buik openen: Roman*, Haarlem: In de Knipscheer; also at: http://www.quest-journal.net/shikanda/literary/Buik%20Openen.pdf

van Binsbergen, Wim M.J., 1999a, Culturen bestaan niet: Het onderzoek van interculturaliteit als een openbreken van vanzelfsprekenheden, oratie, Faculteit Wijsbegeerte, Rotterdam, Erasmus Universiteit; also at: http://www.quest-journal.net/shikanda/general/gen3/cultbest.htm

van Binsbergen, Wim M.J., 1999b, 'Cultures do not exist', Exploding self-evidences in the investigation of interculturality', *Quest, An African Journal of Philosophy*, 13, 1-2, Special Issue: Language & Culture, pp. 37-114, also at: http://www.quest-journal.net/Quest_1999_PDF_articles/Quest_13_vanbinsbergen.pdf; also incorporated as ch. 15 in: van Binsbergen, Wim M.J., 2003, *Intercultural encounters: African and anthropological towards a philosophy of interculturality*, Berlin / Boston / Muenster: LIT, pp. 459-522

van Binsbergen, Wim M.J., 2000, 'Sensus communis or sensus particularis? A social-science comment', in: Kimmerle, H., & Oosterling, H., 2000, eds., *Sensus communis in multi- and intercultural perspective: On the possibility of common judgments in arts and politics*, Würzburg: Königshausen & Neumann, pp. 113-128; reprinted in 2003: 317-332.

van Binsbergen, Wim M.J., 2003, *Intercultural encounters: African and anthropological towards a philosophy of interculturality*, Berlin / Boston / Muenster: LIT; also at: also at: http://quest-journal.net/shikanda/intercultural_encounters/index.htm

van Binsbergen, Wim M.J., 2010, 'The continuity of African and Eurasian mythologies: General theoretical models, and detailed comparative discussion of the case of Nkoya mythology from Zambia, South Central Africa', in: van Binsbergen, Wim M.J., & Venbrux, Eric, eds. *New Perspectives on Myth: Proceedings of the Second Annual Conference of the International Association for Comparative Mythology, Ravenstein (the Netherlands), 19-21 August, 2008*, Haarlem: Papers in Intercultural Philosophy and Transcontinental Comparative Studies, pp. 143-225, also at: http://www.quest-journal.net/PIP/New_Perspectives_On_Myth_2010/New_Perspectives_on_Myth_Chapter9.pdf

van Binsbergen, Wim M.J., 2012, *Before the Presocratics: Cyclicity, transformation, and element* cosmology: The case of transcontinental pre- or protohistoric cosmological substrates linking Africa, Eurasia and North America, special issue, QUEST: An African Journal of Philosophy/Revue Africaine de Philosophie, Vol. XXIII-XXIV, No. 1-2, 2009-2010, pp. 1-398, book version: Haarlem: Shikanda; also at: http://www.quest-journal.net/2009-2010.htm

van Binsbergen, Wim M.J., 2015, Vicarious reflections: African explorations in empirically-grounded intercultural philosophy, Haarlem: PIP-TraCS - Papers in Intercultural Philosophy and Transcontinental Comparative Studies - No. 17, also at: http://www.quest-journal.net/shikanda/topicalities/vicarious/vicariou.htm

van Binsbergen, Wim M.J., 2017, Religion as a social construct: African, Asian, comparative and theoretical excursions in the social science of religion, Haarlem: Shikanda, Papers in Intercultural Philosophy and Transcontinental Comparative Studies, No. 22; at: ee: http://www.quest-journal.net/shikanda/topicalities/rel%20bk%20for%20web/webpage%20relbk.htm).

van Binsbergen, Wim M.J., & Woudhuizen, Fred, C., 2011, Ethnicity in Mediterranean Protohistory, British Archaeological Reports (BAR) International Series No. 2256, Oxford: Archaeopress, also at: http://www.quest-journal.net/shikanda/ethnicity_mediterranean_protohistory/ethnicit.htm

Van Binsbergen's 'Cultures do not exist' revisited

by Pieter Boele van Hensbroek

Abstract: When Wim van Binsbergen's *Cultures Do Not Exist' Exploding Self-Evidences In The Investigation Of Interculturality* appeared in the journal *QUEST* in 1999 (Volume XIII No. 1-2) it was a bold statement, like a battle cry, trying to shake off forever the 'sticky' culturalist talk about 'cultures' that has been repeating itself a thousand times for at least the whole twentieth century. This culture talk, this way of speaking about cultures, claimed to respect and honour undervalued cultural expressions and traditions such as those of Africa, but at the price of reducing a great wealth of cultural diversities to rather flat and worn-out stereotypes such as 'communal Africans versus individualist 'Westerners' or 'spiritual Africa versus the materialist 'West' '- how flat the work can be! Almost twenty years after van Binsbergen's 'Cultures do not exist' appeared, it is good to evaluate his argument and evaluate how the message has been received (and ignored) in writing about Africa and African Philosophy.

Key Words: Intercultural Philosophy, Anthropological theory, Philosophy, Concept of Culture, Sangoma

Introduction: 'Cultures Do Not Exist' 20 years after

In 1999 Wim van Binsbergen offered the journal *QUEST – An African Journal of Philosophy / Revue Africaine de Philosophie* a welcome gift by submitting his landmark article 'Cultures do not exist' to its special issue on Language & Culture (Quest, volume XIII, no 1-2). It was in fact more than an article. The 78 pages long treatise contains van Binsbergen's complete analysis of the deeper challenges of elaborating a practice of Intercultural Philosophy; a programmatic framework for this young academic field exposed initially in his inaugural speech after being appointed to the professorial chair of Intercultural Philosophy at the Faculty of Philosophy of the Erasmus University, Rotterdam. How should this new branch of academic activity be positioned relative to the disciplines of Anthropology and of Philosophy? How should it position itself relative to major intellectual trends such as structuralism, post-modernism and post-colonial studies, as well as to questions of method and epistemology? But also, and courageously in the university world, how should the practice of Intercultural Philosophy fruitfully and authentically liaise with the existential personal cross-cultural experiences and engagements of the academic investigator himself or herself?

Thus, van Binsbergen set a very broad agenda for one text, an almost impossible agenda. But the author could refer to many texts in his own extensive corpus to substantiate many of his points. Also, an agenda that can only be addressed by somebody with great depth of scholarship in a range of disciplines, and van Binsbergen commands this agenda. He easily manoeuvres in the jungle of deeply philosophical questions in academia, plus has a broad overview of the relevant empirical disciplines like Anthropology and History (the treatise includes almost 16 pages of works referenced). Mobilising these extensive resources, the text carves out a clear path for Intercultural Philosophy to move forward.

'Cultures do not exist' was as provocative and challenging as its title is. It expressed a great dissatisfaction with Intercultural Philosophy as practiced to date, but an equally massive dissatisfaction with many aspects of the discipline of Anthropology. The article could be read as a harsh criticism, but also as an invitation to a new practice of 'Intercultural Philosophy', which has to be written in parentheses because his argument itself is built upon misgivings about the term Intercultural Philosophy, suggesting that it would be better to speak of 'investigations of interculturality'. Almost twenty years later we can ask the question what remains relevant of van Binsbergen's treatise and how Intercultural Philosophy changed under his influence. For that purpose, this article gives van Binsbergen's arguments central place, extracting its most challenging and remarkable points, and discussing some of its salient points.

Before tracing van Binsbergen's argument, the style and type of the text deserves attention. With its 78 pages in length it is much more than an article, but it is certainly not a booklet. It is one long essay, an argument that adds ever new dimensions to the question what Intercultural Philosophy should be. And a personal argument in the sense that it testified his long personal 'Werdegang', his struggles with theoretical, methodological and philosophical questions, as well as his diverse field work and life experiences. It is also not lightly written but loaded with long sentences and formulations with much 'gravity' which gives the text a somewhat baroque feel. Given its basic messages to philosophers and anthropologists, it is an inherently polemic work. It is a programmatic essay written in a serious tone, and if it is not taken up today in various academic circles as much as it deserves, this may in part be due to the fact that readers may be overwhelmed and somewhat intimidated by its style.

The main ingredients of 'Cultures do not exist' come with van Binsbergen's 'hand luggage', as he calls it, namely his comprehensive knowledge of empirical research in the study of cultural phenomena and his command of theoretical and philosophical issues in Anthropology and Philosophy. Both are vital for framing the outlines of the new discipline because they help to criticize the major weaknesses in available approaches. Philosophers, also many African philosophers, when discussing issues of culture show a serious lack of awareness of the present state of knowledge in Anthropology and History about cultural phenomena; and Anthropologists fail to give proper thought to the philosophical questions of scientific methodology and the theoretical history of the social sciences. So the essay has to fight an academic battle on two fronts as I will review below.

A third front is not academic but that of public talk and journalistic writing about cultural issues. The rather unfortunate 'culture talk' of the last decades which has frozen and absolutized cultural differences in such a way that it suggests that society is a 'snake pit' of contradictory 'cultures', as van Binsbergen observes. Anthropology itself has long said farewell to such a hypostatized view of a culture as a clear-cut entity, but the idea is virulent in society. Van Binsbergen: 'This is the idea that a person, not by her own free choice but by a determination in his innermost essence and totality, represents not only a universal but also a specific (notably *cultural*, or 'ethnic') mode of being human, a mode that she has in common with only a (usually quite small) *subsection* of humanity ...' (van Binsbergen 1999: 40) Such a concept of culture 'combines claims of totality, unicity, integration, boundedness, and non-performativity' (van Binsbergen 1999: 39). As often in 'Cultures do not exist', the analysis does not stop at this point, this, in fact, is general knowledge to most academics in the field. It adds some further interesting reflections on why such mistaken analyses of culture are so popular, relating these again to more fundamental dilemmas. The fixed idea of culture in fact serves some purpose in (post-) modern society. It 'is a thought machine designed to subjectively turn the fragmentation, disintegration and performativity of the modern experience, into unity, coherence, and authenticity' (van Binsbergen 1999: 42). And for this trick to work, the constructive labour in culturalist fabrication and in the running of the 'thought machine' (the performativity of the self-identification) needs to be hidden and forgotten in order to give the identity derived from this 'culture' its self-evidence, as if it had the status of a natural fact. Of course van Binsbergen's argument could here be expanded (and he hints at such an expansion at several points) to an analysis of the political will that may also foster the talk in terms of cultural essences and intrinsically different 'cultural identities'. So-called 'identity politics' is virulent in many countries today, in non-western as well as in western countries. Pumping up cultural differences and identities, is one of the most popular tools to play the game of politics – both by elites claiming to represent marginal groups and by elites trying to cement the established order and exclude minorities, immigrants and deviant sub-groups and persons from it.

That troubling concept of 'culture'

Van Binsbergen's argument continues by tracing the framing of the concept of culture within the disciplines of Cultural Anthropology and of Philosophy. Originally the concept of culture as a bounded and holistic unit was at the basis of Cultural Anthropology, it is also at the basis of the classical format of anthropological monographs that would each focus on a specific 'culture', claiming to picture in one monograph the results of the study of a people or culture as 'a total, bounded, integrated and non-performative form of human existence' (van Binsbergen 1999: 44). Whereas anthropologists gradually distanced themselves from such a hypostatized idea of cultures, this outdated idea of culture was rather uncritically accepted by prominent early scholars in Intercultural

Philosophy. R.A. Mall explicitly connects to it by stating that he takes 'as ... point of departure the existence, side by side, of a plurality of distinct 'cultures''(van Binsbergen 1999: 46).

Van Binsbergen suggests that this attention of philosophers to the idea of distinct 'cultures' has several sources, such as in attempts to conceptualise and criticize the problem of the Eurocentric character of academic Philosophy, and in the attention for 'otherness' in order to show respect for the cultural uniqueness of non-western 'others'. However, the price in terms of inhibiting an understanding of real cultural and intercultural dynamics is high. Therefore, van Binsbergen proposes to speak not of 'cultures' but of 'cultural orientations' as always shifting and refiguring' forms of programming of human representations and behaviour' (van Binsbergen 1999: 52). Using this term allows to do justice to 'the situationality, multiplicity, and performativity' of culture; 'the cultural is a form of programming' (van Binsbergen 1999: 52) and can be analysed also at the levels of specific groups, professions, classes, and religious communities. In that sense 'every human being finds himself at the intersection of *a number* of different cultural orientations, between which there is no systematic connection' (van Binsbergen 1999: 53). For practical purposes we can still use the concept of 'culture' 'as long as we admit the situationality, multiplicity, and performativity of 'culture''. Therefore, in principle 'a constellation (of cultural orientations) ... might be loosely described as 'Islamic culture' or Christian culture'', however in today's interacting world these are hardly useful notions. ''Cultures' in the holistic sense do not exist unless as the illusions of the participants' (van Binsbergen 1999: 55). These may be powerful illusions and hard facts in a social sense, but they do not prove that cultures 'exist'; they are 'Unreal in existence, real in effect'.

Viewed from the perspective of cultural orientations, the study of interculturality is a matter of hermeneutic scholarship; it is a form of empirical study that should take into consideration the methodological requirements of such study. This is where the tradition of Intercultural Philosophy has not been very strong because many practitioners of Intercultural Philosophy tended to 'rush to the description of African philosophies the way these are manifested in myths, proverbs, and oral pronouncements of contemporary thinkers' without giving attention to questions of reliability of sources and of methods. 'For Intercultural Philosophy it is of the greatest importance to realise that rendering the thought of another thinker or of a tradition of thought is an empirical activity with all attending demands of method' (van Binsbergen 1999: 61); a 'southern birth-right to a monopoly of valid knowledge production about the South' is not enough (van Binsbergen 1999: 79).

But van Binsbergen hastens to add that cultural anthropology itself cannot be the solution because of the tendency of many practitioners of cultural anthropology to focus on field work and naïve empiricism, neglecting conceptual and epistemological complications. Van Binsbergen complains for instance that 'hardly any traces can be found today of what was Wittgenstein's gift to anthropology: the promising discussions, as of the late 1950s, of rationality, magic, and the recognition of the truth problem...' (van Binsbergen 1999: 74). Van Binbergen also observes a limited reception of phenomenology and a quite late reception of Foucault.

Intellectual fireworks

The real intellectual firework starts in section 9 of van Binsbergen's essay. Titled 'From ethnography to Intercultural Philosophy: Beyond the ethnographic epistemology', this section attacks the asymmetry in intercultural communication involved in anthropological studies. Anthropologists are much concerned with providing an *emic* rendering of the world view and life form on those studied, but they stop at the point where the possibility of the correctness of indigenous views may arise. In the practice of anthropologists 'the participants' statements to the effect that their collective representations are a valid description of reality' always remain to be judged by the researcher's own, scientific view of what is real and what is not. Anthropologists do not even give the collective representations of the people they study the 'benefit of the doubt' – while the correctness of the North Atlantic scientific conceptual frameworks is never doubted. For van Binsbergen this issue of ignoring the truth claim of indigenous representation became an existential problem during his field work practice in various parts of the world. As an anthropologist one fully participates in all aspects of life of the people studied and thus as anthropologist of religion one participates fully in ceremonies, dances, trances, prayers. But in the scientific rendering of all this in academic texts, the content, the beliefs inherent in these practices are judged automatically from the point of view of

North Atlantic world views. For van Binsbergen this practice of 'joining them in the field and betraying them outside of the field' felt more and more as a professional hypocrisy derived from a hegemonic world view. Thus, as he recounts, during field work in Guinea Bissau in 1983 he first decided to become a real patient of an oracular healer and later, in Francistown, Botswana, he decided to 'liberate' himself from the limitations imposed by the scientific discipline and actually went through all the learning and initiations to himself become an authorized *Sangoma* healer. This means that he was no longer discarding the truth claim inherent in *Sangoma* healing, and thus became a nomad 'between a plurality of cultures, (rather) than ... self-imposed prisoner of a smug Eurocentrism' (van Binsbergen 1999: 82). Such transgression of rules of behaviour of academic Anthropology (of course looked at very negatively by his professional colleagues) is for van Binsbergen an essential step towards the possibility of Intercultural Philosophy. Intercultural Philosophy means taking the cultural constellations involved in intercultural encounters as in principle of equal status as those current in the academic community. They are forms of life, or language games, on an equal footing where there is no simple way to decide which one is correct, and probably no way at all to do that, as Winch, following Wittgenstein, argued.

Towards the end of 'Cultures do not exist' van Binsbergen elaborates how Intercultural Philosophy can be practiced by building upon such a view of egalitarian encounter. This requires a paradigm of communication where the intercultural philosopher is a mediator, not just translating one language into another, but creating 'an interface, a plateau, from which access to both cultural orientations may be gained' (van Binsbergen 1999: 89). Such a communication and mediation unavoidably cannot be an exercise of creating an objective picture of the ideas of any of the parties, but it is itself a constructive exercise. He refers to the theory of hermeneutics here: 'for a long time already recognized by modern hermeneutics ... (it is) producing not a faithful representation of the original, but an innovative novel creation ...' (van Binsbergen 1999: 89). He contrasts this view with Kwasi Wiredu's argument that the simple fact that understanding between people from different cultures are possible proves that there is a universal component in all cultures. Van Binsbergen suggests a different view both of the units (cultures) and the interactions involved. The units are not holistic and bounded cultures, but 'every human situation always involves a variety of cultural orientation, between which there is a constant interplay, both within one person with his many, varied, and other contradictory roles, and between a number of persons in their interaction with each other'(van Binsbergen 1999: 91). Thus communication and dealing with differences and translations is the rule; in a way, 'communication ... produces cultural difference instead if a pre-existing cultural difference engendering, secondarily, specific forms of intercultural communication' (van Binsbergen 1999: 93). In the alternative that van Binsbergen hints at, becoming aware of cultural differences and engaging to overcoming these is part of the human condition rather than the exceptional case of the meeting of 'cultures'.

From local to global cultural patterns

Interestingly, van Binsbergen's perception of Intercultural Philosophy or philosophy of interculturality, in terms of a constant interaction, becoming aware of differences and sameness individually and collectively has its parallel in his interesting discoveries of historical -cultural interactions between Africa and other parts of the world, and already in early periods of human cultural history. One would have expected that van Binsbergen, by diving into the world of *sangoma* and thereby becoming a 'local' in a much deeper and more existential sense than normal anthropology would allow him, would be induced to underscore the idea of cultural differences and the deeply unique nature of *sangoma* thought and belief. However, in van Binsbergen's biography it is exactly this local experience which led him to see connections previously unnoticed between Africa and other continents. The key to this discovery is the mathematical pattern of sixteen different possible combinations underlying the *sangoma* oracle. This same pattern can also be found in Arabian magic, in Chinese *I Ching*, in South Asia, and in ancient Babylonia. 'The parallels were so striking, so detailed, that I had to seriously consider the possibility of cultural diffusion' (van Binsbergen 1999: 84). Van Binsbergen describes this as a 'head-on collision with the central theory of classical anthropology since the 1930s: the historical and cultural specificity of distinct, for instance African, societies' (van Binsbergen 1999: 84). This discovery led him to partly reaffirm Afrocentric positions from Cheick Anta Diop in Senegal to Molefi Kete Asante in the USA and to an extensive study of Martin Bernal's book *Black Athena: The Afroasiatic Roots of Classical Civilisation*. Van Binsbergen traces among others the detectable historical interactions, especially between Egypt and the rest of Africa; Africa 'having first served as *a* (not: *the*) major source and subsequently as principal recipient of Ancient Egyptian civilization' (van Binsbergen 1999: 86). These claims synchronise with Afrocentric analyses and go counter the more accepted view among students of Africa that the actual diversity of African cultural orientations and traditions is overwhelming. In van Binsbergen's view diversities are great but at a deeper level there are astonishing cultural continuities over extensive historical periods and over great distances within the African continent as well as with other parts of the world.

Reflections: 20 years after

After this extensive, but still very incomplete, overview of van Binsbergen's essay we can ask the question where it lands us. Which parts of the argument convince? And did it actually convince its audiences, in particular that of intercultural philosophers? The answer to this last point does not seem positive; Intercultural Philosophy did not go into new directions and is still in its infancy. In fact, it may be that the project of Intercultural Philosophy is being overtaken by various more specific exercises in a globalized practice of philosophy, like the cosmopolitan approach advocated by Ganeri of New York University who advocates a philosophical practice of 'binocularity' that freely shifts between views deriving from a diversity of cultural and philosophical traditions. Another promising approach is that of Comparative Political Theory (CPT) which studies political ideas the world over in their interaction, specificity, and fruitful engagement.

If we look at the broad range of arguments included in 'Cultures do not exist', then an evaluation of each single point is difficult. However, I want to discuss three points. The first one points to the question of the critique of the definition of culture as an integrated whole, with internal consistency and bounded externally. Van Binsbergen's argument is particularly strong here and in academic literature we only find confirmations of it, from Amartya Sen's book *The Argumentative Indian* to Roger Brubaker's *Ethnicity without Groups*. However it is doubtful if outside the academia, in politics and journalism for instance, much changed. There is a strong will to divide up the word, if not in 'races', then in 'cultures'. The mission of 'Cultures do not exist' to rid ourselves of misleading illusions and 'culture clashes' when speaking about cultural phenomena is still as urgent as twenty years ago. This is not to blame van Binsbergen's text, but noting that texts may not do the job in eliminating illusions where there is a strong political will to foster such illusions for purposes of creating divisions, fear and fostering power.

I am challenged by two arguments of van Binsbergen which I will here discuss or at least try to carry further. First of all van Binsbergen's suggestion that there are hidden patterns of thought , ritual and belief that run through African cultural complexes, connect them with Ancient Egypt and even far beyond. Van Binsbergen argues that this is reason to support for instance the Afrocentric claims about a common source of African cultures in ancient Egypt and a basic commonality of African cultural complexes. This is an interesting argument in particular because van Binsbergen founds this Afrocentrism in empirical research rather than in an assumed, quasi racially grounded, spiritual unity of Africans, or in a shared African personality or African soul, or even in an African race. Making his case, van Binsbergen criticizes, for instance, Kwame Anthony Appiah's picture of cultural diversity on the African continent. However, it may be questioned if van Binsbergen's findings of hidden cultural continuities need to be taken as refuting the prevalence of cultural differences. There is no reason why not both positions may largely go together. Despite sharing certain patterns in divination practices deriving from cultural contacts in times long past, it may be that societies and cultural patterns still developed largely on different tracks, in different directions. For most practical purposes diversity may have to be reckoned with. When shaping political arguments, this may be an important point. There may still be very convincing reasons not to try to ground arguments for African unity, for instance, in a shared cultural or racial 'African' substratum, as Appiah forcefully argues, but ground it in the political need for such unity deriving from a shared recent history of colonialism and a shared marginalized position in the present world system.

Another, and final, point to discuss here about van Binsbergen's arguments concerns his challenging statement that the potential truth value of indigenous traditions is not acknowledged in anthropological studies – the 'joining them in the field and betraying them outside the field' argument explained above. What is the core of the problem here? Of course there is an element of dishonesty here towards the persons studied. Secondly, there is an asymmetry in that the standards used for deciding about a possible truth value of any statement are derived one-sidedly from that of the regular scientific world view. Such asymmetry is very obvious among classical authors in Anthropology such as Lucien Lévy-Bruhl in his writings on 'primitive mentality', but also in E.E. Evans Pritchard's more dry observation that many of the views of the peoples he studied (famous are the Zande) are simply not correct. One can even say that the famous anthropologist Robin Horton, when criticizing Evans Pritchard for mistakenly assuming that science can claim to make true statements, did not solve the problem of asymmetry because by proposing criteria such as 'openness' of a knowledge system (rather than the always contestable empirical correctness of it) he still used criteria derived from the scientific knowledge process in order to judge others. However, van Binsbergen's option of embracing the possible truth value of indigenous knowledge and absorbing for instance *sangoma* divination (or being absorbed in it) may not be the only way to achieve symmetry. One could also start arguing from the other end, so to say, namely not from divination but from our understanding of science (and van Binsbergen would probably not object to this approach). At least as problematic as beforehand excluding the possible correctness of indigenous knowledge systems is the attitude of unfounded belief in science as a truth-producing or universalist, rational machine. It may be better to restore symmetry in the study of knowledge systems by maintaining a highly sceptical attitude to knowledge production on both sides. From contemporary empirical Science Studies to Michel Foucault we are invited to view science itself as a social-cultural process tied with a thousand strings to the society and history that produces it. It is possible to achieve the symmetry that van Binsbergen demands by not necessarily elevating divination to the status of a truth-producing activity (which apparently frightened many of his colleagues), but of developing a higher level of self-reflexivity in any scientific work, resulting in a much more modest view of the game of science that we are involved in. When 'Cultures do not exist' is read in this light, it may be that section 12, titled 'To Intercultural Philosophy as a medium', which tries to develop a dialogical process of gaining intercultural understanding, may be appreciated even by those who resist that we should believe in divination practices.

References cited

Appiah, K.A., 1992, *In My Father's House: Africa in the Philosophy of Culture* New York and London: Oxford University Press.

Asante, M.K., 1990, Kemet, afrocentricity and culture, Trenton: Africa World Press.

Bernal, M., 1987, Black Athena: The Afroasiatic Roots of Classical Civilisation, London: Free Association Press.

Brubaker, R., 2004, *Ethnicity without Groups*, Cambridge Mass. & London: Harvard University Press.

Diop, C. A., 1959, L'unité culturelle Africaine, *Présence Africaine*, 24(25), 60–65.

Evans Pritchard, E.E., 1965, Theories *of Primitive Religion*, London: Oxford University Press.

Ganeri, J., 2016, Why Philosophy Must Go Global: A Manifesto. https://www.academia.edu/25799760/Why_Philosophy_Must_Go_Global_A_Manifesto_201 6_

Horton, R., 1967, African traditional thought and Western science, *Africa, 34*, 50–71, 155–187.

Horton, R., 1982, Tradition and modernity revisited. In M. Hollis & S. Lukes, eds., 1982, *Rationality and relativism*, pp. 201–260, Oxford: Basil Blackwell.

Lévy-Bruhl, L., 1992, *La mentalité primitive,* Paris: Alcan.

Sen, A., 2006, The *Argumentative Indian*: Writings on Indian History, Culture and Identity, London: Allen Lane, Social Studies of Science London: Sage.

van Binsbergen, W., 1999, Cultures Do Not Exist. In: *Quest: Philosophical Discussions*, vol XIII, no 1-2. p. 37-114; revised and expanded version reprinted in van Binsbergen 2003: *Intercultural Encounters*, Munster / Berlin: LIT.

Wiredu, K., 1990, Are there cultural universals? In: *Quest: Philosophical Discussions*. Vol. IV, no 2, p. 4-19

Les défies de la participation africaine au monde africaniste occidental

par Julie Ndaya Tshiteku

Resume: Recueillir les données auprès des populations sujettes de sa recherche est un des défis majeurs pour tout ethnographe. Dans sa pratique, il utilise des méthodes et des outils de la recherche de terrain, l'observation participante, l'interview, les entretiens, les questionnaires. Ses méthodes et outils de recueil des données, conçues dans un monde totalement étranger des populations étudiées, sont considérées comme la norme garantissant la qualité du travail scientifique. En recourant à ces techniques, les chercheurs, en majorité occidentaux, ont ramené des éléments des cultures pour eux inconnues, différentes de l'environnement dans lequel ils sont nés et où ils ont grandi. Mais bien que les populations concernées par les enquêtes ethnographiques se reconnaissent dans le savoir fourni sur elles, les moyens que les ethnographes utilisent pour y accéder sont quelques fois vécus par leurs interlocuteurs comme une imposition, une violence sur leur mode de communication, une intrusion dans la vie des sujets de l'étude qui réveille en outre des attentes jamais satisfaites.

Cette contribution entame, comme une ethnographe soi, parmi ses compatriotes et partageant la réalité sociale étudiée, un regard sur les questions des outils utilisées en Anthropologie pour étudier l'autre. L'Anthropologie promouvant le relativisme culturel, nous proposons, dans le cas des recherches parmi les populations de Congo Kinshasa, les *masolo*, une méthode conversationnelle simple, très personnelle, qui se déroule en plein lieu de la vie sociale. Les *masolo* se caractérisent par la non neutralité, l'égalité de chercheur et son empathie avec les sujets de recherche.

Le chapitre examine aussi quels sont les dangers des *masolo*? Cette méthode peut-elle résoudre le dilemme du chercheur africain de vouloir être accepté par les établissements scientifiques intercontinentaux eten même temps, et leur permettre de rester fidèles aux compatriotes.

Introduction

Au mois de mai 2016, un séminaire intitulé *'Where is the African in 'African' Studies?* avait lieu au Centre d'études africaines à Leiden-Pays Bas. La question de la rencontre provenait d'un constat de la prédominance des scientifiques non africains dans les études africanistes. Et le thème de la rencontre réveillait dans ma conscience les obstacles que j'avais éprouvées en 1999, lorsque, grâce à la rencontre avec le professeur Wim van Binsbergen, l'occasion se présentait de faire des recherches de doctorat en Anthropologie dans un cadre scientifique occidental, comme décrit dans le *post scriptum* de mon ouvrage intitulé *Prendre le bic* (Ndaya 2008: 194-203).

La recherche anthropologique implique la production des connaissances au delà de limites sociales culturelles grâce au travail de terrain. Dans cette discipline, les africanistes des pays occidentaux sont numériquement plus nombreux[1] Et bien que depuis la moitié du XXème siècle, le relativisme culturel a traité les dilemmes dans l'interaction nord sud, les outils utilisés dans le travail de terrain et développés par les occidentaux comme c'est le cas de l'observation «neutre» sont prisés comme garant de la qualité du travail scientifique. Le chercheur est sensé recourir aux techniques prescrites qui lui permettent d'accéder à la compréhension de la vie sociale des autres, en articulant les référents empiriques aux interprétations théoriques.[2]

Ainsi les chercheurs, en majorité occidentaux, ont ramené des éléments des cultures pour eux inconnues, différentes des environnements où ils sont nés et où ils ont grandi. Bien que les populations concernées par les enquêtes se reconnaissent quelques fois dans le savoir produit, les moyens utilisés par les ethnographes pour y accéder sont parfois vécus par leurs concernés comme une violence sur leur mode de communication. Il y a donc peu d'équilibre, peu de poids africains dans les méthodes de la science en Afrique. C'est une des raisons

---

[1] C'est grâce aux ressources économiques que les africanistes de l'Atlantique du nord sont en majorité et ont aussi un rôle dominant. Par ce fait, ils réduisent ceux qui sont dépendants de leurs ressources à néant, les mettant dans l'impossibilité de se définir soi même, de définir leurs recherches, leurs priorités ainsi que leurs méthodes et techniques pour les réaliser.

[2] Voir entre autres : Blanchet 1985; Olivier de Sardan 1995; Grawitz 1990; Fabian 1983, 1999; Malinowski 1989.

pour lesquelles Wim van Binsbergen quitta sa chair de professeur en Anthropologie pour devenir philosophe interculturel (van Binsbergen 1999). Le développement d'une perspective anti hégémonique sur le savoir africain est devenu un thème qui traverse les débats sur l'afrocentricité, avec une opinion exprimée sur les méthodes.

Ce texte est un essai servant à introduire, en tant qu'ethnographe parmi ses compatriotes, un regard réflexif sur les questions de la neutralité dans le travail de terrain. Tout en participant au débat sur l'impact des méthodes de recherche en sciences sociales et en l'élargissant, [3]on se propose de «reprendre», en ouvrant une fenêtre sur la réalité congolaise, notre pratique et notre vécu de ce qui nous a semblé difficile à être considéré comme une évidence: les outils de recueil des données proposés par les Occidentaux classiques. La question centrale est: comment construire un savoir valide, comme *insider* désirant rester fidèle et impliqué parmi sa propre population, et répondre aux exigences académiques du nord de la validité de l'enquête?

Après un rappel du contexte dans lequel la recherche a eu lieu, les discours dominants sur les méthodes ethnologiques et les problèmes qu'elles impliquent seront examinés. Ayant été dans la position de l'enquêtée, notre expérience personnelle fera partie de cette approche réflexive, interrogeant la démarche de l'acquisition des récits de vie des autres. Ensuite nous étudierons notre propre pratique, le recours aux *masolo* [4](Ndaya 2008:10-14). Le terme lingala *masolo*, qui se traduit en français par les causeries, renvoi à ce qu'Olivier de Sardan (2008: 64) nomme la conversation naturelle, ordinaire, qui se déroule en pleine vie sociale. En recourant à cet outil de collecte de données, nous avons cherché à rester proches du mode de communication naturel des Congolais, caractérisé par des conversations informelles. A cette occasion la question sera posée, comme réflexion à suivre, de savoir si causer peut résoudre les problèmes des attentes et de la hiérarchie dans la rencontre entre l'anthropologue et ses interlocuteurs. Pour terminer, quelques avantages et limites liés à cette technique seront présentés.

---

[3] Voir entre autres : Leservoisier 2005; Malinowski 1985; Copans 1974; Leiris 1969; Mudimbe 1979, 1982, 1988, Ela 1994, Ki-Zerbo 1980, Rabinow 1988.

[4] Prononciation: *massolo*.

Contexte

C'est en 1999, lorsque l'occasion se présenta de faire une thèse de doctorat, que nous avons pensé à mener une enquête ethnographique parmi les femmes congolaises chrétiennes combattantes aux Pays-Bas, en Belgique et au Congo.[5] Les Combattantes (masc. Combattant) sont les adeptes des groupes charismatiques pentecôtistes «Le Combat Spirituel» (*Etumba ya molimu* en lingala, en raccourci «Le Combat»). Bien que le terme «combat spirituel» soit utilisé par presque tous les nouveaux mouvements religieux congolais, afin de désigner leur pratique rituelle, l'exorcisme ou la chasse aux esprits[6]; ce mot est surtout revendiqué par le groupe fondé à Kinshasa par une femme congolaise, Elisabeth Olangi Wosho, que les adeptes nomment maman Olangi.

Le but de l'étude était de connaitre les raisons de l'émergence du Combat, les motifs d'adhésion – des femmes en l'occurrence – et l'impact socio-culturel de ce mouvement pour ceux qui y adhèrent.

Le Combat, appelé aussi «ministère de la délivrance», fut fondé vers la fin des années 1980 au Congo. Dans un premier temps, ce groupe était surtout connu par l'appellation Communauté Internationale des Femmes Messagères en Christ (CIFMC) - parce que les femmes constituaient la principale cible des activités du groupe. Mais aujourd'hui ces activités se sont étendues aussi aux hommes et aux enfants (Ndaya *op. cit.*: 84-97). Ses différentes branches et dépendances transnationales font partie de la Fondation Olangi Wosho (FOW), qui a comme objectif:

> 'La réalisation d'une délivrance spirituelle, sentimentale, émotionnelle et socio-économico-culturelle de l'être humain, par la destruction des liens tissés par la coutume, la famille et la pression sociale, ainsi que sa formation pour favoriser son

---

[5] Nous optons pour continuer à utiliser le terme Combattant(e)s pour désigner les membres de ce groupe, au lieu d'Olangiste (voir Meiers 2013). Ceci parce que, dans les conversations récentes avec eux à Kinshasa, ils (elles) pensent que ceux qui les appellent « les Olangistes » sont souvent des personnes qui veulent saboter leur mouvement ou qui y seraient opposées. Il est bon de noter que le terme Combattant est aussi utilisé pour désigner les Congolais militants activistes de la diaspora qui se nomment « les Combattants de la liberté ». Les Combattants de la liberté combattent les résultats des élections présidentielles de 2012 et l'ingérence étrangère au Congo. l'ingérence étrangère au Congo (Ndaya 2002, 2005, 2007, 2008, 2012, 2013).

[6] Dans la pensée congolaise, le terme esprit est un euphémisme qui désigne les idées, les pensées, les pratiques et comportements acquis.

épanouissement Spirituel, moral et matériel conformément à l'évangile du Christ dans sa plénitude' (Ndaya 2008: 93).

La délivrance en question est illustrée à travers le slogan du Combat: «Nous informons, nous formons et nous transformons». C'est en ce sens que le mot lingala *kobonguana* (guérir), très courant dans le parler des Combattants, désigne la rupture avec les modèles de référence ou les images idéalisées dans le passé. Commencé à Kinshasa, les activités de la Fondation Olangi Wosho (FOW) se sont répandues suivant la mouvance de la migration congolaise, avec des extensions dans le monde (voir aussi Meiers 2013).

Pour les Combattants dont il est difficile de connaitre avec précision le nombre dans le monde, l'idée de la fidélité à la Bible, illustrée par l'expérience concrète de vie de la fondatrice est centrale. Suivant le témoignage de la pionnière, elle avait reçu la mission d'aider les femmes rencontrant des difficultés conjugales, après avoir résolu les siennes et ainsi sauver son mariage grâce aux instructions bibliques reçues de Dieu (Ndaya op cit. 85-97).

L'appellation même « Le Combat » a une double signification. Elle désigne d'abord les problèmes successifs (*mikakatano en lingala*) interprétés comme les attaques visibles d'un agresseur invisible logé dans le corps de la personne souffrante. Pour cela l'état de possession est désigné par des expressions 'Bakangingainzoto' (on m'a fermé le corps), *malady yamabokoyabatu*(maladie provenant de la main des personnes). L'agresseur est nommé tour à tour *ndoki* (sorcier), *mobaliyabutu* (le mari de nuit), etc...

Parmi les compatriotes établis aux Pays-Bas, en Belgique et à Kinshasa où nos enquêtes ont eu lieu, les principaux motifs de consultation sont: le célibat prolongé, les problèmes sexuels, les difficultés conjugales, la stérilité, les échecs multiples, la maladie, la mort, le permis de séjour qui tarde, etc...La guérison n'est possible que lorsqu'on a trouvé l'origine de ses problèmes, identifié l'agresseur et qu'on l'a anéanti.

En général l'avènement du problème, son origine, ses causes, son développement sont imputés aux liens familiaux. La famille biologique, les influences de la modernité et autres, subies par l'individu souffrant sont considérées comme origine du mauvais sort (Ndaya 2012: 415-437). Pour le découvrir, il est obligatoire de participer aux classes d'initiation à la recherche étiologique (Ndaya 2008: 132-153). C'est pour cela que le terme *Combat* désigne aussi l'art psychothérapeutique du groupe. Celui-ci est un rituel d'affliction d'inspiration biblique. Il se déroule sous la forme de rite de passage avec la triple procédure de séparation, de mise en marge provisoire et d'agrégation dans une nouvelle communauté (van Gennep 1981; Turner 1990). La personne souffrante est seule au centre de cette démarche. Il est interdit d'intégrer la famille à la thérapie. De plus, il y a dans la thérapie une partie qui est publique et une autre partie qui est privée. L'initiation à la recherche de l'origine du problème ainsi que la nouvelle socialisation se déroule dans les classes d'enseignement (*mateya*) nommés « les affermissements ». L'établissement du diagnostic et la purification du corps (*kopetola*) (Ndaya op cit. 2012), acte symbolique de la rupture avec les anciens systèmes sociaux et l'incorporation de l'esprit saint (*molimo Mosanto*) comme signe du salut par la fidélité aux Ecritures, se déroule en réclusion. Après ce passage du rituel, en occurrence la purification du corps, il est conseillé de rester membre du groupe.

Les Combattants sont organisés en groupes appelés « bergerie », « filiale » ou « site ». Il s'agit de groupes fermés. Leur fréquentation est généralement limitée aux membres de la communauté internationale de la Fondation Olangi Wosho. Une bergerie n'est pas une structure autonome, elle dépend du siège qui se trouve à Kinshasa. A partir de son groupe particulier, on devient membre de la communauté internationale des Combattants, avec des parents, des frères et sœurs de religion.

Les membres d'une bergerie ne résident pas ensemble. Mais ils se voient régulièrement, du fait des obligations et devoirs des uns envers les autres, ainsi que des activités organisées par leurs responsables.

Bien que cela soit en train de changer, les femmes qui cherchent l'aide du Combat sont généralement des femmes associées aux femmes modernes, elles sont éduquées et citadines. C'est aussi grâce à ces positions, qu'elles peuvent entreprendre des voyages: par exemple, venir en l'Europe par les propres moyens. Elles donnent par ce fait l'image de la femme émancipée à l'occidentale. Elles ne sont pas toujours perçues de manière positive dans le milieu congolais. Leur image de femmes modernes va à l'encontre de la mentalité. Suivant les recherches faites par Biaya (1992:89) sur la mentalité congolaise, la femme moderne, indépendante et instruite est considérée comme une femme *ndumba*, de peu de moralité. Ce qui est par conséquent un handicap pour pouvoir se marier «convenablement». Une telle femme, dans le contexte congolais, n'est pas l'épouse idéale. Elle est supposée s'être libérée des cinq vertus essentielles de la femme congolaise: à savoir la soumission au mari, la fidélité, la décence en parole et en actes et l'attention aux soins domestiques. Elle engendre l'insécurité pour l'homme qui craint l'enrayement des représentations traditionnelles au sujet de la relation entre la femme et l'homme. Mais une Congolaise n'est pas volontairement célibataire. Le célibat est vécu comme un échec social. Les termes utilisés en milieu congolais *etula* (vieille fille) et «jardins sans clôture» sont des expressions dénigrantes utilisées à l'égard des femmes qui ne répondent pas à l'idéal de la femme. Les expressions:

Basi ya poto bamona trop clair
Les femmes de l'Europe ont vu trop clair,
Mabala ya poto esilaka na aéroport ya Ndjili
Les mariages de l'Europe se terminent à l'aéroport de Ndjili[7]

montrent la méfiance des hommes à l'encontre de ces femmes supposées être contaminées par les idées d'émancipation à l'occidentale. Pour les Congolais, il est question de distinguer les femmes 'à marier' et les femmes avec lesquelles « il faut vivre ». Les femmes avec lesquelles les hommes « vivent » comme deuxième « bureau »,[8] ils les recherchent parmi les femmes autonomes pour vivre l'amour romantique «du chéri» comme diffusé sur les écrans de télévision, mais dans la certitude d'un mariage stable avec une épouse « pure », fidèle à la tradition et qui n'est pas contaminée par l'Occident. Mais ces femmes avec lesquelles les hommes veulent « vivre » aspirent elles aussi à un mariage « normal ». La bureaugamie ne pouvant garantir le modèle idéal du mariage.[9] Afin de désigner ces femmes nous avons utilisé une métaphore symboliquement chargée 'mi-figue mi-raisin' (Ndaya 2003) qui illustre leur oscillation entre différentes identités. Représentantes du Congo moderne, elles veulent vivre leur

---

[7] N'djili est le nom de l'aéroport internationale congolais. Il se trouve à Kinshasa.

[8] Le terme bureau s'étend à toutes les femmes qui vivent avec les hommes mariés.

Cette nomination vient du prétexte que les hommes donnaient à leurs épouses pour expliquer leur arrivée tardive au foyer après le travail. Ces unions peuvent être régularisées ou non par la dot. En outre le deuxième bureau.

[9] On appelle bureaugamie la polygamie moderne. On dit aussi deuxième bureau. Elle peut être connue ou non par les premières épouses. Les « deuxièmes bureaux » sont généralement recherchés parmi les femmes intellectuelles.

indépendance mais désirent terriblement avant tout être «femmes congolaises».

Confrontée dans notre recherche à cette difficile condition humaine nous ressentions la gêne que vivaient les compatriotes parce que cette gêne a été aussi la nôtre. Et c'est à ce moment que la question du choix méthodologique s'est posée.

Quels outils utiliser pour rendre compte de la douleur sans trahir nos origines?

Les techniques d'enquête: Les discours dominants

Au moment où on allait commencer le recueil des données, les discours dominants au sujet des enquêtes anthropologiques, du moins dans la littérature à notre disposition étaient presque exclusivement centrés sur le dialogue, les entretiens et les interviews avec les interlocuteurs. Johannes Fabian, qui a effectué différents travaux en Afrique et notamment au Congo Kinshasa, insiste, dans la majeure partie de ses travaux, sur la primauté du dialogue:

> 'Je reviens sur ma réflexion convaincante que l'ethnologie est essentiellement et non accidentellement communicative et dialogique, la conversation et non l'observation doit être le moyen de conceptualiser la production des connaissances ethnologiques'.
> Fabian (1990b: 4)

Le dialogue, et surtout l'entretien neutre était l'outil de qualité permettant de recueillir parole des acteurs, des autochtones dans leur rôle de fournir l'information aux chercheurs qui la fixent dans son journal de terrain. En effet, les spécialistes des méthodes de recherche en sciences sociales ont pendant longtemps produit des réflexions sur les méthodologies à utiliser pour étudier les autres cultures. Ils ont systématisé les techniques, ont organisé des enseignements sur ces techniques tant et si bien que les termes mêmes d'«entretien», «interview», «empathie», «écoute neutre»sont devenus des modèles nobles. Mais malgré la symbolique publication posthume du journal de terrain de Malinowski (1985), dévoilant l'illusion de la neutralité de l'ethnographe par rapport à son « terrain »;les méthodes léguées par les ancêtres de la discipline constituent encore jusqu'aujourd'hui la tradition de référence. Le relativisme culturel, cher à l'anthropologie est souvent incantatoire lorsqu'il s'agit du recueil des données auprès des interlocuteurs africains.

Et pourtant le terrain des recherches n'est pas un champ vide. C'est un lieu où vivent des humains qui existaient avant l'anthropologue et qui existera après son passage. Et de ce fait, ces populations ont aussi leur propre manière de communiquer. En imposant une autre peut provoquer des situations inattendues. Malgré cela, les spécialistes sont surtout engagés dans une quête, depuis l'origine de la discipline, du bon lieu et des bonnes manières d'être, qui permettrait à l'enquêteur d'être neutre, de prendre de la distance par rapport à lui-même, de permettre la parole de l'autre, de la libérer. On cherche à savoir comment créer un climat de confiance avec l'enquêté; clé supposée de l'interaction. L'attention est donnée au lieu de l'entretien, afin que l'enquêté se livre mieux. Dans tout cela très peu d'efforts sont fournis pour répondre aux préoccupations énoncées par différents auteurs au sujet de la hiérarchie dans la profession anthropologique. Jean Copans (1974) par exemple a mis en lumière le lien entre la pratique anthropologique et l'impérialisme. Et dans la même ligne, différents penseurs africains,[10] ont critiqué l'hégémonie de la pensée occidentale dans la construction des méthodologies pour étudier l'Afrique.

Valentin Mudimbe, dans ses récusations amorcées dans son ouvrage *L'autre face du royaume* et poursuivies aussi bien dans *L'odeur du père* que dans 'The Invention of Africa' présente l'ethnologie comme une science coloniale, née et au service de la colonisation. Ces échos se retrouvent aussi chez J.M. Ela qui dénonce l'aliénation et l'étroitesse des méthodes utilisées par les sciences sociales occidentales pour rendre compte des formations sociales de l'Afrique. Mudimbe demande aux Africains d'analyser les appuis contingents aux énonciations et leurs lieux et de chercher à savoir quelles directions proposer pour que nos discours nous justifient dans nos existences singulières dans une histoire singulière. Tous appellent de leurs vœux un changement de discours, pour reprendre l'expression de Kizerbo, un changement de l'instrument linguistique de la production de la connaissance scientifique.

En même temps, nous étions confrontée à une carence quasi criante des méthodes de recherche en sciences sociales développées par les Africains eux-mêmes. On se retrouvait de fait devant un dilemme: se lancer dans le chemin battu, en recourant aux techniques telles qu'utilisées par la majorité des cher-

---

[10] Voir entre autres : Ela 1994; Kizerbo 1980; Mudimbe 1979, 1982, 1988.

cheurs en sciences sociales, ou marquer une différence, en restant proche du mode de communication de l'environnement dans lequel nous avons grandi, et être reconnue par la communauté scientifique. On a essayé d'exprimer cette tension vécue dans un article intitulé 'Entre le marteau et l'enclume. La dialectique être proche et faire des analyses dans la recherche du terrain' (Ndaya: 2005). On y confessait notre embarras d'être prise entre différentes positions.

La question au centre de notre exercice fut de savoir quel outil de recueil des données utiliser comme anthropologue native pour témoigner de l'être proche? Comment rendre compte de la vie des autres comme si on n'avait pas de compte à régler soi-même? Comment demander aux autres des choses de leur vie comme si la nôtre était exemplaire? Si on demande quelque chose à quelqu'un ou si on écoute quelqu'un, ne doit-t-il pas aussi être possible que cette chose qu'on demande soit aussi quelque chose que l'autre puisse vous demander? Pourquoi la relation avec les enquêtés ne peut-elle pas être, pour reprendre Bourdieu (1993) «une relation sociale dans laquelle l'enquêteur assiste l'enquête, ou on s'assiste mutuellement dans l'effort douloureux et gratifiant à la fois, pour mettre à jour les déterminants sociaux de ses opinions et de ses pratiques dans ce qu'elles peuvent avoir de plus difficile à avoir et à assumer»? Les entretiens, les interviews neutres sont des techniques qui nous sont connues, non seulement pour y avoir été formée, mais aussi pour avoir été interlocutrice des chercheurs. Et cette expérience n'a jamais été édifiante. Elle soulève plusieurs problèmes. D'abord l'imposition de la problématique. En effet, le fait autour duquel la conversation va se dérouler, c'est le chercheur qui l'impose. La rencontre ne se déroule pas autour d'un thème à propos duquel chacun des interlocuteurs trouve son compte, s'exprime en liberté et autonomie. De plus, l'enquêté ne connait pas les origines sociales de celui qui est en face de lui. Généralement l'interlocuteur rencontré n'attend pas que l'enquêteur vienne dialoguer avec lui. Les thèmes proposés ne sont pas toujours au centre des preoccupations de l'enquêté. L'enquêteur livre le sujet de discussion. Et l'enquêté se trouve souvent comme sommé d'échanger sur la problématique.

Ayant été parfois dans la position d'enquêtée, on s'est souvent sentie crispée, craignant de ne pas pouvoir bien faire, de ne pas être à la hauteur, d'être discréditée. On s'est parfois sentie comme prise dans une situation scolaire, à la manière d'un étudiant à la dérive devant un examen, en face d'un enquêteur qui joue de l'argument d'autorité, du fait que c'est lui qui pose les questions. On n'ose pas avancer un «je ne sais pas» ou un «je ne comprends pas ce que vous recherchez», «je ne comprends pas de quoi vous parlez».

Ensuite, il y a les attentes de l'enquête face à l'enquêteur. Qu'y-a-t-il au bout de la chaine des entretiens? Quel est la plus-value sociale de l'enquête pour la personne concernée?

Il y a plusieurs années, on a participé, enthousiaste, à une enquête très vaste d'une chercheure ressortissante du Nord sur la planification des naissances. Nous nous y sommes investies, corps et âme, dans l'espoir qu'allaitent être créées des structures de limitation des naissances qui permettraient de contrôler la fécondité. On a attendu, en vain.

La parole des autres participe à la promotion de l'enquêteur. De plus, les matériaux recueillis sont souvent discutés ailleurs que parmi les gens qui les ont produits.[11] Les interlocuteurs investissent dans la recherche en fonction de ce qu'ils sont et en fonction des enjeux qu'ils pensent être liés à cette recherche. On donne à l'enquêteur le don du vécu, en explicitant les expériences. On n'ose pas refuser, de peur de manquer probablement un avantage, dans l'idée que l'enquêteur constituerait un capital social. Et cette idée qu'il y aura quelque chose au bout de la chaine est renforcée par les annotations dans le cahier, la multiplication des entretiens, les demandes d'éclaircissements, la reformulation de ce qui a été dit. Que l'enquêteur le veuille ou non, l'enquêté a une représentation de l'enquêteur qui influe sur la relation et la parole produite. Pierre Bourdieu signale que l'enquêté entre en contact en fonction des représentations qu'il se fait de l'univers de la recherche de l'enquêteur et ceci est particulièrement le cas dans la rencontre entre l'Occidental et l'Africain. En effet, seul l'Occident a su si bien vendre son image de philanthropie, d'être là pour résoudre les problèmes des autres. Paternalisme, condescendance avec les personnes interviewées font que les images mentales, incorporées dans les contacts historiques assignent un sens à la relation entre les ressortissants nord-atlantiques et les Africains. Comme nous l'avons écrit (Ndaya 2008: 130), sans que celui qui interviewe s'en rende compte, il laisse flâner l'impression d'être

---

[11] La carence de la littérature en général et de la littérature sur l'Afrique en particulier, dans les universités africaines, n'est plus à démontrer. Et ce malgré les recherches faites dans différents pays de ce continent. En outre, les ouvrages écrits sur les recherches faites sont quelque fois dans des langues que les concernés (leurs enfants et petits-enfants) ne maitrisent guère.

investi d'une mission de redresser la situation de ceux qu'il interroge, un peu avec l'air de *'je vais vous débrouiller ça'*. Il y a des attentes vis-à-vis de l'Européen qui est considéré comme celui qui peut améliorer les conditions de vie des gens. Comme exprimé par Pierre-Joseph Laurent, dans sa communication orale lors de la conférence IAACHOS en octobre 2013,*«on nous assigne sur le terrain des rôles que nous ne saurons assumer»*.

Et un autre problème soulevé par les outils de recherche est l'imposition d'une manière de communiquer. L'entretien, et plus exactement l'écoute neutre, ressemble bien plus à un interrogatoire, ce qui donne l'impression d'être au tribunal. Ceci fait que l'entretien de l'enquête devient un lieu d'agressions symboliques. Ce n'est pas une situation de dialogue, mais un rapport de force. La rencontre semble en effet un jeu « question / réponse » durant lequel l'enquêteur gère une relation où l'enquêté n'est pas toujours dans une position de lui poser aussi ses propres questions. Les questions « pourquoi» par exemple, très normales dans la communication en milieu occidental, implicite dans le milieu congolais, le doute sur la véracité de ce qui est dit, une contestation. Le terme « pourquoi » met l'interlocuteur sur la défensive.

Et enfin, il faut souvent fixer des rendez-vous. Ce qui donne un caractère formel à la rencontre renvoyant implicitement à la question de la hiérarchie. Par le rendez-vous, l'enquêteur réveille la conscience de l'enquête. Celui-ci ne parlera pas avec l'enquêteur comme s'il parlait avec d'autres personnes, il est oblige de structurer ses propos.

Nous ressentions le malaise de devoir utiliser les outils de recherche consacrés par des diplômes, en niant par la même occasion notre capacité à communiquer, qui nous a permis de mobiliser les ressources de notre environnement. Pour cela nous sommes restée *insider*, en recourant dans le travail de terrain aux causeries (*masolo*[12], en lingala).

Causer, les causeries ou masolo

Causer ou *kosolola* en lingala, c'est être en plein dans la réalité culturelle congo-

---

[12] Ce terme se retrouve dans différentes langues congolaises: *kuyukila* en tshiluba.

laise (Ndaya 2008: 13). C'est une pratique proche de celle qu'Olivier de Sardan (2008: 64) nomme une conversation ordinaire, en situation naturelle, qui se déroule de manière banale, détendue et en assumant le risque que les autres personnes se mêlent à cette conversation. Le sujet de la conversation y est impromptu, spontané, non sollicité a priori par le chercheur.

Nos *masolo* ont eu lieu apartir de notre expérience de membre d'une « Bergerie »[13]. A partir de cette posture, nous remplissions les différents devoirs et obligations d'une chrétienne combattante au sein de la Communauté internationale des Combattants (CIFMC). Grâce aux activités organisées dans ce cadre, les relations se nouaient (ou dénouaient) avec « les copines ». C'est ainsi que nous avons recueilli les données auprès de plusieurs d'entre elles. Kombi Charlotte était l'une d'elles. [14] La relation avec Kombi s'est développée à partir de la participation aux enseignements bibliques donnés par la Bergerie de Bruxelles. Pour cette raison, on logeait chez elle lors de nos séjours dans la capitale belge.

A Charlotte nous avions raconté les raisons de notre présence aux affermissements: recueillir les données pour écrire une thèse de doctorat sur le Combat. Ce motif était considéré par elle noble: il fallait écrire, pour faire connaitre l'œuvre de Dieu dans la vie des femmes. Elle-même était venue au « Combat », comme elle le disait, pour « arracher son mariage». Car elle avait 38 ans mais ne portait pas encore la bague au doigt. Elle avait été bloquée dans son évolution personnelle par les pratiques qui avaient été exécutées sur sa mère lors du décès de leur père. Ses oncles s'étaient accaparés spirituellement de la famille. Son mariage était confisqué, afin qu'elle reste célibataire pour servir les intérêts de ses frères, qu'elle avait beaucoup aidé pour avoir été « bureau » de différents hommes d'affaires congolais. Nous avons dit à Charlotte que c'est en effet un grand problème pour les femmes qui ont fait des études d'avoir un mari. Mais en même temps elles sont utilisées comme des « atouts » par certains membres de leurs familles. Charlotte acquiesçait.

Elle disait que les problèmes proviennent du fait qu'on reste liées aux coutumes. Charlotte disait qu'elle voulait téléphoner à sa mère pour lui demander là

---

[13] Partiellement. Il y a eu une porte que nous ne voulions pas franchir.

[14] Pour l'histoire plus détaillée de Charlotte, voir Ndaya 2008 : 110.

où avait été enterré son cordon ombilical[15] afin qu'elle aille le déterrer. Nous lui avons dit de ne pas le faire. Ce type d'accusation crée des divisions.

Nos conversations avec Charlotte se déclenchaient sur la base de thèmes abordés dans les prêches. Le thème «liens de servitude» était favori. Charlotte disait être poursuivie par la servitude. Un jour, pendant que nous faisions la vaisselle, elle se mit à écraser les cancrelats qui fourmillaient dans ses armoires, disant vouloir les brûler car il s'agissait d'espions, que certains membres de sa parenté s'étaient métamorphosés en ces bêtes afin de surveiller ses mouvements. Nous avons dit à Charlotte qu'on pouvait chercher un produit pour lutter contre ces insects.[16] Charlotte disait que les pièges des démons sont nombreux et qu'il faut être vigilant.

Nos *masolo* prenaient différentes couleurs. Il est difficile de tout reproduire sur papier avec des mots. Car on ne peut pas voir comment on claquait nos mains l'une contre l'autre, on se touchait, on produisait un bruit en poussant la langue contre le palais pour marquer la désapprobation, les rigolades et même les pas de danse de triomphe, en chantonnant ensemble les chansons religieuses lors des repas, les déplacements, en faisant la cuisine, lors de tressage des cheveux, de la couture etc...C'étaient des moments d'interactions qui n'avaient pas beaucoup à voir avec un entretien caractérisé par une écoute neutre.

En outre, si seulement l'interaction avec Charlotte est ici racontée, on n'était pas toujours seule avec elle. Des visiteuses entraient pour vendre les marchandises et se mêlaient à nos causeries, la télévision était ouverte. Il arrivait qu'à partir des rencontres chez notre interlocutrice on nouait une amitié avec une visiteuse qui elle aussi nous faisait le don de son récit.

Nous avons donc résolu le dilemme qui s'est posé à nous comme congolaise,

---

[15] La société congolaise est parsemée de rites qui se déroulent autour de la personne, dès sa naissance jusqu' à sa mort. C'est le cas de l'activité autour du cordon ombilical (*mututu, motolu*) lors de la naissance. Chez différents peuples du Congo, le bout sec du cordon ombilical qui tombe du bébé après quelques jours de vie reçoit des soins particuliers. Suivant les ethnies, on l'enterre ou on le conserve. Aux Kasaï les mères l'enterrent sous un bananier dans le village résidentiel.

[16] Lorsque nous lui avons dit qu'il existait des insecticides pour traiter les armoires dans les supermarchés, Charlotte ne pouvait pas croire que les Blancs pouvaient avoir des insectes, comme les cancrelats dans leur maison.

dans un cadre outre-atlantique, au sein desquels les outils de recherche classiques sont présentés comme garant de la qualité du travail scientifique en restant proche de notre *background* culturel. Nous avons opté pour les *masolo*, afin que nos échanges s'éloignent de l'entretien-interrogatoire neutre. On ne notait pas, on n'enregistrait pas en présence des copines, sauf lorsque les circonstances de la participation à l'initiation l'exigeaient. On a recueilli les matériaux de la recherche en s'assistant mutuellement. Cette assistance devenait la plus-value mentale;«le contre don» face à l'information reçue.

Il n'y avait au départ pas d'imposition du sujet. Le sujet était secondaire à la rencontre mais devenait une cause commune. Ainsi les conversations se caractérisaient par l'inversion. Comme enquêtrice on n'avait pas le monopole de poser les questions, ni de l'orientation de l'exploration. On n'était pas en retrait de la problématique. On s'est aussi livrée. Bien entendu nous recueillions des données au sujet du motif d'adhésion de Charlotte, comme pour de tant d'autres femmes congolaises à devenir membres du Combat. Nous nous immixtions dans sa vie mais on l'a laissée aussi s'immiscer dans la nôtre. A chaque fois que des éléments d'identification survenaient, il y avait une ouverture. On n'a pas été inactive dans l'interaction. On donnait à voir et à entendre aussi. On avait de quoi parler. On est intervenu activement, voire on apporté un jugement ou réagi explicitement à la demande d'avis, tel qu'on sentait sur le coup dans l'action, à la manière de la «réflexivité reflexe» (Bourdieu) comme la capacité de l'enquêteur à intervenir au bon moment. Les *masolo* rejoignent cela. Elles faisaient de la rencontre des sessions thérapeutiques « à la congolaise » avec le soulagement mental que cela procure, d'avoir raconté, d'avoir échangé, d'avoir ouvert des fenêtres.

Mais malgré ces avantages, les *masolo* comme outils d'enquête ethnocentriste presentent aussi des désavantages et ceux-ci sont à explorer. On peut citer l'investissement de la mémoire qu'elle exige, du fait de ne pas prendre des notes. Comment retenir l'information reçue? Il faut dès lors mettre sa créativité en marche en cherchant différents « trucs » pour fixer l'information. En outre, elle demande la maitrise de la langue, de la communication verbale et non verbale. L'implication dans les relations, savoir les créer, s'y engager et les maintenir.

Conclusion

Dans ce texte nous avons voulu partager et transmettre notre expérience, basée sur comment résoudre le dilemme qui se posait à nous: faire une enquête anthropologique chez soi. Sur base des expériences passées, on ne s'est engagée dans les sentiers battus, en utilisant l'entretien comme outil de recueil. Quelle que soit sa forme, l'entretien n'est jamais une conversation ordinaire. Il s'écarte parfois du mode de communication des autres et installe un sentiment de dette. Causer fut libérateur. En s'immergeant dans le groupe et dans les relations sociales, la causerie est une technique de recueil des données qui s'écarte probablement de celles qu'approuvent les méthodologies d'enquête ethnologiques. Mais si l'anthropologie est la rencontre avec l'autre, le relativisme culturel nous interpelle. On doit tenir compte de l'autre, qui est en face de soi comme individu programmé, avec ses spécificités culturelles, comme dans sa manière de communiquer. Nous ne pensons pas que livrer son expérience, parler de soi, c'est heurter l'enquêté. Mais c'est plutôt introduire un changement dans la nature de l'interaction par la rupture des hiérarchies. L'enquêteur ne peut pas faire comme s'il annihilait ce qu'il est socialement. Les questions, les expériences personnelles échangées font du travail de terrain une relation d'assistance et d'enrichissement mutuel. Autrement dit, la rencontre devient une réelle interaction. Les interventions de l'enquêteur sur les dons qui lui sont faits peuvent aider l'enquêté à ouvrir ses horizons en le dotant de certaines ressources.

Cet essai est une proposition d'engager une réflexion sur les outils d'enquête. «Le terrain» de recherche n'est pas un champ vide. Même si la rencontre vaut le temps de l'enquête, il est nécessaire que l'enquêteur aussi se livre. Ceci est aujourd'hui plus que jamais nécessaire. L'intérêt sans précédent dans l'histoire de la discipline pour les études anthropologiques dans les Universités africaines, remarquable par la création des départements et des centres des recherches ethnologiques, exige la révision des méthodes de recherche qualitative.

Bibliographie

Biaya Tshikalak, 1992, « Femmes, possession et christianisme au Zaïre. Analyse diachronique des productions de la spiritualité chrétienne africaine»,Thèse de doctorat, Québec, Université Laval, 494p.

Bimuenyi Kweshi O., 1978, « *Religions africaines, un lieu de la théologie chrétienne africaine*»,Cahiers des religions africaines, n°24, pp. 159-225.

Blanchet A. et alii, 1985, «*L'entretien dans les sciences sociales*», Paris, Dunod.

Bourdieu P., 1993, (dir.)«*La misère du monde* » Paris, Seuil.

Breda C. 2013, «La modernité insécurisée. Anthropologie des conséquences de la mondialisation», Academia-L'Harmattan.

Briggs, D. Warren, 1986, «Learning how to ask. A socio-linguistic Appraisal of the role of the Interview in social science research», Cambridge, Cambridge University Press.

Copans J., 1974, «Critiques et politiques de l'anthropologie», Paris, Maspero.

Ela, J.M., 1994, «Restituer l'histoire aux sociétés africaines. Promouvoir les sciences sociales en Afrique», Paris, L'Harmattan.

Grawitz M., 1964, «*Méthodes des sciences sociales*», Paris, Dalloz.

Hermesse J. &alii, 2011, « Investigations d'Anthropologie prospective. Implications et explorations éthiques en anthropologie»,Academia-L'Harmattan.

Ki-Zerbo J., 1980, «De l'Afrique ustensile à l'Afrique partenaire' » in: *Les Dépendances de l'Afrique et les moyens d'y remédier*, Paris: Berger-Levrault, pp. 42-55.

Leiris M., 1969, «L'ethnographe devant le colonialisme»,*Les Temps Modernes*, 58(1950), reprinted in Cinq études d'ethnologie, Paris, Gonthier-Denoël, pp. 83-112.

Malinowski B., 1985,«*Journal d'ethnographe*»,Trad. fr., Seuil.

Malinowski B., 1989, «*Les argonautes du Pacifique Occidental*»,Trad. fr.,Paris, Gallimard.

Meiers B., 2013, « Le Dieu de Maman Olangi Ethnographie d'un combat spirituel transnational»,Academia-L'Harmattan.

Mudimbe V.Y., 1979, «*L'écart*», Présence africaine. Paris.

Mudimbe V.Y., 1982, «L'odeur du père. Essai sur les limites de la science et de vie en Afrique noire », Présence africaine, Paris.

Mudimbe V.Y., 1988, «The Invention of Africa. Gnosis, philosophy and the order of knowledge», Bloomington: Indiana University Press.

Ndaya J., 2003, « Mi figues mi raisins. l'oscillation des femmes congolaises entre les positions traditionnelles et les positions cosmopolites », Leiden, Pays-Bas, Centre d'études africaines.

Ndaya J., 2005a, «Entre le marteau et l'enclume. Ou la dialectique être proche / faire des analyses dans la recherche de terrain», In: *Quest: An African Journal of Philosophy / Revue africaine de Philosophie* 17: 125-140.

Ndaya J., 2007, « Les enveloppes pour papa Daniel: la transformation des relations domestiques dans les ménages congolais de la diaspora », In: M. de Bruijn, R. van Dijk, J.B Gewald (ed.) *Strength beyond structure. Social and historical trajectories of agency in Africa*. Leiden, Brill: 144-162.

Ndaya J., 2008 « Prendre le Bic. Le Combat Spirituel congolais et les transformations sociales», Leiden: AfricanStudies Centre.

Ndaya J., 2011, «A la recherche d'un autre système de valeur: une approche anthropologique d'un mouvement religieux congolais », In: Kalamba, N. &Bilolo M. (sous la dir. de), *Héritage du discours théologique négro-africain Mélanges en l'honneur du prof.dr. Bimuenyi Kweshi*. Munich, Freising, Kinshasa: African University Studies.

Ndaya J., 2012, « Purification in the healing rituals in the Congolese migrant church Le Combat Spirituel», In: *How purity is made*. Ed. Petra Rosch and Udo Simon. Leipzig: Harassowitz, pp 415-437.

Ndaya J., 2013, '« Take your pen»,Self-divination on the Congolese diaspora', in van Beek W. & Peek M.P. (eds.): *Reviewing reality. Dynamics of African divination*, pp. 257-271.

Olivier DE Sardan J.P., 1995, « La Politique du terrain. Sur la production des données en an-thropologie », *Enquête*, n°1, p.71-107.

Olivier de Sardan J.P., 2000, « Le 'je'méthodologique»,*Revue française de sociologie*, 41-3, pp. 417-445.

Olivier de Sardan J.P., 2008, « *La rigueur du qualitatif. Les contraintes empiriques de l'interprétation socio-anthropologique»*, Anthropologie prospective, Louvain-La -Neuve, Academiano 3, pp. 62-65.

Rabinow P., 1988, «Un ethnologue au Maroc. Réflexion sur une enquête de terrain», trad. Fr., Paris, Hachette.

Turner V., 1990, « Le phénomènerituel, Structure et Contre-structure », Paris: PUF.

van Binsbergen, W., 1999a «Culturen bestaan niet'. Het onderzoek van interculturaliteit als een openbreken van vanzelfsprekendheden». Rede in verkorte vorm uitgesproken bij de aan-vaarding van het ambt van bijzonder hoogleraar grondslagen van interculturele filosofie, Rotterdam: Faculty of Philosophy, Erasmus University, Rotterdamse filosofische studies XXIV.

Van Gennep A., 1981, « *Les rites de passage* », Paris: A. et J. Picard éd.

The Belly Open: Fieldwork, Defecation and Literature with a Capital L

by Sjaak van der Geest

Abstract: This contribution takes the opening scene of Wim van Binsbergen's fieldwork novel EenBuikOpenen (Opening a Belly) as a starting point for a reflection on the hidden worries, shame and discomfort around defecation among anthropological fieldworkers in other cultures. The essay focuses on the question of why this part of life remains largely untold in the many fieldwork accounts that are now being published, and what role literary work can play in the articulation of personal emotion in anthropological fieldwork.

Keywords: fieldwork, anxiety, defecation, Literature

Introduction

'He strained as if his very intestines were in need of being expelled and moaned softly, without conviction, for his mother. All that came out was some yellowish-brown foam and mucus. The sharp-edged lumps that Pieter felt in his abdomen were nothing but cramps. Squatting down as far as he could, he supported himself with one arm on his thigh and clasped the roll of toilet paper, reduced by half since the day before. With the other hand he held his pants away from whatever might come out of his body and whatever was already flowing beneath him. Dizzy, nauseated, he wondered how he had caught this diarrhoea, and if this was how he would be spending the entire seven months of his research time in this North African village' (van Binsbergen 1988: 9; translation Nancy Forest-Flier).

Twice (if I remember well) Wim and I corresponded orally or in written form about an anthropological issue that we seemed to disagree on. Once was through three brief articles on ethical concerns during fieldwork that appeared in *Human Organization* and to which Wim refers in his *Vicarious Reflections.* [1] The other time was after I had read his stories *Zusters, Dochters* (Sisters, Daughters). He gave me a copy of the book and scribbled a friendly kinship term on the front page. I was impressed by the way he captured 'the existential thrust' (Wim's term, thirty years later) of the fieldwork experience. I told him that his literary work was perhaps even more effective ethnography than his elaborate anthropological essays. He did not appreciate the comment and in reaction used a term that I cannot repeat in a contribution to a book of friends. But I wonder how he would react today. This is one of the reasons why I have chosen to renew our too brief debate of about 25 years ago.

Let me first give an example to explain why I was impressed. In the first story, *God prutst maar wat bij het scheppen* (God is messing around with creation), a Zambian woman, Pauline, tells her friend (the writer) the story of her love for a man, Patrick, who was very different from all those men who wanted to have sex with her. She tells the story while she and the friends spend the night together. The setting of the night the mud house, the simple room, and the very bed that plays a role in her story captures the double meanings and intimate contradictions of love and sex; more than I have found in the desperate anthropological attempts to dissect the 'mechanisms' of love. Little details in her narrative about her daily life and work in a shop provide the 'stuff' that stories are made of but would need long explanations in an anthropological account. The point of the love story is that Patrick respected her and treated her as a 'sister'. Her love for him would last as long as they remained 'brother and sister', until

---

[1] My thanks go to Wim van Binsbergen, who over the past 35 years has been a unique and outstanding colleague in anthropology in the broadest sense of the term. I don't need to explain – certainly not to Wim – that the term 'unique' has many shades of meaning. Personally, I had the privilege of receiving from him numerous stimulating comments on my own work, which began with my PhD thesis. When I finally defended that thesis in the Dutch ceremonial fashion, Wim was prepared to fulfil the humble task of 'paranymph', which is generally regarded as a token of friendship. More recently, our paths have gone in different directions, but this fact increases my gratitude for having been invited to contribute to this *Festschrift* and being given the opportunity to re-ignite an old fire. I apologize for the self-plagiarism in this text (Van der Geest 2007); it is better to admit this before you are 'accused' of it. I thank Nancy Forest-Flier for translating the fragment from Wim's novel into Literary English, and Zoe Goldstein for editing the final text.

the night that Patrick persuaded her to have sex with him, and then it was over. This point is alluded to in the title of the story and in a traditional song that Pauline and her friend had heard that very evening: God created people and trees, but 'Why did he create my sister Shongo so beautiful that I wish she was-not my sister?'

The rapprochement between anthropology and literary text can be viewed from two perspectives: the anthropological character of the literary text and the lit-erary style of the anthropological text. The 'discovery' that Literature such as novels and poems contains ethnography is related to the search for an 'experi-ence-near' anthropology. The story provides the 'true-to-lifeness' that the an-thropologist is looking for and thus offers an attractive format for the presentation of research data. Ironically, the authority and authenticity of a story – that apparently does not try to prove anything is greater than that of an academic text that wants to bring insight to the reader. This latter task is at-tempted by following various anthropological conventions, including theoreti-cal digressions, which may rather disturb the ethnography.

Literary writers are better equipped to reveal the more hidden experiences and thoughts of their interlocutors than anthropologists. They do not have to draw a representative sample and are not dependent on painstaking interviews with informants who are unwilling to answer impertinent questions. Literary writers often know what they write about from personal experience. Things they have picked up along the way. Their 'research method' is more natural participant observation than that of the anthropologist. They have seen and heard it all and speak from experience. They can close their eyes and explore their memory.

Fieldwork, anxiety and diarrhoea

My contribution will be a reaction to Wim's remark about a contradiction in his work:

> 'What was effectively expressed in the routinised, globalised discourse of professional anthropology... on second thought turned out not to capture the existential thrust of the fieldwork encounters on which it was based, and what was even more regrettable, did not make any sense to my original fieldwork hosts, and could hardly be a source of pride and identity to them. And what came closer to the latter (*e.g.* my 1988 novel in Dutch, *Een Buik Openen* [Opening a Belly], on my first fieldwork in North Africa, 1968;

and many of my poems) was, with some exceptions, considered irrelevant to the furtherance of anthropology' (van Binsbergen 2015: 3).

The quote that opened this essay is the first paragraph of the novel he mentions in the citation above. The novel is about an anthropology student doing fieldwork in Tunisia, unmistakably a story full of autobiography. The defecation scene sets the tone for a whole series of anxieties that I will not be able to elaborate upon here. Diarrhoea as a metonym for the insecurity, fears, frustrations and unfulfilled desires of anthropological fieldwork suffices for the purpose of my contribution. It enables me to present some of my thoughts about Literature and fieldwork, as well as about dirt, defecation and intimacy. Colleagues who have supervised students travelling abroad to do research in harsh conditions know that fears about defecation are as prominent as fears about food, health, privacy and loneliness. But we hardly ever hear or read about defecatory anxiety; the topic seems more at home in novels than in anthropological publications.

Growing academic reflexivity has treated us to a wave of publications in which the personal anxieties of the author in the field are presented and discussed, sometimes in intimate detail. Surprisingly, however, one of the main worries of fieldwork defecation remains conspicuously absent. Miller (1997: 22) praises the bravery of anthropologists who have 'endured life without toilet paper', but how and if they defecated remains a mystery. Van der Veer (1988: 21), who is one of those 'brave anthropologists', writes that 'the symphony of the bowels' dominates the diaries of anthropologists in the field, but can rarely be heard in their academic publications. He is undoubtedly also speaking of his own experience. The diarrhoea of the diary turns into constipation at the threshold of civilization. Sometimes it does not even enter the diary. Malinowski's strictest diary never mentions this most mundane, drab, everyday activity. Seeing his tent pitched on the shore in one of the photographs of his *Argonauts*, one cannot help being curious. It is ironic, to say the least, that he cancelled out his own defecation while preaching his creed of 'biopsychofunctionalism'.

Thinking of the 'horror' of my own toilet experience on my first morning in the field in Kwahu, Ghana, and the events that followed, I wonder how one can cut out such incidents from reflexive contemplation. I have described my experiences elsewhere (Van der Geest 1998) and it would become a monotonous symphony to repeat the story here. It suffices to note that it was not only the rebellion of four of my five senses (fortunately taste was not involved) that made me run away from that filthy public toilet. The absence of privacy was equally decisive for my fear of the situation. Feeling the eyes of the squatting figures on me though nobody looked at me directly I found it impossible to squat between them, incapable of coping with the technical and social ineptness of handling my own dirt and the dirt around me.

Relating this incident to the rest of my fieldwork, as a reflexive anthropologist should do, I can see one major implication. My running away from that place and my subsequent almost continuous avoidance of local public toilets have made me aware of a serious shortcoming in my participation in the daily life of the community. If toilet training constitutes the entrance to culture, my truant reaction made me lose this essential opportunity. How can I write intelligently – as I have tried to do – about dirt and cleanliness in Kwahu society if I failed to attend the initiation where the principles of purity and danger are taught?

Assuming that many of my colleagues, in similar circumstances, did the same, I suggest that this omission can be an important motivation for silence. Not speaking the local language and failing the toilet test are two awkward shortcomings in anthropological fieldwork. Both are usually concealed. Without directly lying about it, anthropologists tend to give an impression of language capacity by liberally using vernacular quotes. About defecation, they just hold their tongue, as they should in the civilized world of academic discourse. Yet even if we feel uncomfortable about the topic in our own ethnographic work, should we not be more open about it for the sake of our students? Several of my colleagues who have been involved in the supervision of students' fieldwork have told me about their students' fears of defecation in the field. One told me that he could read the emotional burden of fieldwork in his students "infantile obsession with their own defecation". This silence reminds me of the secrecy surrounding initiation rituals. When many years ago Freilich (1977) called fieldwork an initiation rite, he was more right than we realized at the time.

This is not to say that all fieldworkers are always silent about the subject. Some have made one or two remarks about their experiences, keeping it decent and limited. Dentan (1970), who did research in Malaysia, writes that he always had company when he went to relieve himself:

I found it hard to adapt to the fact that going to the river to defecate meant answering cries of 'Where are you going?' The evasive answer, 'To the river,' merely led people to ask, 'Why are you going to the river?' A mumbled 'To defecate' brought a reply of either 'Have a good defecation' or, sometimes, if the speaker was a man, 'Hang on, I'll come with you.'

Evans-Pritchard also complained about the lack of privacy and found it increasingly difficult to defecate before the eyes of his Nuer public (I never found the exact quote). Goodenough (1992) provides a more relaxed picture of his toilet use on one of the Gilbert Islands in the Pacific. He was the only person using the outhouse on the beach; the children used the place to fish and to play. Whenever he needed to go there the children politely gave him passage. On his return they would ask him the traditional question: 'Did you?' The reply was a joyful 'I did'.

A few anthropologists volunteered to tell me about their uncomfortable (or peaceful) toilet experiences in the field. Irene Agyepon, a medical doctor from Ghana with anthropological talents, wrote to me that she could not stay overnight in a fishing village because of the toilet conditions. Defecation had to be done in the bush and the faeces were immediately consumed by pigs. This was too much for her. Peter Ventevogel, anthropologist and psychiatrist, sent me a paragraph from his personal diary, also in Ghana:

> "Been to the toilet. A ditch of one by ten metres, three metres deep. My diarrhoea is back. While the yellow strings fall down an old man is hunching at the other side, in his hand an empty cornhusk to clean his buttocks. My God, everything goes wrong.... I must give up all ambitions. I will never become a medical anthropologist" (diary 17 October 1991).

Ivo Strecker and Jean Lydall (1079) wrote an extensive diary (three volumes) about their fieldwork among the Hamar people in Ethiopia. There is very little in it about defecation but in an email message (May 2003) Strecker summarized their experiences as follows:

We found it enchanting to go as the Hamar do into the bush and relieve ourselves there in the heart of nature, surrounded by plants, birds and insects crawling on the ground who would turn our faeces to dust in no time. During the morning hours the air would still be cool and the world would still be fresh, during midday one would search for a shady place and at night we would walk carefully to avoid getting scratched by the thorny bush, and not to disturb and get bitten by a snake.... The plant we preferred as 'toilet paper' was *baraza* (*grewiamollis*). It is used in countless rituals of the Hamar. There are several entries in the work journal where we mention how we got sick and how this brought us close to the Hamar.

His remark about sickness is significant. Falling sick and defecating (the two are not unrelated) are intense examples of sharing life conditions, of being, after all, of the same species. They constitute crucial elements in the experience of participatory fieldwork.

Defecation as a literary subject

Together with sex and death, defecation has proved the most frequent reason to use euphemisms. The need to avoid the topic, however innocent and natural it may seem, occurs worldwide. This avoidance is also noted in anthropology. Rachel Lea (2001:51) rightly remarked that defecation 'was ignored in ethnography just as it is ignored in daily life'. Not writing about faeces seems part of a general complex of avoiding the issue.

One academic explanation for the near absence of defecation in anthropological writing is the claim that defecation, like sleeping, is a non-issue, an activity that is non-social and non-cultural because it takes place in a social and cultural vacuum. Defecation may be relevant for biology, the medical sciences and psychoanalysis, but not for social scientists, as it lacks any social dimension (Lea 2001: 8-9). My point is that the widespread concern about privacy surrounding the topic rather constitutes evidence of its high social and cultural relevance. The anthropological silence is directly related to this social and cultural relevance (read: embarrassment, unease).

A more plausible 'theory' for the absence of defecation in ethnography and in anthropological reflection is the disgust of this 'matter out of place', to use Mary Douglas' (1966) famous definition of dirt. If speaking, let alone writing, about shit – to call the substance by (one of) its name(s) – is improper, an anthropology of defecation would be equally improper. It does what it claims is 'not done'. If shit is dirt, the anthropologist will become dirty by association, an example of bad taste, or worse, a childish or psychiatric character, or a case of 'narcissistic epistemology' (Quigley, cited in Lea 2001: 14). As the Ghanaian proverb goes, 'If you talk about shit, the smell clings to you'.

Writing, like speaking, is a metonymic act of making present. Writing about defecation takes the activity out of its hiding place and shows it in public. The impropriety of defecating in public extends itself to rules of not speaking about it or referring to it in any other sense, including academic writing. It is true that there are certain situations in which the topic can be discussed, where it is 'framed' or 'bracketed off', as Lea calls it. These are mainly medical contexts and temporary rites of inversion, such as during carnivals and other folk festivities. Anthropological literature does not belong to these free havens of defecatory talk.

My explanation of the anthropological avoidance of defecation, in spite of its high cultural and social relevance, is both embarrassing and ironic. It shows how much anthropologists remain trapped in the rules and conventions of their own culture. I call this 'ironic' since anthropologists claim to take distance from their own culture. They love to justify their ethnographic work as cultural critique, a contribution to defamiliarization by cross-cultural juxtaposition (Marcus & Fischer 1986). Artists, like literary authors, seem more inclined and able to break conventions and broach topics about which we have learnt to keep silent.

When half a century ago Laura Bohannan wanted to let readers look behind the scenes of fieldwork, she opted to write a novel and a pseudonym, as she feared that the novel would damage her scholarly reputation. Her novel *Return to laughter* (Bowen 1964) provided a more effective tool for writing about the day-to-day affairs and hiccups of anthropological fieldwork in a colonial African society. I do not remember that there was anything about defecation in the novel, but the choice for a literary form was significant. It allowed for more personal anxiety and other normal human sentiments, without risking being criticized for exhibitionism or making an inappropriate display of emotion and personal drama.

Another Dutch novel, *Allemaal projectie* (All projection) by Gerrit Jan Zwier (1980), tells the story of an anthropologist who is so scared of doing fieldwork that he secretly stays home and employs a Moroccan assistant to do the work. When his fraud is threatened with discovery, he flies to Morocco and takes up residence in a hotel. His assistant is as fraudulent as his employer, and fills out the tests and questionnaires himself. The anthropologist's boss is very pleased with the outcome of the research. I suppose that the author, who is also a geographer and anthropologist, did not write an autobiographical story, but the novel certainly reveals real emotions and practices that occur in fieldwork.

Wim's graphic description of diarrhoea and angst in his novel would probably not have been appreciated, let alone accepted for publication, in a conventional anthropological account. It would have been rejected as what I have just called exhibitionism and an inappropriate display of emotion and intimacy. The novel, however, allows him to reveal this very personal part of his fieldwork. But is it 'irrelevant to the furtherance of anthropology', as he worried about in the quote above? I do not think so; the tandem of academic and literary work that he has pursued in his career presents a challenge for anthropology as scholarly tradition. Both writing options still have to come to terms with one another. Discussions about subjectivity, narrative, auto-ethnography, aesthetics, intersubjectivity, introspection and serendipity are indications of a future anthropology in which boundaries and overlaps between these various perspectives will be more profoundly and eagerly examined.

References cited

Bowen, E.S. 1964, *Return to laughter*. Garden City NY: Doubleday.

Dentan, R.K. 1970, 'Living and working with the Semai'. In: G. D. Spindler (ed) *Being an Anthropologist*. New York: Holt, Rinehart & Winston, pp. 85-112.

Douglas, M. 1970, *Purity and Danger: An analysis of concepts of pollution and taboo*. Harmondsworth: Penguin.

Freilich, M.1977, *Field work: An introduction*. In: M. Freilich (ed) *Marginal natives: Anthropologists at work*. New York: Harper & Row, pp. 1-37.

Goodenough, W.H., 1992, 'Did You?' In: P. R. Devita (ed) *The naked anthropologist: Tales from around the World*. Belmont, CA: Wadsworth, pp. 112-115.

Lydall, J. & I. Strecker1979, *The Hamar of Southern Ethiopia*. Vol. I: Work journal. Hohenschaftlam: Renner Verlag.

Marcus, G. & M. Fischer 1986, *Anthropology as cultural critique: An experimental movement in the human sciences*. Chicago: University of Chicago Press.

Miller, I. 1997, *The anatomy of disgust*. Cambridge, MA: Harvard University Press.

van Binsbergen, W., 1984, *Zusters, dochters: Afrikaanse verhalen*[Sisters, daughters: Africanstories]. Haarlem: In de Knipscheer.

van Binsbergen, W., 1988, *Een buik openen [Opening a belly]*. Haarlem: In de Knipscheer.

van Binsbergen, W., 2016, *Vicarious reflections: African explorations in empirically grounded intercultural philosophy*. Haarlem: Shikanda Press.

Van der Geest, S., 1998, Akan shit: Gettingrid of dirt in Ghana. *Anthropology Today* 14 (3): 8-12.

Van der Geest, S., 2007, Not knowing about defecation. In: R. Littlewood (ed) *On knowing & not knowing in the anthropology of medicine*. Oxford: Berg, pp. 75 –86.

Van der Veer, P. 1988, De hurkende mens: Een essay over etnografische verbeelding [Squatting man: An essay aboutethnographicimagination]. *Hollands Maandblad* 30, no. 491: 21.

Zwier, G.J. 1980, *Allemaal projectie [All projection]*. Amsterdam/Brussel: Elsevier Manteau.

Becoming Chinese Indonesians

An affirmative approach to social identity

by Stephanus Djunatan

Abstract: A review of becoming Indonesian Chinese should deliberate a philosophical perspective beside the empirical approach to a production of social identity. This article would begin with an exploration that a formation of social identity could not be detached from the power transaction among group identities. This means a struggle for maintaining collective identity by conducting power in dealing with the supposedly major group identity. I do not mean to review the historical record of the political struggle on how Chinese in Indonesia acquire their political identity. Instead, this paper will explore both the internal signification and the external identification of becoming Chinese Indonesians. The exploration concentrates on the experiential-epistemic approach of having a meaningful social identity both for internal and external groups. This concentration is necessary since an interpretation of a meaningful social identity does not ignore an experience of violence in cultural context. Following an elaboration of cultural violence conducted by Johan Galtung, a social recognition of identity could render such experience about agendas of systemised violence. The level of cultural violence per se is a realisation of the principle of negativity and that of the exclusion of the middle. The application of such principles of thought defines the epistemic side of the experiential-expistemic approach. Thus, one should consider an alternative of experiencing and thinking about social identity in a non-violent way. This article will consider an alternative account argued by Wim van Binsbergen. He elaborates the identification of 'sensus communis' and sensus particularis' while he discusses his critique to Kantian argument of '*sensus communis*'. The interplay between becoming '*communis*' and '*particularis*', implies a principle of thinking which intensifies both methods of identifications: they are becoming *communis* and *particularis*. I would suggest this alternative approach as the principle of affirmation. The identification of Chinese Indonesians as social identity should be deliberated through this approach. The complex identification of their identity will continue to be problematic unless they take into account the strategy to keep the interplay betweenbeing the inherent part of the national identity and being the unique ethnic groups within the context of intercultural interaction in the Archipelago.

Keywords: principle of negativity, the principle of affirmation, *sensus communis, sensus particularis,* Chinese Indonesians, cultural violence, social identity identification, intercultural interaction

Introduction

One can consider becoming an Indonesian Chinese from a philosophical perspective, apart from the interdisciplinary approach to the recognition of social identity.[1] This article would begin with an exploration of social identity as an experience of making one's communal identity meaningful. This perspective refers to the internal signification of one's social identity and the external recognition for her communal identity. The former is an effort to define social identity amid many other ones. This signification is not as simple as answering-the question 'who am I?' in order to know herself.[2] The more one tries to respond to that question the more she realises that her attachment to the communal identity to which she belongs could not be detached from the power transaction among group identities. This means that the definition of identity for the sake of national recognition of one's social identity inevitably includes a struggle for maintaining power transaction with the supposedly major group identity.

This article does not cover in detail the political struggle on how Chinese descendants in Indonesia acquire their political transaction for the sake of their group identity. Instead, it will dig deeper into both the internal signification on how Indonesian Chinese understand their collective identity; and the external

---

[1] Some published anthologies on interdisciplinary approaches to Chinese Indonesians' social identity are Mackie (ed.) 1976; Tan (ed.) 1979; Coppel 2006; Suryadinata (ed.) 2008; Wibowo & JuLan (eds.) 2010. Some regional journals are also dedicated to the Chinese Indonesians' social identity: *Journal of Asian Survey* vol. 42 (Jul.- Aug. 2002), *The legacy of violence in Indonesia; Journal of Genocide Research* vol. 11 no. 4 (2009). For individual research and empirical exploration concerning Chinese Indonesians' social identity see Schwarz 1999; Freedman 2000; Bertrand 2004; Wijayakusuma 2005; Purdey 2006; Sindhunata 2006; Setiono 2008; Dawis 2009; Tong 2010; Hardiman 2010; Rahoyo 2010; Afif 2012, Hoon 2012; Setiawan 2012; Santosa 2012; Gondomono 2013; Daradjadi 2013; Lan 2013.

[2] *Cf.* Jenkins 2008: chapter 1, esp., p. 5*ff.*

recognition of becoming Chinese in Indonesia -- on how politics, Indonesian laws of citizenship, and public policies define who these Chinese descendants are amongst local identities. This exploration concentrates on how logical pre-supposition of having an identity provides the apprehension of self-identity meaningful both in personal and collective ways. This concentration is neces-sary if one takes into account the struggle to give meaning to the collective identity. This struggle for recognition inevitably insinuates violence in the cul-tural context. In other words, a study of reconstruction of a meaningful identity involves systematised agendas, programmes or policies that promote agendas and practices of violence.

Chinese Indonesians and cultural violence

Cultural violence, according to Johan Galtung is symbolical legitimation of agendas, programs and policies as well as direct actions that harm or eliminate a considered enemy group. Galtung continues his argument by taking into account that violence is justified even when it becomes something adjusted to perpetrators and their victims. A person's mind and attitude can adapt to the fact that violence becomes inherent part of every day life both as person and society. Moreover, Galtung enhances the idea that people is used to legitimise that violence is accepted as part of moral principles. It changes what is wrong intosomething right in moral domain.[3] In other words, at the 'cultural' level, people appropriately adjust themselves to understand violence as something 'normal' for their mind and behaviour. People are getting used to approve – even to back up – the agenda, programmes, public policies that prioritise vio-lent attitude and behaviour towards an enemy group. Thus, desocialisation, repression, categorisation of secondary citizens, expulsion, marginalisation, segmentation, fragmentation between 'our group' against 'the enemy group', and the alienation of the enemy group, even killing them are common agendas and practices for that society.[4] Thus, how can one identify the way of thinking and living that approve such agendas and practices of violence? Galtung has

---

[3] *Cf.* Galtung 2013: 42.

[4] *Cf.* Galtung 2013:43.

traced some areas of life that accommodate legitimatisation of violence against an enemy group. They are religion, language, empirical and formal sciences, and cosmology. Those render people into thinking that violence against the enemy becomes (sacred, popular, national, anthropocentrical) duty for the in-groups to protect themselves against the out- groups).[5]

Following an elaboration of types of violence conducted by Galtung, the recognition of social identities could render implicitly the agenda, programmes or policies of systemized violence besides explicitly destructive actions. The structural and direct violence is the arena where power transaction between social identities occurs. One can trace the arena through its realisation in language, religion, formal and empirical sciences in the case of violence against Chinese Indonesians. The struggle for the recognition of social identity as Chinese group in the Archipelago has had this historical records of power transaction since the 18th century.[6]

The remaining debate of how Chinese Indonesians should perform themselves amidst native identities points out the polarisation of arguments. Chinese Indonesians faced a bipolar position: on the one hand they were obliged to follow the government's programmes of assimilation, and one the other, they proposed the discourse of the integration of Chinese Indonesians as an independent 'ethnic group' among the native ones.[7] Chinese Indonesians had become divided into two bipolar discourses since this country claimed the independence in 1945 to the fall of New Order era in 1998. This debate, moreover, has been associated with the issue of nationalism. The proponents of assimilation group alleged the integrationists as not being fully supportive to the republic. In these exhaustive argumentations the integrationists were alleged less nationalist to the extent that they only maintained their own interest as Chinese

---

[5] Cf. Galtung 2013: 49-56.

[6] Based on historical records, for example the detailed compilation of historical notes made by Setiono (2008), the historical research of the riot of Chinese migrants allied with Javanese forces against VOC by Wijayakusuma, 2005 and Daradjadi2013, I agree to mention that the struggle for the recognition of Chinese people as ethnic identity can be dated in 18th century. The political struggle for this identity recognition is symbolised by the riot of Chinese migrants in Batavia against the VOC paramilitary in 1740.

[7] Cf. Coppel 1976; Purdey 2003; Hoon 2006; Lembong 2008; Setiono 2008; Winata 2008; Afif 2012; Setiawan 2012.

descendants in the Archipelago. The loyalists to the integration programme blamed the assimilationists for their rejection to the fact that they were Chinese by birth. In this sense, the supporters of Indonesian nationalism tried to remove the traditional ties with their ancestors. Therefore, the assimilationists intended to break up their generational relationship with China nationalism.

Later, the younger generations of Chinese Indonesians have no longer been the proponents of either group since the fall of the New Order regime. Despite this emergence of the young's interest, this debate uncovers the fact that the presence of Chinese Indonesians has been problematic due to the way of thinking and living of the Chinese Indonesians themselves and the native ethnic groups. This presupposes the realisation of cultural violence. The Chinese Indonesians still consider their ambiguous presence among the natives. The first sense of this ambiguity can be formulated as: 'we are not really part of this nation; somehow we belong to the identity of our ancestors whose motherland was the southern part of China nowadays; therefore, we are temporarily staying in this country'. The second sense is that 'we are now the citizen of the republic; we should proclaim ourselves as Indonesians, and leave behind our identity as Chinese; therefore, we deserve equal recognition as a unique ethnic group that inhabits the islands within the republic'. The paradox of 'being alien' *vis-à-vis* 'becoming the citizen' refers to the incessant dilemma of the social identity of Chinese Indonesians with which the natives also agree. In other words, the tension of being fully Indonesians or being fully Chinese still maintains.

Meanwhile on the side of the natives, they tend to consider ambiguous attitudes to their Chinese counterparts in many aspects of life. On the one hand, the natives acknowledge that they have equal position with the Chinese Indonesians. Thus the Chinese Indonesians deserve indiscriminative attitude to access public services and facilities. On the other, the natives somehow sense that these Chinese Indonesians still have more privileges for these social facilities and services due to their better economic and educational status. The supposition that Chinese Indonesians are being more privileged people is the main stereotype of the group. Besides the higher economic status, the natives also recognise the Chinese Indonesians as strangers due to the religious convictions this ethnic group embraces. The majority of Chinese Indonesians are Confucians, Taoists, Buddhists and Christians. To the extent that the Chinese Indonesians have different religious convictions, some local social scholars advice even encourage that Chinese Indonesians should deliberately think of their religions. It is better for them if they convert into Islam. The conversion of religion can be the necessary condition for the natives so that they are easier to acknowledge the Chinese Indonesians as the 'genuine' citizens of the Indonesian people.[8] This admonition of religious conversion implies the practice of penetration of the major group into the marginal one. In this case, one can identify the sense of cultural violence.

Above is the realization of cultural violence in the way of thinking and living in the area of public policies, governmental programmes and religion for both Chinese Indonesians and the natives. In the area of language, one could find out the application of some words that contain discriminative meanings.[9] Chinese Indonesians and the natives have applied these words even since the time of colonisation of the country by the Kingdom of The Netherlands. These words are 'pribumi' ('the natives' in English), 'non-pribumi' (the incomers from mainland China), 'totok' (the pure descendants of Chinese), 'peranakan' (the mixed descendants of Chinese). These words signify classification among Chinese Indonesians as if by appearance only they can separate who has the pure blood of Chinese, the 'totok', from who has not, the 'peranakan'; also segregation of the natives, 'pribumi' from the Chinese Indonesians 'non-pribumi' in social and economic domains. These words connote some important signification such as 'being loyal to the republic' against 'being an opportunistic group that intends to keep their own interest than to serve the welfare of the republic' 'being respectful and upholding the ancestral tradition' against 'being impertinent and careless to generational customs and traditions'. Other connotations of these words are associated with the intergenerational issue of authentic identity of being Chinese among the Chinese Indonesians. The allegation of 'mix blood' remains a serious impediment in the internal relationship among the Chinese Indonesians. For example, a Chinese Indonesian parent always poses a question concerning the ethnic identity to their son or daughter when their children have a girlfriend or a boyfriend. They want to know whether their children's partners are pure Chinese descendants or mix-blood or having the

---

[8] See for example a research on Chinese Indonesians' social identity from the perspective of social psychology Afif 2012; Setiawan 2012.

[9] See Djunatan & Setiawan 2013: 77-88. cf. a philosophical reflection on language and violence on Ricoeur 1998.

identity as an ethnic group outside the Chinese Indonesians.

These words explicitly realise the thinking and attitude of violence within the group of Chinese Indonesians and that among the ethnic groups in the Archipelago. Moreover, these indicate that violence in the archipelago is legalised and morally correct. The oriented way of living of both Chinese Indonesians and the natives approves and encourages the actualisation of violence at all levels; from the level of adjusted consciousness and mentality, structural agendas and public policies, to that of direct actions.

An inevitable question appears after one realises the deep source of violence. It is 'what does the kind of reasons encourage morally authorised violence?' This question will evoke one's mind that even at the level of consciousness violence is permissible, and to that extent, it is morally correct. Moreover, the question brings us to deliberate the epistemological reason. Therefore, one could not disregard a philosophical elaboration in order to take into account epistemological reasons to explicate the question above. By suggesting the elaboration of epistemological reasons, one also arrives at an explanation to how identity formations, both in personal and collective domain, necessarily presuppose this epistemic root of the knowledge of self.

The epistemological root of violence

The previous part of this topic explains that one necessarily looks upon epistemological reasons in order to understand the intention of having a violent mind and that of conducting violence in action. Such fundamental reasons also realises an epistemic process of identity formation both for individual and communal knowledge of oneself. In other words, it is indispensable to analyse deeply into an epistemological root of this political struggle for identity.

The analysis of the epistemological root begins with the capacity of our mind to yield a proposition by means of combining two concepts or terms (as an expression of concept).[10] The capacities of our mind can determine the correla-

---

[10] See Maritain, 1946: 2-3, 5-6. On these pages Maritain delineates how our mind constructs verbal proposition by means of correlating two concepts. The construction of a verbal proposi-

tion between terms (as subject and predicate) by composing and dividing them. The operation of composing and that of dividing two terms apply connector words or the *copula*. It follows that there are only two connectors of the subject and predicate: the positive and negative copula. In this sense then, one can also define the composition of two terms as the positive operation and the division of subject and predicate as the negative operation.

This is the basic application of a logical operation by means of which our mind expresses our propositions of what we perceive, and conceive the knowledge that we want to deliver to someone else.[11] At the next level, our mind applies these logical operations to yield new propositions or inferences out of one or some propositions.[12] Thus, the epistemological root in this elaboration is pertinent to an identification of logical operation of positivity and negativity. Indeed, together with the positive operation, the negative is the key operation in defining our understanding of reality. One could argue that there are three conditions of knowledge production.

Both logical operations provide clarity and distinction for our knowledge of reality so as to define what we perceive and conceive in mind. Pursuing the clarification and distinction of the knowledge of reality is a categorical condition that our mind has to maintain. This is the first condition of knowledge production.

Second condition of the application of the key logical operation is pertinent to the identity formation. Especially one applies the negative operation in order to clarify unique identity of oneself. Personal identity in this matter requires specific description concerning 'what kind of person am I?' A person then asserts several mental traits to be her main character. The assertion of the unique set of mental traits is necessary due to the realization of negative operation. One is unique to the extent that she is not the same as the others. Her personality in this sense appears irreplaceable because her characters are not a copy of others'. Her individuality is clear and distinct due to her dissimilarity with the

---

tion belongs to the second capability of our mind to compose knowledge of reality.

[11] *Cf.* The natural function of language as media for human or social interaction in Winch 1990: 44.

[12] *Cf.* Maritain 1946: 1-2, 7-8. These pages explain the third capability of mind that compose an inference out of combining two or more propositions.

others. In other words, the logical operations of positivity and negativity only assert one's subjectivity by establishing the individuality of oneself. She does not necessarily share her identity to others in order to define her subjectivity. It is because of the independence of herself that she claims her identity. Both operations necessarily ignore the subjectivity of others. At the level of individual confirmation of identity both operation cannot be unacceptable. An individual necessarily performs as a subject in her understanding of reality and the existence of herself. This Aristotelian principle of thinking[13] and the Cartesian argument of clear and distinct apprehension of reality render one's identity authenticated and bounded.

The next condition requires more complex understanding than the first two. This condition is concerning social interaction between individuals. One indicates this operation of negativity within the function of language as a tool for social interaction. While referring to Wittgenstein, Peter Winch underlines that the significant function of language through its meaning suggests the necessary interconnection of interacting subjects.[14] In the context of how language influences social interactivity, the key logical operations give direction to our mind in such a way that one can express what in his mind during interaction. The usage of language here is not limited to a tool of communication. One can distinguish the dissimilarity of persons who are involved in an interactive moment. It is precisely the more one indicates dissimilarities among the interactive subjects, the more one realises that application of a negative operation yields a problematic matter concerning one's understanding of the others. This problematic matter has something to do with the way of thinking that applies violence against the others.

In the context of the third condition of the application of the negative operation in a complex social relationship, an Indonesian philosopher, F. Budi Hardiman argues that one should not disregard the existence and presence of the others in social interaction.[15] The presence of others in social interaction appears at least in three statuses. The first status is the others who are very the

---

[13] See. Stebbing 1961: 146.

[14] *Cf.* Winch 1990: 24, 35 – 39.

[15] Hardiman 2010: xix *ff.*

same to oneself. The second one is they who are less identical. And third one is they who are very dissimilar to oneself.[16] Hardiman continues his argument by saying that one can express friendly attitude towards the person who belongs to the first status. One could compromise the presence of the less identical others. It is supposed that one finds difficulty if she deals with others who have no similar identification of identity at all. The existence of very different others in this sense insinuates anxiety even fear especially in moments of social interactions.[17] This inconvenient thought and feeling towards the very different others occurs because one fails to settle common ground for the divergent aspects of identity such as cultural appearances, performances, language, stereotypes, interests, *etc.* This is the reason why one classifies the very different others as 'they' or 'them'. Meanwhile one succeeds to define the middle ground for the first two statuses above so that she categorises them as 'we' or 'us'.[18] By classifying the different others as 'they' or 'them', Hardiman argues that the anxiety and fear of the distinct others stimulate antipathy against the different others or 'the heterophobia'.

In the social realm, this heterophobia can be a reason of why one applies negative operation in order to understand their presence. Hardiman suggests that at least three models of negativity applied in social interaction with the different others.

For the sake of protecting oneself from any inconvenient situations appeared in interactive relation with the different others, one treats the different others as 'the alien'. This issue of survival insinuates the incapability to accept others as fellow human being. They are not the subject of interaction; they are merely alien objects who live in the middle of us. Their presence is a menace to the group of us. To that extent, we can exploit them for the sake of our interest; alternatively, it is morally correct if we can harm them; so that they can not endanger the existence of our group.[19] In other words, so as to keep the interest of survival, the different others deserve to get domination, repression, discrim-

---

[16]Hardiman 2010: 7*ff.*

[17] Hardiman 2010: 6-9.

[18] Hardiman 2010:10-13.

[19] Hardiman 2010: 14-19.

ination and segregation. Hardiman argues that the reason for survival then encourages the incapability to appreciate any forms of differences in the middle of society. Everything that is different is then perceived as the enemy of the group. This is the first negative operation in social interactivity. This negativity causes failure to construct solidarity among different self-representations within a society.[20]

If a society destroys the appreciation of differences and fails to construct solidarity in it, this society is totalising the sameness in the social interactivity. This society is programmed to appear as singular representation, and it demands absolute submission of its citizen. There is no more capability of a citizen but following obediently the ruler's programmes and agendas. The negativity shuts down the capability to express differences in the public domain. The programme then terrorises a subject if she persists to appreciate the difference in herself or in others around her. In short, there is no longer the capability of self-determination. The absolute control over citizens in society is the main goal of the second model of negativity. In turn, this submissive society can follow any order to destroy the alien others who are present within or outside the society. In this context of absolute submission of individuals in society, it is impossible to realise freedom to express different voices and self-representation and public deliberation based on that freedom of expression. The totalitarian ruler then transforms society into an obedient entity.[21]

This obedient entity then appears as ethnic monism. Individuals in this ethnic monism have no name. Everyone is anonym in this submissive society. The uniqueness of each individual is reduced into just a mass. The ruler easily mobilises this mass in order to direct them to act aggressively and destructively against the different others who stay within the society. This is the third model of negativity which reduces uniqueness of individuality into merely a uniform entity and mass. This uniform mass which is controlled and mobilised by the ruler then becomes the end of negativity as such, that is, 'nihility'. In other words, under the absolute authority of the ruler, the monistic mass is not the perpetrator. They are the victim together with the alien others who are the

---

[20] Hardiman 2010: 37.

[21] Hardiman 2010: 42-43.

object of massive destruction aggressively conducted by this anonym mass.[22]

As in the case of Chinese Indonesians, one can detect the actualization of negativity in the middle of the Indonesian society. The presence of Chinese Indonesians would not have been a chronic problem if the authoritarian ruler had not implemented the totalitarian agenda to exploit and mobilise the submissive monistic mass in order to serve the interest of profit by means of occupying the archipelago. The relationship between the Chinese and the local natives had occurred in trading, cultural exchange for ages.[23] Historical records show that if an authoritarian ruler occupies a region or most region of the archipelago, the relationship between Chinese descendants and local people becomes disparate. It means that this authoritarian ruler uses Chinese people to his profit advantage. This kind of ruler manipulates the Chinese community in a region so that the Chinese people exploit the local natives to accumulate profit from that region. The historical analyses also display that this authoritarian ruler could be manifested in anyone of any ethnic backgrounds.[24]

These authoritarian rulers have implemented the negative operation by means of putting Chinese people on the position of complete strangers (the middlemen) whose interest is to gain profit and accumulation as if the accumulation of profit was only things matter for the group's interest. By doing so, the authoritarian ruler stimulates the natives in such a way that they could no longer perceive that the Chinese people is an inherent part of the local community. The natives can impossibly apply the metaphorical 'we' and 'us' while perceiving the presence of Chinese amid them. In other words, the policies, bills are so much discriminative and exploitative that the Chinese people are regarded as a singular entity against the local people whose representation is also constructed as the singular performance for Chinese community. The authoritarian rulers have reconstructed each hostile group as ethnic monism and they have controlled these submissive groups. The ruler can maximise the exploitation of

---

[22] Hardiman 2010: 44-46.

[23] See the historical records of interaction of Chinese and local natives before the era of colonization of the archipelago took place in De Graff et al., 1984; Lombard 1990 vol.II; Stuart-Fox 2003.

[24] See the monumental works on a historical records of Chinese Indonesians in Setiono, 2008, *Tionghoadalam Pusaran Politik* (*The Chinese Indonesian in the vortex of Politics*) cf. De Graff et al. 1984; see also a study of oligarchy power in the Archipelago in Robinson & Vedi 2004.

both by insinuating hatred even resentment between those groups so that the nature of the relationship between two ethnic identities gradually shifted from relatively equal interaction to hostile one. The usage of the terminology 'pribumi' (the native) and 'non-pribumi' (or the opportunistic stranger) undermines intergenerational correlation between the Chinese and the natives (this was common in the Southeast Asia region).[25]

The authoritarian ruler implements the negativity by deconstructing the internal group cohesion among Chinese Indonesians. It strengthens the sensitive arguments on purity and authenticity of the Chinese which I have discussed above. The ruler also maintains the internal opposition on the political affiliation of the integrationists against that of the assimilationists; on the demand of the autonomous ethnic group whose presence does not melt into local community against the admonition to disperse the Chinese identity into the national one through taking over the local identity, such as the religious conviction, for the sake of the new identity for the Chinese Indonesians.[26] The end of this agenda is to undermine Chinese people's solidarity and their respective capability to represent themselves as a unique individuals and community. By doing so this agenda of the ruler imposes the Chinese Indonesians to be submissive so that they can easily be manipulated to harm each other; in turn, the ruler can employ the Chinese to execute exploitation and repression to the native.

A question hints in the application of negative operation in social interaction of individuals. The question is 'how does one avoid the realisation of negativity either it is imposed by the ruler or it is applied automatically by the ethnic groups' thinking and attitude?' The discussion on this question requires an analysis on how our mind apprehends things and individuals. This apprehension of reality does not detach to valuation of our perception and conception about reality: things and individuals. The valuation inevitably renders determination whether the perception and conception of reality are true or false. If one applies this determination of truth value in social relationship, one

---

[25] Cf. Stuart-Fox 2003: chapt. 3-5.

[26] See the insistence of having independent group identity vis-à-vis the argument of melting identity of Chinese Indonesians amid local groups on Purdey 2003; Anthony Reid 2009; Rahoyo 2010; Afif 2012; Setiawan 2012.

can realise that persons who are involved in an interaction inevitably decide whether their counterpart is implicitly categorised in a true or false position. By this categorisation then, a subject decides on what position these dissimilar others hold. Each position then means something for both interactive subjects. Within this context of signification of the presence of subject in social interaction then we should continue with the elaboration of the alternative model of recognition of identity.

The alternative model of recognition of meaningful identity

Is there an alternative for our mind to apprehend the reality of the different others' existence and their presence in the middle of our society? Besides the logical operation of negativity, one's thinking capability is able to operate positive operation of apprehending things and individuals. To that extent the positive operation is complementary to negative one. Both operations enable our thinking process to determine the existence of a subject by clarifying a definitive understanding of oneself (this is the function of the positive operation) and distinguishing the existence of oneself from the one of the others (this is that of negative one). The latter distinction implies the elimination of the non-identical others. In other words, the logical operations of positivity and negativity are complementary to the extent that both provide foundation to apprehend things and individuals. Such foundations are either the sameness or the difference. Our mind knows something by means of taking a position as the subject applying these two logical operations. Besides these two logical operations, the knowing subject must follow the third principle, that is, the exclusion of the third possibility to know things and individuals (or *principium tertii exclusi*). This exclusion will keep knowledge definitive; meaning that the clear existence of things is obtained according to the function of certain logical operations and the determination of the truth values of the knowledge.[27]

---

[27] *Cf.* Jacques Maritain, 1946: 128-136. See also L. Susan Stebbing 1961: 30-33. In logic a truth value is determined out of correlation of two different propositions. The differences of the proposition are either on the quantitative form or on the qualitative one. The result of the determination appears in three values: true, false and undetermined.

A Japanese philosopher, Nishida Kitaro argues against this model of the principle of thought. Instead, he provides another logical operation which he calls 'the absolute affirmation *qua* negation'.[28] He explains that our mind establishes the idea of things by disposing the real thing to the position of an object. The first operation is then to affirm the identical explanation to the thing one perceives by virtue of negating the object in the perceiving process. The second phase of this logical operation is to negate the perceiver, the subject of knowing process. In other words, these double negations assert the awareness of the perceiver so as to affirm the presence of the object, the dissimilar counterpart in knowing process. If one considers this application of the absolute negation to be mutual determination of the presence of the perceiver and the perceived in knowing process in the social interaction, one determines both presences of 'I' and 'the Others' being true.

Through this argument, Kitaro argues that if one perceives and conceives reality, she does not negate the existence of reality. Suppose then that it is a flower. The end of this negation is to conceive the idea of flower out of the existing flower. She also negates her position as the knower in order to experience the existing flower, and not only the image of one. Thus, the combination of double negation, that is the negation of the position of perceiver and that of the position of existing perceived thing per se, affirms the existing identity of both sides.[29]Considering Nishida arguments of affirmation of the perceiver and the perceived, the more one deals with the nature of social interaction between individuals, the more one realises limitation of the function of the logical operation of single negativity in directing the determination of truth in a social

---

[28] See Kitaro 1970: 29-30, 240; 1987: 73. On the first book Kitaro argues for the negation of subjective position, the perceiver and the negative one, the perceived. The negation to both positions affirms the determination that they are present. Therefore, he writes 'But since absolute negation is an affirmation in the world of the present, which is the place of the mutual determination of individuals, this world of present is both determined by and in turn moves individuals.' While on the latter book Nishida explain as follows: *'the dynamic world always expresses itself within itself through its negation; at the same time, it forms itself within itself and become creative through its own self-affirmation. (the negation of its own negation).'*

[29] See Kitaro 1970: 245. Here Kitaro writes 'For things which are considered to exist in the actual present are both subjective and objective. Even in the case of perceptual objects, they both transcend us and yet are our own sensations'. The double negation is concerning of the position of subject and object in a knowing process. Then he continues '(P)ursuing this direction to its logical conclusion, everything subjective must be negated; the subjective-objective world of reality must be negated.'

relationship. Moreover, if one also considers the application of the principle of the excluded middle in the relationship of individuals, one indicates that this principle only supports the determination of the identical identity of individual in a social interaction. It does not provide any reasons to recognise the presence of the dissimilar others to the extent that these others cannot hold equal or even similar positions as subjects. One of the individuals will assume the position as a subject, and the others as objects. Once the position of the subject is taken, the subject claims a substantial status; subsequently, the others have subordinate positions. Leaving this operation of single negativity behind is to assert that there is the capability to claim the third position that is the recognition of 'similar' status in between the subject and its objects. In other words, one can infer that the position subject opens the possibility to treat the others' equally of having the similar recognition. This means that the subject abandons the manipulation and exploitation of the object for the sake of its interest of knowledge. This manipulation and exploitation in this sense then refer back to the realisation of violence at the structural level and that of direct acts of aggression against the dissimilar others.

In order to develop the alternative principle of thought which Kitaro proposes above, it is indispensable to take into account the argument of 'sensus communis' and 'sensus particularis' as elaborated by Wim van Binsbergen (a Dutch intercultural philosopher) for Nkoya Identity in Zambia. The argument reflects how one can understand the recognition of social identity, the Nkoya identity, in the Kazanga Festival.[30] Indeed, the reconstruction and reproduction of the identity of Nkoya comprise political and economic interests, commodification, and political agendas.[31] These interests may keep the latent manifestation of inter-ethnic and cultural violence. However, in my argument, this power exchange for the sake of identity recognition offers an alternative mode of producing and reconstructing a group identity.

These identifications of becoming 'communis' and 'particularis' imply a principle of thinking which intensifies simultaneously both logical operations of positivity and negativity in social domain. This principle of thinking brings to

---

[30] See van Binsbergen 2003: chapt. 9.

[31] See van Binsbergen 2003: 328ff.

light a paradoxical mode of thinking. The simultaneous operations of positivity and negativity become the complementary paradox in apprehending the reality. I would suggest this principle of thinking as that of affirmation. In this sense, the principle of affirmation is the third function for logical operations beside the positivity and negativity. This third function keeps the interaction of knowing the subject and its perceived things. This interaction can be said as the inclusion of both sides in an event of apprehension. The perceiving and perceived sides necessarily retain the presence of each other.[32] In other words, this mode of thinking insists the position of neither subject nor object. It provides that apprehension of reality constitutes the side of perceiving and conceiving and the one of the perceived and conceived. The determination of the truth value then comes after the correlation of both sides entails a comprehensive picture of one's experience of knowing. The evaluation does not reside in the output of knower, that is, the idea of reality. It is through experience of knowing, that is, to be the perceiver and the perceived, the evaluation of knowledge then occurs. The result of this mode of thinking is an experience to acknowledge the existence of both sides without trapped into an elimination of the subordinated position.

In this sense, one can consider the hint of the principle of affirmation in the dynamic of identification, the interplay of social identity between the sense of 'communis' and that of 'particularis', which is argued in van Binsbergen's philosophical reflection on the recognition of Nkoya identity in Kazanga Festival above. The sense of communis (the sense of 'we-us-ness') explicates how the Nkoya people become inclusive to the extent that they include the others as inherent participants of the celebration. By doing so they are incorporated into a larger supposedly national identity; and they experience a sense of belonging to the same, late-modern humanity.[33] Meanwhile, the sense of 'particularis' (the sense of we-them-ness) enhances the idea of being particular ethnic group, a sense of belonging to the unique representation of ethnic group, as the Nkoya

---

[32] Cf. Chinese philosophy symbolises the third principle of knowing, that is, the comprehensive understanding of reality that comprises of two sides, in the Yin Yang diagram. Some philosophers then consider this holistic understanding of reality the correlative thought. See. Joseph Needham 1956/2005 vol. 2, chapt. 13 section ff; Ames & Hall 2003: 11-54; Williams 2006: 433-436.

[33] See van Binsbergen 2003: 326-327.

people.[34] These two poles, the sense of *communis* and that of *particularis* reveal a negotiation of power transaction for the sake of experiencing social identity meaningful, both for the individuals and the collective. These poles, indeed, entail the tension of becoming a group identity. On the one hand, it is tempting to insist on one pole and to ignore the other one in order to maintain an authentic and bounded identity. On the other, one faces a danger of losing authenticity or even the integrated identity group if one deals with too much negotiation in power transaction in order to regain recognition from the others. This tension, however, necessarily occurs to the extent that this identity politics provides meaning for one's presence and existence as a unique ethnic group. Thus, the solution to tension of the temptation of being bounded by identity and that of being inauthentic, as the output of negotiating identity in accordance with the globally formatted one, is to weave both poles into the experience of affirmation of the meaningful group identity.[35]

In addition to my exploration to the social recognition of Chinese Indonesians, these modes of identity identification collectively and nationally become significant. Furthermore, both modes of becoming *the communis* and *the particularis* hint a shift from a merely epistemic level of comprehension of group identity into an experiential-epistemic one. The experiential-epistemic level of determination of group identity means that the understanding of social identity for Chinese Indonesians discovers complex aspects that reconstruct complementary sets of identification. Those sets are the experience of the emergence of '*communis*', being the perceiver who share the national identity of the Indonesians with other ethnic groups in a plural society in the archipelago; and that of the emergence of '*particularis*' or being the perceived whose presence is defined according to the Chinese Indonesians themselves and to the other ethnic groups in the archipelago. In other words, the emergence of *communis* explicates the plurality of ethnic groups in the archipelago who shares the same national identity. The acknowledgement of plurality avoids any insistent programmes and agendas to reconstruct the uniform identity for Indonesian peo-

---

[34] See van Binsbergen 2003: 327.

[35] *Cf.* van Binsbergen 2003: 332. He concludes that 'Acknowledging the tension amounts to an admission of the legitimacy of both the polar positions between which the tension is generated; that surely is a more promising strategy than seeking to resolve the tension by privileging one of the constituent poles by the application of rigid binary oppositions.'

ple (in accordance with a certain ideology or a religious conviction). Meanwhile, emerging *particularis* signifies that every ethnic group in the archipelago is irreplaceable, independent. By virtue of this independency, each of them deserves to live and maintain their respective customs, rituals, conviction, traditions, way of life, politics, local laws, including their occupation of the land, *etc.* This acknowledgement of being specific social identity then fortifies the assertion of plurality in the Indonesian society. In this sense one realises the complementary correlation between the emerging of *communis* and that of *particularis* to the extent that these emergences continue a dynamic and complex correlation among social identities.

The interplay of emerging the '*communis*' and the '*particularis*' above reveals that one can not ignore one of this pair of emergence if one deliberates the formal recognition of social identity for each ethnic group in Indonesia, including the Chinese Indonesians. If one only determines identity by recognising one polar, and ignoring the other, one is entrapped in a preferential option to violence. The Chinese Indonesians should take into account the interplay of these emergences unless they intend to be entrapped in the vicious circle of violence.

If one falls into the trap of violence, it is hard to draw a line between perpetrators *vis-à-vis* victims in all levels of violence. The power transaction between groups only strengthens this ambiguous position. My exploration to the violence against Chinese Indonesians as the social identity indicates this confusion. Chinese Indonesians' position can be either as perpetrators, or at least active or passive proponents of violence agenda of violence against the rival group, or they are exposed or suppressed victims of any agenda of cultural violence. In this sense, it is problematic to determine which parties are enacted as the 'occupying knower, the powerful subject, or the manipulated known, or the weak object.[36]

---

[36] See Djunatan & Setiawan 2013: 78-88. In these pages we explain the implication of the usage of discriminative terminologies such as the native (*pribumi*) and the strangers (*non-pribumi*) and the pure *vis-à-vis* the mix blood. Such implementation of the terminologies in social interactions between ethnic groups yields the confusion of which parties deserve to be the victims (metaphorically describe as the slave group) against the perpetrators (the master group) within the context of violence.

Persistency on one side of the emergence means another trap, that is, the double dilemma of expressing the social identity for Chinese Indonesians. In this case I develop an explanation of the dilemma of being Chinese Indonesians suggested by Susan Giblin.[37] One should take into account seriously this double dilemma since this is also an output of cultural violence. The dilemma is as follows.

First, on one side, the involvement of Chinese Indonesians into the national or local domain of politics, economy, law, and culture indicate their intention to be involved in the development programmes of the Republic of Indonesia. They take seriously the social issue of poverty, social injustice, poor education, social discrimination and they want to solve this problem together with other people of Indonesia. Therefore, the Chinese establish social foundations, non-governmental groups, social charity programmes for the sake of supporting the national development programme. Yet, for the sake of this good intention some Chinese deliberately abandon 'Chinese-ness';in turn, they replace their Chinese heritage with local identification such as converting into the local religious convictions, speaking the local language and actualizing the local way of life. On the other, this resolution is understandable to the extent that everyone can choose their respective horizontal affiliation. Yet this resolution to renounce the Chinese-ness is unacceptable due to the demand to be faithful to the cultural heritage for the sake of the appreciation of basic rights that everyone can inherit their respective group identity to the extent that accepting the heritage means to declare the right to express unique and irreplaceable social identity. Moreover, the good intention of involvement in social arena does not automatically means good acceptance before the local identity. The local identity still questions the loyalty of the Chinese Indonesians. They still sceptically consider that this good intention is purely for the sake of the national welfare. Their scepticism is due to the historical record that indicates another agenda in the Chinese's involvement, that is, their hidden intention to merely gain profit

---

[37] See Giblin 2003: 345; Djunatan & Setiawan 2012: 57; cf. studies on defining meaningful the social identity for Chinese Indonesians within the context of violence against this group identity in McTKahin 1946; Wertheim 1956; Anderson 1983; Suryadinata 1976, 1978, 1982, 2001, 2008, 2010; Aguilar 2001; Cribb 2002, 2009; Kusno 2003; Purdey 2003; Chua 2004, 2008; Winarta 2008; Soebagjo 2008; Lembong 2008; Panggabean & Smith 2008; Mackie 2008; Knorr 2009; Reid 2009; Leggett 2010; Rahoyo 2010; Wibowo & JuLan 2010; Hoon 2006, 2012; Santosa 2012; Setiawan 2012; Lan 2013.

from the situations.

Second, on the one hand the revival of Chinese heritages by virtue of cultural celebration, rituals, application of mandarin in daily conversations and mass media, culinary, songs and dances among the local groups in the archipelago is an appreciation of the right of expressing the unique social identity, being Chinese. The Chinese Indonesians deserve to have social acknowledgement for their social identity. The expression of cultural heritage will support the plural performance of the republic. Moreover, the revival signifies the respect to human rights and equality among ethnic groups. On the other, the appreciation of cultural rights does not motivate Chinese Indonesians in order to have commitment to be involved in the national agenda. This revival only entraps Chinese into segregation from the Indonesian society. They become a more exclusive group identity among the local ones by virtue of the revival. The suspicion and scepticism of the local groups remain since the proposition of respecting human rights and equality is only a rhetorical commitment of the group. The daily interaction still indicates that the practice of subordination is occurred between the Chinese against the locals.

The first dilemma above in my argument expresses the choice of keeping the first emergence of *communis* as if it were only the answer to resolve the problematic presence of the Chinese Indonesian among the local groups. The second dilemma signifies the insistence to preserving the emergence of *particularis*, as Chinese Indonesians. This preservation of becoming *particularis* does not provide the answer to the question of meaningful presence of the Chinese Indonesian both for the signification of presence interpreted by the internal group and by the external group, the other ethnic groups. The dilemmas in turn only keep the practice and systemised program of cultural violence occurs.

In other words, keeping alert from an inclination of entrapping oneself into the endless cycle of cultural violence, one should begin to consider a non-violence approach to an experience of interweaving modes of realisation of group identities. This non-violence approach starts with accepting both the emergence of *communis* and that of *particularis*. These emergences, like I have discussed above, deliberate these emergences an alternative experiential-epistemic understanding of a group identity in a social interaction. This is a worth strategy to be taken into account both by Chinese Indonesians and other ethnic groups in the archipelago.

Concluding remark

The experience of intercultural encounters within Indonesian archipelago has a long history of peaceful and warring times. There have been conflicting occurrencesbetween ethnic identities besides fruitful interaction of ethnic groups. The power transaction among ethnic groups reminds us the danger of incessant violence at all levels of life if one can prefer the paradigm of either negativity or positivity respectively to the third possibility that provides meaningful presence for each ethnic group. The former hints the preferential option to violence, while the latter concerns with the non-violence approach for the sake of a resolution of latent conflicts of social identity identification.

One can analyse some verbal expressions as far as this inclination of violence is concerned. This verbal expression implies the preference to opt for violence in the deep level of understanding. The third method of affirmation to both the negative and positive operations at the epistemic level encourages us to cease the expression of the violent-contained thinking paradigm. The abandonment of this verbal expression also strengthens the awareness to avoid the preferential option to cultural violence. This is an effort to continuously choose some sets of constructive expression both in verbal and gestural ways. These affirmative sets of verbal expression bring the idea that identity identification in power transaction reveals necessity of emergenceofthe 'communis' as well as the 'particularis' simultaneously. In other words, becoming Chinese Indonesian means to accept an interplay between becoming the 'communis', as an inherent part of national identity in the archipelago and becoming the 'particularis' as a unique identity representation in accordance with its cultural heritage.

This interplay of emergences also occurs inside the group identity. Becoming the pure descendants and becoming the mix blood display an interactive transaction in defining substantial expression of what Chinese-ness is in the context of diaspora. In this sense, becoming Chinese Indonesians implies a reconstruction of a bounded identity. Following Amin Maalouf, identity identification constitutes two associative attachments, the vertical and the horizontal ones.[38] Firstly, one inherits her vertical line, the legacy of cultural tradition or heritage; secondly one can choose any kind of horizontal affiliation to any kinds of identification such as religious, political and economical one. These associative attachments in my opinion represent explicitly the interplay of becoming *'communis'* and *'particularis'* in the sense of identity identification for Chinese Indonesians.

References cited

Afif, Afthonul., 2012, *Identitas Tionghoa Muslim Indonesia: PergulatanMencariJatiDiri*, Depok: PenerbitKepik.

Aguilar, Filomeno V., 2001, 'Citizenship, Inheritance, and the Indigenizing of 'Orang Chinese' in Indonesia' *Positions: east Asia cultures critique*, Volume 9, Number 3, Winter 2001, pp. 501-533, Duke University Press.

Ames, Roger T., & Hall, David L., 2003, *Daodejing: 'Making This Life Significant', A Philosophical Translation*, New York: Ballantine Books.

Anderson, Benedict R. O'G., 1983, 'Old State, New Society: Indonesia's New Order in Comparative Historical Perspective' *The Journal of Asian Studies*, Vol. 42, No. 3 (May, 1983), pp. 477-496, Association for Asian Studies.

Budi Hardiman, Frans., 2010, *Massa, Terrordan Trauma: MenggeledahNegativitasMasyarakat Kita*, Maumere: LedalerodanLamalera, 2010.

Chua, Christian., 2004, 'Defining Indonesian Chineseness under the New Order', *Journal of Contemporary Asia*, 34: 4, 465-479.

Chua, Christian, 2008, 'The conglomerates in crisis: Indonesia, 1997-1998', *Journal of the Asia Pacific Economy*, 13: 1, 107-127.

Coppel, Charles A., 1976, 'Patterns of Chinese Political Activity in Indonesia,' in *The Chinese in Indonesia: Five Essays*, ed. J.A.C.Mackie, Melbourne: Thomas Nelson, in association with the Australian Institute of International Affairs, 1976, pp. 19-76.

Coppel, Charles A., ed., 2006, *Violent Conflict in Indonesia, analysis, representation, resolution*, London & New York: Routledge.

Coppel, Charles., A., 2008, 'Anti-Chinese Violence in Indonesia after Soeharto' Suryadinata, Leo (ed.), 2008, *Ethnic Chinese in Contemporary Indonesia*, Singapore: Institute of Southeast Asian Studies.

Cribb, Robert., 2002, 'Unresolved Problems in the Indonesian Killings of 1965-1966' *Asian Survey*, Vol. 42, No. 4, *The Legacy of Violence in Indonesia* (Jul. - Aug., 2002),pp. 550-563, University of California Press.

Cribb, Robert & Coppel, Charles A., 2009, 'A genocide that never was: explaining the myth of anti-

---

[38] Maalouf 2003: 102*ff.*

Chinese massacres in Indonesia, 1965-66', *Journal of Genocide Research*, 11: 4, 447-465.

de Graaf H.J., *et al*., 1984, *Chinese Muslims in Java in the 15th & 16th centuries, the Malay annals of Semarang and Cerbon*, Merle Ricklefs.

Daradjadi, 2013, *Geger Pacinan, Persekutuan Tionghoa-JawaMelawan VOC*, Jakarta: Kompas.

Dawis, Aimee., 2009, *The Chinese of Indonesia and Their Search for Identity: The Relationship Between Collective Memory and the Media. Translated into Bahasa Indonesia in 2009*, Orang Indonesia MencariIdentitas, Jakarta: GramediaPustakaUtama.

Djunatan, Stephanus & F. X. Rudi Setiawan, 2013, *Eksplorasi Paradigma negativities sebagai-akarkekerasankultural, pendekatan hermeneutic atasKajianKekerasanmassalterhadap Orang Tionghoa Indonesia*, Bandung: LPPM UNPAR; this research report can be accessed at http://journal.unpar.ac.id/index.php/Sosial/article/view/243

Freedman, Amy L., 2000, *Political Participation and Ethnic Minorities, Chinese overseas in Malaysia, Indonesia and the United States*, London & NY: Routledge.

Freedman, Amy L., 2003, 'Political Institutions and Ethnic Chinese Identity in Indonesia', *Asian Ethnicity*, 4: 3, 439-452.

Galtung, Johan & Dietrich Fischer., 2013, *Johan Galtung, Pioneer of Peace Research*, Heidelberg: Springer.

Giblin, Susan., 2003, 'Civil Society Groups Overcoming Stereotypes? Chinese Indonesian Civil Society Groups in Post-Suharto Indonesia', *Asian Ethnicity*, 4: 3, 353-368

Gondomono., 2013, *ManusiadanKebudayaan Han*, Jakarta: Kompas.

Hoon, Chang-Yau, 2006, 'Assimilation, multiculturalism, hybridity: The dilemmas of the ethnic Chinese in post-suharto Indonesia', *Asian Ethnicity*, 7: 2, 149-166.

Hoon, Chang-Yau., 2012, *IdentitasTionghoa, Pasca-Suharto, Budaya, Politik, dan Media*, Jakarta: Yayasan Nabil & LP3ES.

Jenkins, Richard., 2008, *Social Identity, key ideas*, NY: Routledge.

Joe Lan, Nio, 2013, *PeradabanTionghoa, Selayang Pandang*, Jakarta: KepustakaanPopulerGramedia.

Kahin, George McT., 1946, 'The Chinese in Indonesia', *Far Eastern Survey*, Vol. 15, No. 21 (Oct. 23, 1946), pp. 326-329

Kitaro, Nishida., 1970, *Fundamental Problems of Philosophy, the World of Action and the Dialectical World*, transl. David A. Dilworth, Tokyo: Sophia University.

Kitaro Nishida., 1987, *Last Writings, Nothingness and the Religious Worldview*, Honolulu: University of Hawaii Press.

Knörr, Jacqueline., 2009, 'Free the dragon' versus 'Becoming Betawi': Chinese identity in contemporary Jakarta', *Asian Ethnicity*, 10: 1, 71-90.

Kusno, Abidin., 2003, 'Remembering/Forgetting the May Riots: Architecture, Violence, and the Making of 'Chinese Cultures' in Post-1998 Jakarta' *Public Culture*, Volume 15, Number 1, Winter 2003, pp. 149-177 (Article), Duke University Press.

Leggett, William H., 2010, 'Institutionalising the Colonial Imagination: Chinese Middlemen and the Transnational Corporate Office in Jakarta, Indonesia', *Journal of Ethnic and Migration Studies*, 36: 8, 1265-1278, First published on: 28 June 2010 (iFirst)

Lembong, Eddie., 2008, 'Indonesian Government Policies and the ethnic Chinese: some recent developments', Suryadinata, Leo (ed.), 2008, *Ethnic Chinese in Contemporary Indonesia*,

Singapore: Institute of Southeast Asian Studies.

Lombard, Denys., 1990, *Le Carrefour Javanais, Essai d'histoire globale vol. II Les reseaux asiatiques*, Paris: École des HautesÉtudes en Sciences Sociales. Translated into Bahasa Indonesia in 1996/2005, *Nusa Jawa: SilangBudayaJaringan Asia*, volume II: Jaringan Asia, Jakarta: Penerbit Gramedia Pustaka Umum & Forum Jakarta – Paris, École Francais d'Extrême – Orient.

Maalouf, Amin., 1996, *In the Name of Identity: violence and the need to belong*, New York: Penguin Books.

Mackie, Jamie A.C., ed., 1976, *The Chinese in Indonesia: Five Essays*, Melbourne: Thomas Nelson, in association with the Australian Institute of International Affairs.

Mackie, Jamie., 2008, 'Is there a Future for Chinse Indonesians?' Suryadinata, Leo (ed.), 2008, *Ethnic Chinese in Contemporary Indonesia*, Singapore: Institute of Southeast Asian Studies.

Maritain, Jacques., 1946, *Formal logic*, revised ed., translated from French by Imelda Choquette, NY: Sheed& Ward,

Needham Joseph., 1956/2005, *Science and Civilization in China* vol. 2, 'History of Scientific Thought', Cambridge: Cambridge University Press.

Panggabean, Samsu Rizal & Smith, Benjamin., 2008, 'Explaining Anti-Chinese Riots in Late 20th Century Indonesia' Forthcoming in *World Development*, February 2011 Peck Yang, Twang, 1998, *Chinese Business Elite in Indonesia and the transition to Independence 1940 – 1950*. Kuala Lumpur: Oxford University Press.

Purdey, Jemma, 2002., 'Problematizing the Place of Victims in Reformasi Indonesia: A Contested Truth about the May 1998 Violence' *Asian Survey*, Vol. 42, No. 4, *The Legacy of Violence in Indonesia* (Jul. - Aug., 2002), pp. 605-622, University of California Press.

Purdey, Jemma, 2003., 'Political Change Reopening the AsimilasivsIntegrasi Debate: Ethnic Chinese Identity in Post-Suharto Indonesia', *Asian Ethnicity*, 4: 3, 421-43.

Purdey, Jemma, 2006, *Anti-Chinese Violence in Indonesia, 1996-1999*, Singapore: Singapore University Press in association with Asian Studies Association of Australia.

Rahoyo, Stefanus., 2010, *DilemaTionghoaMiskin*, Yogyakarta: Tiara Wacana.

Reid, Anthony., 2009, 'Escaping the burdens of Chineseness', *Asian Ethnicity*, 10: 3, 285-296.

Ricoeur, Paul., 1998, 'Violence and Language' in Bulletin *de la Société Americaine de philosophie de langue Française*, Vol. 10, Issue 2, Fall (1988).

Robinson, Richard & Vedi, R. Hadiz., 2004, *Reorganizing Power in Indonesia, the politics of oligarchy in an age of markets*, London & NY: Routledge.

Santosa, Iwan, 2012, *Peranakan Tionghoa di Nusantara, CatatanPerjalanandari Barat keTimur*. Jakarta: Kompas.

Schwarz Adam, 1999, A *Nation in Waiting*, NSW: Allen &Unwim.

Setiawan, Teguh., 2012, *Tionghoa Indonesia, Cina Muslim dan Runtuhnya Republik Bisnis*, Jakarta: Republika.

Soebagjo, Natalia., 2008, 'Ethnic Chinese and Ethnic Indonesians: A love-hate relationship' Suryadinata, Leo (ed.), 2008, *Ethnic Chinese in Contemporary Indonesia*, Singapore: Institute of Southeast Asian Studies.

Setiono, Benny G., 2008, *Tionghoadalam Pusaran Politik*, Jakarta: Trans media pustaka.

Sindhunata, G., 2006, *KambingHitam, teori René Girard*, Jakarta: PT. Gramedia.

Suryadinata., 1976, 'Indonesian Policies toward the Chinese minority under the New Order', *Asian Survey*, Vol. 16, No. 8 (Aug., 1976), pp. 770-787, University of California Press.

Suryadinata, Leo., 1978, *The Chinese Minority in Indonesia 7 Papers*, Singapore: Chopmen Enterprises.

Suryadinata, Leo & Sharon Siddique, 1982, 'Bumiputra and Pribumi: Economic Nationalism (Indiginism) in Malaysia and Indonesia' *Pacific Affairs*, Vol. 54, No. 4 (Winter, 1981-1982), pp. 662-687, Pacific Affairs, University of British Columbia.

Suryadinata, Leo., 2001, 'Chinese Politics in Post-Suharto's Indonesia: Beyond the Ethnic Approach?' *Asian Survey*, Vol. 41, No. 3 (May - Jun., 2001), pp. 502-524, University of California Press.

Suryadinata, Leo., ed., 2008, *Ethnic Chinese in Contemporary Indonesia*, Singapore: Institute of Southeast Asian Studies.

Suryadinata, Leo., 2010, 'AkhirnyaDiakui, agama khonghucudan agama buddhapasca-Suharto', Wibowo, I., & Ju Lan, Thung (eds.), 2010, *Setelah Air Mata Kering, MasyarakatTionghoa-Pasca- peristiwa Mei 1998*, Jakarta: Kompas.

Stebbing, L. Susan., 1961, *A Modern Elementary Logic*, New York: Barnes & Noble.

Stuart-Fox, Martin., 2003, A Short History of China and South East Asia, tribute, trade and influence, NSW: Allen &Unwin.

Tan, Mely G., 1979 (ed.), *Golongan Etnis Tionghoa di Indonesia, suatumasalahpembinaankesatuanbangsa*. Jakarta: LEKNAS – LIPI &YayasanObor Indonesia.

Tong, CheeKiong., 2010, *Identity and Ethnic Relations in Southeast Asia, RacializingChineseness*, Dordrecht: Springer.

van Binsbergen, W.M.J., 2003, *Intercultural Encounters*, Berlin: LIT.

Wertheim, W.F., 1956, *Indonesian Society in Transition*, The Hague: Van Hoeve.

Wibowo, I., & Ju Lan, Thung., 2010, *Setelah Air Mata Kering, MasyarakatTionghoaPasca-peristiwa Mei 1998*, Jakarta: Kompas.

Winch, P., 1990, *The idea of Social Science and its relation to Philosophy*, 2nd edition, London: Routledge.

Winarta, Frans H., 2008, 'No More Discrimination against the Chinese', Suryadinata, Leo (ed.), 2008, *Ethnic Chinese in Contemporary Indonesia*, Singapore: Institute of Southeast Asian Studies.

Williams, C.A.S., 2006, *Chinese Symbolism and Art Motifs, A Comprehensive Handbook on Symbolism in Chinese Art through the Ages*, 4th revised, Singapore: Charles E. Tuttle Co., Inc.

Wijayakusuma, Hembing, H. M., 2005, *Pembantaian Massal 1740, tragediberdarahangke*, Jakarta: Pustaka Populer Obor.

Facets of the Egyptian Ennead in relation to a posited Indo-European and Chinese Ten-God system

by Emily Lyle

Abstract: The theory of a ten-god system is one I have been exploring for some time. It can be stated most simply in the Chinese case as that of the ten Heavenly Stems, eight of which, represented at the periphery, are equated with the eight trigrams, while the remaining two are centrally located. In Indo-European terms, the posited system deals with four old cosmic gods (the Indian Asuras) and four young social gods (Devas), plus a king of the living (Indra) and a king of the dead (Yama).Since the gods can be regarded as the head terms of an extensive series of correlations, they are related to directions of space, segments of time, colours, elements, *etc.* Taken together the Indo-European and Chinese patterns may serve to suggest a possible preform of the Egyptian ninefold Ennead which, with the living king, Horus, has a tenfold structure.

Keywords: Egyptian cosmology, Indo-European cosmology, Chinese cosmology, pantheon, Horus, trigrams, elements, Presocratic philosophy

Wim van Binsbergen in his book, *Before the Presocratics. Cyclicity, transformation, and element cosmology: The case of transcontinental pre- or protohistoric cosmological substrates linking Africa, Eurasia and North America* (2009-2010), which I shall use as my main point of contact with his extensive work, mentions that he does not always feel equipped to deal with entire correlative systems (2009-2010, 31) and, for his purpose in tracking specific features as they make the transitions between continents, it is not essential that he should be so equipped. The range he exhibits in his interesting explorations of whole networks of cosmological stories is already vast (see, *e.g.*, the useful survey in van Binsbergen, with Isaak 2007), without taking everything on board. However, for many purposes it is valuable to consider cosmologies as wholes and that is what I shall be doing in this article as in a number of previous studies (see, *e.g.*,Lyle 1990,2012abc, 2013, 2014ab, 2015ab). The four elements that van Binsbergen is concerned with can be argued to relate to the four levels of a universe consisting of an upper sky or heaven (corresponding to air), a lower sky (corresponding to fire), earth, and sea (corresponding to water). This sequence differs from that in Aristotle which runs: fire, air, water, earth, but this Aristotelian sequence can be corrected back to a probable pre-scientific cosmological sequence, as listed above, in the light of comparison with the sequence of the humours (Lyle 1990: 149-152; Dijksterhuis 1961: 22-34). Van Binsbergen says (2009-2010: 174):

> [I]t is central to our present argument to ask whether the Presocratic four-element doctrine was a Greek invention, or was merely the surfacing, in Western Eurasia, of a much more widespread and much older cosmological substrate. The latter view, of course, is the one advocated by [this book].

It is also the one endorsed in this article, which understands archaic cosmologies with correspondence systems including elements to have been in existence millennia before the Greek Presocratic philosophers. It can be argued that the main difference that came in with the Presocratics whom van Binsbergen takes as his focus is that they were thinking in terms of a real 'as is' scientifically present world, while the cosmologies can be regarded as fictive 'as if' worlds which hung together as integrated systems but might or might not have been believed as true (Lyle 2013: 95-97).

In the 'as if' world of the proto-Indo-Europeans, we can see everything coming into existence in terms of a theogony with ten gods. I shall use the Greek gods for reference here in the main argument. The first of the gods is the primal goddess, Ge, the earth, who gives birth to Uranus, the heaven. The pair generate Cronus (taken to correspond to lower sky), and, according to the Lyle model though not stated in Hesiod, they also generate Poseidon at this point (sea, water). These are the old gods of the cosmic levels who are to be distinguished from the set of young gods. In Hesiod, the young gods arethe six born to Rhea and swallowed by Cronus, the last of whom is Zeus. Treating the matter more broadly in the Indo-European context, the six young gods can be identified as the king (Zeus or Indra), and his rival, a king of the dead (Typhon or Yama) and another brother, plus his queen and her two brothers (the twin Dioskouroi or Aśvins) (see Lyle 2012a).

As regards the correspondences between Indo-European and Chinese cosmologies, in both cases there is a developmental sequence. In the Chinese case, the supreme pole generates the two exemplars (*e.g. yin* and *yang*) which generate the four images (*e.g.* the seasons) which generate the eight trigrams (Graham 1986: 68), that is to say that there are four stages that correspond to the generations in the less abstract Indo-European formulation. In the Chinese case, equivalents of the Indo-European gods are identifiable in the ten *gan* or Heavenly Stems, the central two of which relate to the centre while the remaining eight relate to the four directions and seasons and to the eight trigrams. The *gan* are known from the time of the Shang in the second millennium BCE (Chang 1978: 13; Keightley 18-19, 25-26), and they occur in yang (bright) and yin (dark) pairs in the order: Jia, Yi; Bing, Ding; Wu, Ji; Geng, Xin; Ren, Gui. A correspondence between the *gan* and the Indo-European gods can be made out on the basis of the structures and a few external indications, as discussed in the quotation below (Lyle 2014a: 213-214):

The central two in the [*gan*] series, Wu and Ji, relate to the centre and correspond to the two kings who have this position. They can be ignored in terms of the four-season cycle, and attention can be paid next to the second-last pair, which is anomalous in two ways. When the *gan* are attached as the posthumous names of the Shang kings, they fall into two sets which are headed by Yi and Ding, except for these two, Geng and Xin, which can belong to either party. Geng and Xin are also absent from the series of legendary predynastic kings which includes the other six in this series (Chang 1978; Chao 1982, pp. 11-18, 111; Keightley 2000, pp. 132-133). In the light of my work on the Indo-European parallel, I explain the different nature of Geng and Xin as a gender one, and interpret them as the two goddesses in the system. It can then be proposed that the three pairs remaining, when this pair is excluded, form a male triad of the kind that Georges Dumézil located in the Indo-European context, which related to: (1) the sacred, (2) physical force, and (3) prosperity and fertility.

The order is largely a hierarchical one and runs 1: Jia, Yi; 2: Bing, Ding; King:

Wu, Counter-king: Ji; Females: Geng, Xin; 3: Ren, Gui. The Indo-European pattern of the ten gods has the correspondences: 1 heaven, priest; 2 lower sky, warrior; king, counter-king; primal goddess, queen; sea, cultivator. There are no ambiguities here, but a straightforward set of parallels. This schema applicable to both Indo-European and Chinese materials has only recently been made available in full (Lyle 2014a, but *cf.* Lyle 1990, 105-155, for a partial treatment), and has yet to be debated by specialists in both subject areas. However, the schema can be taken as a point of departure for comparison with the material from Ancient Egypt, which has the potential to enrich our understanding of all the cosmologies concerned.

In both the Indo-European and Chinese cases, it can be presumed that the old-god sequence begins with the primal goddess who is responsible for the series of births that results firstly in the creation of the cosmic levels of heaven, lower sky, and sea, and secondly of the king and queen and their brothers. In the Chinese case, the creation sequence can be shown in terms of the trigrams, with the three levels (from bottom up) of the yin/yang pairs of above / below, dry / wet, and dark / light (Lyle 1990: 139). The trigrams can be laid out in a way that accords with this sequence of division. This is not found in the established set of trigrams explored by van Binsbergen which is not informative in the same way about the cosmology. From the point of view of tracking developments, however, the fact that the established set of eight is a culturally learned arbitrary one can be an asset, and van Binsbergen has found it useful in linking China and West Asia (2009-2010, 234-252).

In approaching the highly elaborated Egyptian system of thought which had currency over a period of two millennia (Troy 1994, 3), it is particularly important to stress that comparison should not be seen as reductive. Each of the systems under review has its own richness. I shall be comparing one component of the Egyptian system – the Heliopolitan Enneador 'Nine Gods'– with the model based on Indo-European and Chinese sources. We can begin by looking at the shape of the Ennead from origin to culmination (Troy 1994, 5-12). The first five gods represent a cosmology – the primordial substances of creation (Frankfort 1946, 182).

The sequence of the Ennead of Heliopolis is well established. It consists, after the abyss, Nun, of the creator god, Atum. After this androgynous figure masturbates, the twins Shu and Tefnut are brought into existence. Following them are Nut, the female sky, and Geb, the male earth. Geb and Nut are the parents of the young gods who form the gendered pairs: Osiris and Isis, Seth and Nephthys. This sequence establishes a set of nine consisting of five old gods and four young gods, and to the young gods is added Horus, either in the next generation as the young Horus or within the same generation as Osiris as the elder Horus, creating a tenfold set. The diagram by van Binsbergen shows both the Horus alternatives (2009-2010, 130). In either case, Horus is bound to Seth in an oppositional relationship. As Henri Frankfort notes (1946, 21-22), these two gods are partners, who are 'the antagonists per se – the mythological symbols for all conflict'. They mirror the Indo-European concept of a positive divine king of the living opposed by the negative counter-king, who rules the realm of the dead. Frankfort describes the entire Ennead and its functioning in the following way (1946, 183):

Thus the Ennead was formed out of the five cosmic gods and Osiris with the three gods of his circle. Here we hold the clue to its meaning; it was a theological concept which comprised the order of creation as well as the order of society. It is peculiar to the Egyptian concept of kingship that it envisaged the incumbent of that office as part of the world of the gods as well as of the world of men. If Osiris was the son of Geb and Nut, he was also the dead king in Egypt. And if Horus, the living king stood outside the Ennead, he was yet the pivot of this theological construction. Horus lived perennially in each king. Hence each king at death receded before his successor and merged with Osiris, the mythological figure of 'king's father'; the power that had been guiding the state sank back into the earth and from there continued its beneficial care of the community, shown in the abundance of the harvest and the inundation. Egyptian views of great antiquity, rooted in African beliefs, find expression in this conception of the dead king's future life.

There are wide differences from the Indo-European / Chinese pattern, but there are also marked similarities which encourage us to think that there is the same underlying structure and that we can usefully explore it in terms of modifications that were made to it.

The similar facets are the following. In all three cases the king is centrally important and is the key component that comes at the close of the sequence. He is associated with other young gods, one of whom is his antagonist. The set of young gods is preceded by a set of cosmic gods who are generated over two generations from an androgynous primal source. The androgynous source in the Indo-European case is the female, Ge, who bears the male within her as can be seen when she first gives birth and produces the male sky. The complex figure of Atum has female as well as the more overtly present male attributes, as Lana Troy explains (1986, 16):

Atum creates using his phallus. This act of creation has a special male reference. The explicit masculinity of Atum's act does not, however, exclude the feminine element from the birth of the twins. The androgynous role of the god is recognized in the texts as characteristic of the creator. [...] The masculine element of Atum's creative act is clearly indicated in terms of the manner in which it takes place. If however the texts are closely examined, the dualism of male and female are discerned as present in the description of the creation of the twins.

The Egyptian sequence stresses male/female complementarity at every level after Atum: She / Tefnut, Geb / Nut, Osiris / Isis, and Seth / Nephthys, and this gender complementarity can be seen as a specifically Egyptian feature. It can accordingly be suggested that the system was modified in this direction in the Egyptian context, and that the places of some of the male figures in a proto form of the cosmology were taken by females. In the set of old creation gods, we can consider the pair of Shu and Tefnut, of whom Shu (male) is the dry component and Tefnut (female) is the wet one in a dry / wet complementarity that is also typical of Egypt like the male / female one (Troy 1994, 38-39). The Greek god equivalents in the model are Cronus (lower sky) and Poseidon (sea). We can quite readily posit a modification by which the male god of the sea, partnered with lower sky, becomes the wet female partner of Shu, who is also the lower sky or atmosphere since he spans the area between earth and heaven when he holds up the heaven (Troy 1994, 6-9). By these correspondences, in the terms discussed above, Shu would relate to the element fire and Tefnut to the element water, while Nun and Geb would relate to the elements air and earth.

This leads into the question of the identities of earth and heaven, for clearly there is a gender switch with Nun being a female heaven and Geb being a male earth by contrast with the male Uranus as sky and the female Ge as earth in the Greek context. It is worth observing, however, in relation to the story of the separation of earth and sky, that it does not matter which of the two is above and which below. The point seems to be that the generating power of the cosmos comes to an end with the birth of the children who have been conceived before this separation (Lyle 2012a, 64-68, 125 n55; Pinch 2004, 174). In the Indo-European case, there is the violence of a castration (Lyle 2012, 103-106) while in the Egyptian case, Shu simply holds sky and earth apart. The pregnant female then proceeds to the multiple birth of the young gods. When this birth is connected with the five epagomenal days (extraneous to the 360 days of the regular 30-day months) five young gods are born (Lyle 1990, 110-112; Griffiths 1970, 134-135, 291-294) and this is an instance where Horus is added to Osiris and Isis, Seth and Nephthys, so supplying the expanded tenfold set of gods and not simply the nine of the Ennead.

In conclusion, it can be said that the cosmology found in the Ancient Egyptian, Chinese and Indo-European traditions falls into two parts. The first of these covers three generations of cosmic gods and the second treats the young gods centred on a king. The king is experienced as uniquely important. In the Egyptian case, he forms the link between humans and gods by being both the current living king and the god Horus. In the posited Indo-European structure, all ten gods are understood as having human equivalents within a four-generation structure, which presents three generations of ancestors as corresponding to cosmic gods and a final generation as that of the king and his contemporaries. In the Chinese case, the structural components are particularly strongly marked. The yin or dark aspects correspond to the old or cosmic gods. The yang gods are centred on the centrally placed king who has as antagonist a young yin god, these corresponding to the antagonists, Horus and Seth. Despite differences of emphasis, it seems that the Egyptian, Chinese, and Indo-European schemes have structures in common which probably imply common origin, and which can definitely open the way to viable cross-cultural comparison.

References cited

Chang, Kwang-chih, 1978, 'T'ien kan: a key to the history of the Shang', in Roy, David T. and Tsien, Tsuen-hsuin, eds, *Ancient China: Studies in Early Civilization*, Hong Kong: Chinese University Press, pp. 13-42.

Chao, Lin, 1982,*The Socio-Political Systems of the Shang Dynasty*, Monograph Series No. 3, Nankang, Taipei, Taiwan: Institute of the Three Principles of the People, Academia Sinica.

Dijksterhuis, E. J., 1961, *The Mechanization of the World*, Oxford: Clarendon Press.

Frankfort, Henri, 1948,Kingship and the Gods: A Study of Ancient Near Eastern Religion as the Integration of Society and Nature, Chicago: University of Chicago Press.

Graham, A. C., 1986,Yin-Yang and the Nature of Correlative Thinking, Singapore: Institute of East Asian Philosophies, National University of Singapore.

Griffiths, J. Gwyn, ed., 1970, Plutarch's 'De Iside et Osiride', Cardiff: University of Wales Press.

Keightley, David N., 2000,The Ancestral Landscape: Time, Space, and Community in Late Shang China (ca. 1200-1045 B.C.), Berkeley: Institute of East Asian Studies, University of California, Berkeley and Centre for Chinese Studies.

Lyle, Emily, 2010, 'The Cosmological Theory of Myth', in: van Binsbergen, Wim M. J. and Venbrux, Eric, eds, New Perspectives on Myth. Proceedings of the Second Annual Conference of the International Association for Comparative Mythology, Ravenstein (the Netherlands), 19-21 August, 2008,Haarlem: Shikanda, pp. 267-277.

Lyle, Emily, 2012a, Ten Gods: A New Approach to Defining the Mythological Structures of the Indo-Europeans, Newcastle upon Tyne: Cambridge Scholars Publishing.

Lyle, Emily, 2012b, 'Stepping Stones through Time', Oral Tradition, 27, 1: 1-11,electronic.

Lyle, Emily, 2012c, 'Entering the Chimeraland of Indo-European Reconstruction',Retrospective Methods Network Newsletter, . No. 5, December: 6-10,electronic.

Lyle, Emily, 2013, 'Defining the Religion that Lay behind the Self-Colonization of Europe', in: Cox, James L., ed.,Critical Reflections on Indigenous Religions, Farnham, Surrey and Burlington VT: Ashgate, pp. 93-101.

Lyle, Emily, 2014a, 'The Correspondences between Indo-European and Chinese Cosmologies', in: Antoni, Klaus, and Weiss, David, eds, Sources of Mythology: National and International Myths, Berlin and London: LIT Verlag, pp. 209-217.

Lyle, Emily, 2014b, 'The 'Order, Chaos, Order' Theoretical Approach to Reconstructing the Mythology of a Remote Past',CosmosI, 30: 37-48.

Lyle, Emily, 2015a, 'The Hero Who Releases the Waters and Defeats the Flood Dragon', Comparative Mythology 1: 1-12, electronic.

Lyle, Emily, 2015b, 'The Cosmic Connections of the Eight Key Points in the Indo-European Ritual Year',The Ritual Year 10: 9-27.

Lyle, Emily,1990, Archaic Cosmos: Polarity, Space and Time, Edinburgh: Polygon.

Pinch, Geraldine, 2004, Egyptian Mythology: A Guide to the Gods, Goddesses, and Traditions of Ancient Egypt, Oxford: Oxford University Press.

Troy, Lana, 1986, Patterns of Queenship in Ancient Egyptian Myth and History, Boreas 14, Uppsala: Acta Universitatis Upsaliensis.

Troy, Lana, 1994, 'The First Time: Homology and Complementarity as Structural Forces in Ancient Egyptian Cosmology', Cosmos 10: 3-51.

van Binsbergen, Wim M. J., with Mark Isaak, 2007, 'Transcontinental Mythological Patterns in Prehistory',Cosmos 23: 29-80.

van Binsbergen, Wim M. J., 2009-2010, 'Before the Presocratics. Cyclicity, transformation, and element cosmology: The case of transcontinental pre- or protohistoriccosmological substrates linking Africa, Eurasia and North America', Quest: An African Journal of Philosophy / Revue Africaine de Philosophie vols. 23-24, nos 1-2.

The traditions of Gulfeil

Projection of Israelite-Assyrian history on the local conditions

by Dierk Lange

Introduction

Situated on the River Chari about 60 km south of Lake Chad, the small Kotoko town of Gulfeil is blessed with rich traditions pointing to an eventful past. Dealing apparently with the history of the townspeople on the local site, the available oral accounts were provided by three informants from different quarters of the town.[1] Two of the informants were descendants of the primordial ancestor Mamba and priests of the deities Garé and Gara, and the third was an indistinctive old man.[2] From published summaries it appears that the narrative sources are in chronological order and allow us to trace the presumable history of the town from the beginning until recent times. Taken at face value, the reconstructed history refers to the founding of the town by Sao immigrants, the extinction of the first inhabitants of the town, the arrival of a second group of migrants, also belonging to the Sao, the second founding of the town, the later conquest of the town which precipitated the flight of most of its inhabitants,

---

[1] Lebeuf / Rodinson 1948 ; Lebeuf 1969.

[2] Lebeuf / Rodinson 1948.

and the conversion of the remaining inhabitants to Islam, leading to the loss of their identity as Sao.[3]

Far less comprehensive is the Gulfeil King List (henceforth GKL), written in Arabic, which is presently only available in the form of a rough French translation. Offering the names of sixty kings who supposedly ruled over the town before 1900 – including a gap of five missing names following the first king – it is of a non-narrative, somewhat archival character. It is divided into two sections of which the first, extending over the reigns of twenty-two kings but comprising (on account of the omissions) only seventeen names, is said to deal with non-Muslim Sao kings.[4] This first section of the GKL, like that of Kusseri, is separated from the second by some words distinguishing between the 'pagan' and the Muslim kings. In the list of Makari the distinction between pagans and Muslims can only be inferred from oral traditions. In addition, we note that five of the seventeen 'pagan kings' stand out from the others by the epithets 'the great' or 'the small' added either orally in Kotoko as *enduma* and *swcayre*, or written in Arabic as *al-kab3r* and *al-sag3r*.[5] We also observe that three of the pre-Islamic kings are called by the Arabic name Ahmad and that in fact only five of the first thirty-two so-called 'Muslim kings' bear Arabic names.[6] It would therefore appear that the distinction between 'pagan' and 'Muslim' kings cannot possibly reflect the difference between the situation prior to and posterior to the introduction of Islam into the society of Gulfeil, perhaps in the eighteenth century. In spite of the scarcity of information it contains and the seemingly incomprehensive nature of its names, the GKL constitutes an important document for the legitimacy of political rule. In 1948 a descendant of the last nineteenth-century Kotoko ruler produced a version of the list omitting all the kings of the Sao period and also the first non-Sao ruler, Duna (23). Trying to bolster his claims to kingship in Gulfeil by this list, the contender to the throne apparently tried to use non-descent from the Sao kings as a decisive argument

---

[3] Lebeuf / Rodinson 1948; Lebeuf / Masson Detourbet 1950; Griaule / Lebeuf 1951 ; Lebeuf 1969.

[4] Lebeuf / Rodinson 1948 ; Lebeuf, 1969 ; A. Masson Detourbet (later A. Lebeuf) obtained a king list in 1948 which omits all the 22 Sao kings and also Duna (23) and begins with Madu (24) (Lebeuf / Rodinson 1948: 39-40).

[5] Lebeuf / Rodinson 1948 : 36 n. 5, n. 11. Lukas notes *dwmde* 'great' and *tá'bu* 'small' for the dialect of Logone Birni (*Logone-Sprache*, 135, 138).

[6] Lebeuf / Rodinson 1948 : 33-40; Lebeuf 1969 : 78.

for his own claims to succession.[7] However, since, in addition to the Sao kings, he also omitted the first non-Sao king, it seems that he was concerned with more than just the issue of the Sao.

Though transmitted in chronological order, the information included in the oral legendary accounts and the written king list is at first sight difficult to interpret.[8] In the absence of any valid synchronism, even a combination of these different types of evidence would have been of limited use. However, since the material can be shown to reflect precise developments in ancient Near Eastern history, the available data acquire an unexpectedly clear historical meaning. In the light of previous research on other dynastic traditions of Central Sudanic kingdoms, we realize in particular that the traditions of Gulfeil offer an outline of Israelite-Assyrian history beginning with the independent kingdom of Israel, continuing with the Neo-Assyrian domination of the Fertile Crescent and ending with the Babylonian conquest.[9] From these elements it can be inferred that the period of the Sao corresponds to a limited period of ancient Near Eastern history to the exclusion of all others.

First episode: Migration of Israelite patriarchs

In Gulfeil, there are three different traditions of origin which deal with the arrival of the migrants from a distant country in the north or the north-east situated beyond the desert to the region of Lake Chad.[10] According to the first, the migrants were only three brothers, Mamba, Teri and Abrabimo, who were followed by their wives.[11] The second version mentions five migrants, Abra-Shemshem, his two twin sisters, Sodo (Chodo) and Modo, his sister Udme Massassaryo and the latter's daughter Sao Udme Mashilanga.[12] The third version

---

[7] Lebeuf / Rodinson 1948 : 39-40.

[8] Griaule / Lebeuf 1948 : 4-6; Lebeuf 1969 : 67-70.

[9] Lange, 2011: 11-18; 2012:158-164; 2009: 369-375; 2011: 584-591.

[10] Griaule 1943 :86; Boulnois 1943 :99.

[11] Griaule 1943 : 86. Boulnois refers this version to Griaule and Lebeuf ('Migration', 99 n. 2).

[12] The names Abra-Shemshem (French: Abra Chemchem) and Massassaryo are noted as Abra çwmçwm and Abra *chêmchêm* and Maçaçaryo and Massassaryo in Lebeuf / Rodinson 1948: 36,

refers to the migrants as a male and a female hero, Mamba and Sao Udme Mashilanga, and two further male heroes, Sodo and Modo.[13] In the three available publications, these three versions are recorded in varying sequences and obviously the local informants do not know which group arrived first and which last.[14] Moreover, it should be noted that another set of traditions attributes the actual founding of the town either to Mamba or to Sao Udme Mashilanga, the younger sister of Mamba.[15] Probably the accounts reflect the opinions of the putative descendants of the major figures who belong to distinct families living in different quarters of the town.[16]

In the GKL, the first king is called Mamba (Ar. *Mımba*). He is followed by five other figures whose names are said to have been forgotten. The 'forgotten names' may have consisted of those of Mamba's 'brothers', Teri and Abrabimo, and they may have included the name of the important figure of Oumar (4) mentioned in the narrative accounts in the context of the sacrificed and immured daughters.[17] The note concerning the forgotten names clearly shows that the GKL was for some time transmitted orally.

Some light can be thrown on the meaning of the presumed migration of the Sao to the region of Lake Chad by the attempt to identify these figures. The first legend of provenance which came to the note of modern scholars is that of Mamba, his two brothers and their wives, invaders from the far north whose huge bodies are said to have been reduced to normal size by their crossing of the desert.[18] On the basis of the biblical Table of Nations we realize that Mamba (1) is probably a distorted form of the name of Eber, designating the fourteenth

---

and Griaule / Lebeuf 1948:5 n. 1).

[13] The name Sodo (French: Chodo) is noted as *codo* in Lebeuf / Rodinson 1948: 36, and as *sodo* in Griaule/Lebeuf 1948: 5 n. 1, and the name Sao Udme Mashilanga (French: Sao Oudmé Machilanga) as Sao *udwme macilanga* Lebeuf / Rodinson 1948: 35 n. 2, and *machilânnga* in Griaule / Lebeuf 1948: 5 n. 1.

[14] Lebeuf / Rodinson 1948 : 36; Griaule / Lebeuf, 1945 : 5 n. 1; Lebeuf 1969 : 68. Griaule has only the version of Mamba-Abrapémon and Boulnois that of Abrachamcham (Boulnois 1943: 99-100).

[15] Lebeuf 1969 :68.

[16] Griaule / Lebeuf, 1948: 6, 12-13.

[17] Griaule/Lebeuf, 1948 : 6; Lebeuf 1969 : 69.

[18] Griaule 1943 : 86-87; Boulnois 1943 : 99 n. 2.

patriarch after Adam in the biblical counting.[19] As the eponymous ancestor of the Hebrews, Eber/Mamba is perfectly suited to be considered as the leader of the migration of Israelites. His descendants live nowadays in Guéréndouma, the eastern quarter of the town in which is located the king's palace.[20] He is followed by Teri (2) whose descendants live in the two southern quarters of the town and in whom we recognize Terah, the father of Abraham, whose name also designates a locality near Harran on the Balikh River.[21] Abrabimo (3) or Abrapemon whose descendants live in the two northern quarters of Gulfeil seems to be the name of Abraham with a second element, -bimo perhaps related to the second syllable of Abra-ham, distinguishing the name from Abra-Shemshem.[22]

These different Israelite figures are well known from the biblical accounts of Genesis. If we consider the city-state of Gulfeil as a microcosm of ancient Israel, Abra-Shemshem's description as the major founding figure of the polity correlates quite well with the biblical information. As for Abrabimo / Abrapemon his duplication of Abra-Shemshem possibly refers to two different groups of immigrants, both claiming descent from the Israelite patriarch Abraham. Also observable in other kingdoms of the Central Sudan and in particular the written chronicle of the neighbouring Kanem-Bornu, the D3w1n,[23] the projection of Israelite traditions on the local conditions is particularly striking for Gulfeil. It testifies to the importance of Israelite elements among the founders of the founding groups of the small city-state south of Lake Chad.

Yet, Abra-Shemshem, contrary to the dynastic figure Abrabimo, is not an isolated hero. His complex and widely known tradition connects him with several women and thus emphasizes the female factor in Gulfeil history. It has Abra-Shemshem as the great leader of the migration, attributes to him twin sisters, Sodo and Modo, who are sometimes also considered as his wives, and in addition ascribes to him another sister, Udme Massassaryo, and the latter's daugh-

---

[19] Gen 10:24-25; Hess 1898; Freedman 1992: 260; Liverani 2005: 242.

[20] Griaule / Lebeuf 1948: 12-13; Lebeuf / Detourbet 1950 : 79.

[21] Gen 11:24-32; Hess 1898; Liverani 2005: 387-8.

[22] Griaule 1943 : 86; Griaule / Lebeuf 1948 :12-13.

[23] Lange 1977 : 22-23; 65; id., 2009 : 588-597.

ter, Sao Udme Mashilanga.[24] According to a related version of the tradition, Mamba was accompanied by the heroine Sao, also called Udme Mashilanga, and he further had two male companions, Sodo and Modo. In this case the main hero is not Abra-Shemshem but Mamba, who was accompanied by Sao Udme Machilanga and Sodo and Modo, the latter being described as men and not women.[25] Apparently, the third legend corresponds to a combination of the first and the second. The group of migrants is said to have brought with them, or to have found on the local mound, the regalia of the future polity: an iron helmet, an iron missile, a stick of authority, a Quran in a leather bag and a spear.[26]

On account of the five 'forgotten names' at the beginning of the GKL, we cannot expect to find any of these names in the list. Estimating, however, that Mamba and Abra-Shamsham are both figures in the early history of the Gulfeil people, it is certainly not an unreasonable assumption that apart from Mamba (1) there was also Abra-Shemshem or a related name designating the same figure among these names, which the dynastic tradition of Gulfeil failed to transmit.

In Abra-Shemshem and his two twin sisters and wives, Sodo and Modo, we recognize Abraham, his wife Sarah and his concubine Hagar. The first element of the name Abra-Shemshem is apparently an abbreviated form of Abraham – comparable to the ancient form *Abram* mentioned in Gen 17:3. The second element consists perhaps of *Yzm* (Hbr. 'Shem'), designating the eleventh patriarch, Shem the son of Noah, and the third element may be *Shɪm* (Ar. and perhaps Aram.) referring to 'Syria'.[27] Abraham, the 'Father of many nations' (Gen 17:5), the Semite from *Shɪm*/Syria, would indeed be an appropriate designation for the emblematic – though fictive – leader of an exodus from Syria-Palestine comprising members of various national communities. Sodo and Modo, the two sisters of Abra-Shemshem, who are sometimes even considered as the hero's

---

[24] Boulnois 1943 : 99-101. According to other authors, Sodo and Modo were twin sisters but not wives of Abra-Shemshem (Lebeuf / Rodinson 1948: 34-36; Griaule / Lebeuf 1948: 5 n. 1; Lebeuf 1969: 68).

[25] Griaule / Lebeuf 1948 : 5 n. 1; Lebeuf 1969 : 68.

[26] Boulnois 1943 :105-8; Griaule / Lebeuf 1948 : 5, 5 n. 6; Lebeuf 1969 : 273-4.

[27] Wehr 1979: 525.

wives, seem to correspond to Sarah and Hagar, the wife and the concubine of Abraham.[28] Sarah was the cousin of Abraham, but on two different occasions Abraham pretended that Sarah was his sister.[29] Hagar was the handmaid of Sarah who was given by her mistress to Abraham in order to provide him with a child. By their sons, Ishmael and Isaac, the two women became the ancestresses of many nations.[30] In biblical terms it is therefore quite plausible that, besides Abraham, the Gulfeil tradition of provenance also mentions his two wives, and that it emphasizes their close connection by designating them as twins.

In addition to Abra-Shemshem / Abraham, Sodo / Sarah and Modo/Hagar, the tradition mentions Udme Massassaryo as the third sister of Abra-Shemshem, and it attributes to Udme Massassaryo a daughter called Udme Mashilanga.[31] The element Udme in both names being etymologically explained by Kot. *u dw me* (or *u dê me*) – 'daughter of the king',[32] both female figures seem to be associated with male political power. Placed in a somewhat marginal position with respect to the biblical figures of Abraham, Sarah and Hagar, Udme Massassaryo and her daughter Udme Mashilanga do not appear to belong to the biblical tradition.

With respect to Udme Massassaryo, we note that her name seems to be related to that of *'ayyûr* (Hebr.) or *ay-yur* (Akk.) designating the patron deity of the city of Assur, the state deity of Assyria and by extension the Assyrian state itself.[33]To *Ayyur* was added either a prefix *ma-*, having a localizing function in various Semitic languages, or the term *mât* as in Akk. *mât ay-yur* 'land of Ayyur'.[34] Such an assumption is supported by the phonetic closeness of the name Massassaryo to the name of Assur – *Ayyur > (M)assass(a)r(yo* – and by the legend of Semiramis which seems to have been widely known as the founding tradition of

---

[28] *Cf.* Lange 2009: 589-597; 2012: *144-6*.

[29] Gen 12:13 (Pharaoh); Gen 20:2 (Abimelech).

[30] Gen 16:5; E. A. Knauf, 'Hagar', ABD, III, 18-19.

[31] Boulnois 1943 : 99-110; Lebeuf/Rodinson 1948 : 35-36; Griaule / Lebeuf 1948 : 5 n. 1.

[32] Lebeuf / Rodinson 1948 : 35; Griaule / Lebeuf 1948 : 5 n. 2.

[33] Boulnois writes Oudmé Massario and Oudmé Maçario (Boulnois 1943 : 100-1), Lebeuf / Rodinson and Griaule / Lebeuf write Oudmé Maçaçaryo (Lebeuf / Rodinson 1948 : 36) and Massassaryo (Griaule/ Lebeuf 1948 : 6) and Lebeuf writes Oulmé Massassaryo (Lebeuf 1969 : 68).

[34] Gesenius 1842: 149; von Soden 1952: 64.

Assyria.[35] Building on the legend of Semiramis but distinguishing between the founding heroine and her daughter, Semiramis / Sammuramat, Udme Massas-saryo therefore appears to be the female personification of the legendarized national deity Assur. In the Table of Nations we find with respect to the sons of Shem five collaterals, the great-grandfather and four great-uncles of Eber: Elam, Asshur, Lud and Aram were the eponymous ancestors of Elam, Assyria, Egypt and Aram-Damascus, while Arphaxad, mentioned between Asshur, Assyria and Lud, Egypt, was the great-grandfather of Eber, the eponymous ancestor of the Hebrews. An association of Massassaryo / Asshur with biblical patriarchs is accordingly quite in line with Israelite genealogical thinking.

Originally called Sao and thus singled out by the same name as the great ances-tral people of the Kotoko, Udme Mashilanga is one of the key figures in Gulfeil traditions.[36] Sao or Udme Mashilanga (Kot 'daughter of the king', 'never dry') is apparently the eponymous ancestress of the Sao, and therefore the tracing of her prototype might provide important evidence for solving the problem of the Sao. In order to determine her identity, we have first to consider the onomastic context in which she is situated by the different traditions of provenance. In four of the five traditions she is mentioned in connection with biblical figures: she came to the region of the Lake Chad basin together with Mamba, Eber, she is associated with the three patriarchs, Mamba, Eber, Teri, Terah and Abrapemon, Abraham, she is adjoined to her uncle Abra-Shemshem, Abraham, her twin aunts, and her mother Udme Massassaryo, and she is related to three men, Sodo, Sarah, Modo, Hagar and Mamba/Eber.[37] In spite of her marginal position in the traditions of provenance, founding traditions of the town attrib-ute a major role to her: though sometimes in agreement with the GKL they describe Mamba, Eber as the only founder of the town, most traditions depict her as the heroine who planted the spear on the mound near the River Chari and thus single her out as the founding heroine of Gulfeil.[38] Being distinguished from the biblical figures of Shem (11), Eber (14), Terah (19), Abraham (20), Sarah

---

[35] Olmstead 1923: 158-9.

[36] Lebeuf / Rodinson 1948 : 34; Lebeuf / Detourbet 1950 : 45; Griaule / Lebeuf 1948 : 5.

[37] Lebeuf / Rodinson 1948 : 34-36; Griaule / Lebeuf 1948 : 4-5; Lebeuf 1969 : 68.

[38] Griaule 1943 : 86; Boulnois 1943 : 100-1; Lebeuf / Rodinson, 1948 : 36; Lebeuf / Detourbet 1950 : 45; Lebeuf 1969 : 68.

and Hagar by their marginal position, Sao Udme Mashilanga and her mother Udme Massassaryo seem to belong to a non-Israelite ethnic tradition. If it is correct to consider her as the eponymous ancestress of the Sao, then only her own descendants and not those of the biblical patriarchs can properly speaking be designated as Sao.

Sao Udme Mashilanga resembles by her name, her female gender and her description as an African founding heroine, Saw (10), the last pre-Islamic figure in the Kanem-Bornu dynastic tradition.[39] On account of this resemblance and the Israelite-Assyrian onomastic context of her appearance in the Gulfeil traditions she seems to correspond to Semiramis, the legendary founding heroine of the Assyrian Empire.[40] In view of the title Ligonge, Lugal designating the Assyrian kings enthroned in Babylonia in the Makari King List, the name *Mashi-langa* may perhaps be interpreted as a duplicated Hebrew-Sumerian name designating an anointed king (Hebr. *mašîh*, Messiah) and a ruler of Babylonia (Sum. *lugal* 'Lord'). The female character of both figures can be explained either as reflecting the direct repercussions of the legend of Semiramis, Sammuramıt, or as resulting from the influence of the underlying belief in the great female deity.[41]

As founding figures distinct from the biblical patriarchs, Udme Massassaryo and Sao Udme Mashilanga bear witness to the Assyrian encroachments on Israel. Just as Semiramis corresponds to a projection of the historical Sammuramat, the mother Adad-nirari III (810-783), into the distant past of Assyria, the association of Assyrian figures with Abraham may have resulted from the projection into the distant past of the Assyrian encroachments on Israel under Shalmaneser III in 841, Tiglath-pileser III in 734-733 and Sargon II in 722 BCE. Similar to the Sango episode in the dynastic tradition of the Oyo-Yoruba, this association could reflect the early invasion of Israel by Shalmaneser III.[42] Yet, on account of Sao Udme Mashilanga's association with a younger generation than Abra-Shemshem, Sodo and Modu, it seems indispensable to exclude Abraham and his contemporaries from the concept of Sao. Though already

---

[39] Lange, 2011 : 15; id. ; 1977 : 29-30, 67-68.

[40] Diodorus II: 4-6, 13-20; Roux 1992 : 301-2; Pettinato 1991 : 1-23.

[41] Pettinato 1991: 45-57.

[42] Lange 2004: 239-241; *id.*; 2011: 585-6.

mentioned in connection with Abraham, Sao Udme Mashilanga herself should be considered as a projection of the real 'Sao period', when her counterpart, Magira Dama, the sister of Baggari/Tiglath-pileser III, and hence a contemporary of the great conqueror, is supposed to have founded the city-state of Gulfeil for the second time.[43]

Second episode: The kingdom of Israel (884-722)

In certain aspects, the founding legend of Gulfeil is patterned on the legendary history of Israel. The most important figure in this respect is Mamba/Eber, a hero who is generally considered as the primordial ancestor of the kings of Gulfeil.[44] Non-Israelite heroic figures like Udme Massassaryo, and her daughter Sao Udme Mashilanga, may be thought to correspond to posterior juxtapositions resulting from later Assyrian encroachments on Israel. As for the great Israelite kings David and Solomon, their absence from the dynastic records of Gulfeil, as from other Central Sudanic traditions, can be explained by their post-seventh century BCE emergence as rulers of the United Kingdom in the historical tradition of Israel.[45] In their stead we find a figure called Oumar ('Umar) or Guémé Oumar, described as a descendant of Mamba/Eber, in the position of the first historic king of the people. Probably to be regarded as number six after the five forgotten kings omitted from the GKL, between Mamba/Eber (1) and Mashi (7), Oumar occupies an important position in the tradition. A recorded narrative account mentions him as a powerful survivor from the first town of Gulfeil, whose daughters were caught by the second group of conquerors and sacrificed.[46] Most likely he corresponds to Omri (884-873), the earliest important historical king of Israel, who is also mentioned in the tradition of the Oyo-Yoruba.[47] Although the dynasty founded by Omri ruled only until 841 BCE, his name and not that of David or Solomon appears in

---

[43] Lebeuf / Rodinson 1948 : 36; Griaule / Lebeuf 1948 : 5-6; Lebeuf 1969 : 68-69.

[44] Lebeuf / Rodinson 1948 : 34; Lebeuf 1969 : 68, 78.

[45] Liverani 2005 : 92-101; Garbini 1988 : 21-32.

[46] Griaule/Lebeuf 1948 : 5-6; Lebeuf 1969 : 68-69.

[47] Johnson 1921: 151-2; Lange 2011: 585.

Assyrian records as the designation for the Israelite ruling houseup to the end of the Israelite Kingdom in 722 BCE.[48] This outstanding position explains why he is remembered by Oyo-Yoruba and Gulfeil oral traditions as the eldest Israelite king.

In order to be able to construct a town wall as solid as those of the ancient abandoned cities on the Black Sea, far to the north, Abra-Shemshem;Abraham believed that he had to bury his own sister alive in a jar at the bottom of the wall. He took either one of his twin sisters Sodo, Sarah or Modo, Hagar or his sister Massassaryo, Asshur for this sacrifice.[49] After her metamorphosis into a monitor lizard, this animal became the totem animal of the town: no king is elected against the monitor's wish and in order to ensure its goodwill the king has to offer regular sacrifices and protect it against all injuries.[50]

By a strange shift from the sister to the mother, the kings of the town had to kill their mothers at the time of their enthronement. Mashi refused to comply with the age-honoured ritual practice and in consequence all the people of Gulfeil were wiped out by an epidemic and the town became an empty place.[51] Yet, in spite of the sacrilegious act, some versions of the tradition accord him the prestigious name, Mamat/Muhammad, inconceivable for an Assyrian ruler. Figuring in the seventh position of the GKL, Mashi (7) seems to have been a highly respected king on account of his second name and his brave resistance to an inhuman custom.[52] In all likelihood he is identical with the Israelite king Menahem (749-738), the eighth successor of Omri, having ruled longer than one year. In fact, Menahem was a usurper who seized power during the crisis of leadership in 749 and who hastened to pay tribute to Tiglath-pileser III (744-727) in exchange for acknowledgement as vassal.[53] With respect to the missing names at the beginning of the GKL, we may guess that the next three kings

---

[48]Luckenbill 1926-7 ; 1 Kgs 16:21-28; Liverani 2005: 108-9.

[49] Boulnois notes the hesitation of his informants with respect to the buried sister, either Udme Massassaryo or one of the twin sisters (Boulnois 1943: 101). Lebeuf mentions the live burial of one of the hero's sisters and calls the monitor Massassaryo (Lebeuf 1969: 68, 295-6).

[50] Boulnois 1943 : 101-2; Lebeuf 1969 : 69, 246.

[51] Lebeuf / Detourbet 1950 : 46; Griaule / Lebeuf 1948 : 5; Lebeuf 1969 : 68.

[52] Griaule / Lebeuf 1948 : 5-6; Lebeuf 1969 : 68.

[53] 2 Kgs 15: 19-20; Luckenbill 1926-7; Liverani 2005: 144.

after the generally accepted ancestor Mamba (1)/Eber, and before Mashi (7) / Menahem, were Teri(2),Terah, Abrabimo (3)/Abraham and Oumar (4)/Omri, while the two remaining ones were Israelite kings who reigned between Omri (884-873) and Menahem (749-738). The legend's attribution of the desertion of the town to Mashi's sacrilegious act is therefore in accordance with the loss of Israel's independence under Menahem. As the last king of independent Israel, he certainly deserves to be treated with special attention by the tradition of Gulfeil.

Contrary to the narrative account which has Mashi / Menahem as the last hero identifiable as an Israelite figure, the GKL mentions two supplementary kings, Semade (*Swmadw*) (8) and Ahmad (9) before Baggari-the-Great (10), known by oral tradition as the second great newcomer.[54] On account of Mashi's probable equation with Menahem (749-738), it is tempting to see in the two additional figures the last two Israelite kings, Pekah (736-732) and Hoshea (732-722), who ruled before the Assyrian conquest of Israel in 722 BCE. With respect to Pekah his entirely different name may be explained by the assumption of biblical scholars that this king adopted his name from his murdered predecessor, Peka-hiah (738-736), and that he might in fact have originally been known by a different name.[55] Since Pekah's father is said to have been called Remaliah, and considering the possible transformation of Remaliah to Semade, this hypothe-sized name may have been that of his father, or at least it may have comprised the second element *ben Remaliah* 'son of Remaliah': *(R)ema(l)i(ah)* > *(S)wma(d)w*. As for Ahmad, the name is equivalent to Muhammad, and can be considered as an appreciative designation for the last Israelite king, Hoshea, who suspended paying tribute to Assyria, and who was therefore imprisoned. At the same time, Samaria, the Israelite capital, was besieged and finally con-quered by the Assyrian army.[56]

By combining the available information on the history of the pioneer state founders of Gulfeil, we note that both the GKL and the narrative tradition of the town trace the newcomers to Israelite antecedents. Enumerating several

---

[54] Lebeuf / Rodinson 1948 : 36; Lebeuf 1969 : 78.

[55] 2 Kgs 15:23, 27.

[56] 2 Kgs 17:1-6; Liverani 2005: 145.

Israelite kings, and without indicating any geographical location, the king list suggests that the immigrant state builders looked back at the history of the state of Israel, extending from 884 to 722 BCE, as the beginning of their own history. In this respect the notable omission of the legendary kings David and Solomon provides precious evidence for the authenticity and the pre-exilic nature of the dynastic tradition. As for the country of origin of those who gave form to the GKL in the first place, it can hardly be doubted that it was either Israel or a country to which Israelites had been deported by the Assyrians.

Third episode: The 'Sao period' Assyrian rule over the ancient Near East (734-605)

According to the narrative tradition, a second group of migrants arrived in the town at a time when all of its first inhabitants had died. It comprised a woman, the Magira Dama, and four men, Baggari-the-Great, Madao, Alanga and Ald-jauna; Magira Dama was the sister of Baggari and the last two were brothers. The newcomers found that the former inhabitants had left the regalia, which included a Quran and a spear.[57] Magira Dama took the spear and planted it once more at the centre of the town, thus showing that she took possession of the locality in the name of the second group. Then the five leaders built their houses at the four cardinal points: Baggari in the north, Magira Dama in the east, Madao in the west and Alanga and Aldjauna in the south.[58] However, when Baggari wanted to build a bigger wall it fell down at each successive attempt. Taking advantage of the absence of Oumar at his farm, the newcomers seized his three daughters, Gara/Shawe, Garé and Windé, Massassaryo, who still lived in the town (in spite of its general state of abandonment), placed them alive in great jars and immured them successively in the walls of the town, which thereupon no longer collapsed. When Oumar came back from his farm and failed to find his daughters, he wanted to fight the newcomers, but he abstained from attacking them when he saw three monitor lizards who claimed to be his metamorphosed daughters. They were now the 'masters of the

---

[57] For photos of the regalia of Gulfeil, see Boulnois 1943: 106 Pl. I, 108 Pl. II.

[58] Lebeuf / Rodinson 1948 : 36; Lebeuf / Detourbet 1950 : 46; Griaule / Lebeuf 1948 : 5; Lebeuf 1969 : 68.

town'(*madêho*). After having been proclaimed the new king of the town, Baggari built his palace in the centre of Gulfeil.[59]

It should be noted that the Makari narrative tradition distinguishes in a similar way between two phases of the peopling of the town, interrupted by a period of void. In this case the first founding of the town is attributed to Ma Sugu, apparently an Assyrian leader, and the second to Musakala, probably an Israelite leader.[60] In spite of the similarity of the two founding movements separated by an interruption, we note that there is an inversion of the Israelite factor in the two traditions: while in Gulfeil the Israelite figure leads the first movement, in Makari it leads the second. Moreover, in Gulfeil both movements are part of ancient Near Eastern history, while in Makari only the first was an event in the Near East, the second pointing to the great exodus to West Africa and hence to the founding of the local city. In spite of these fundamental differences, the present analysis of the Gulfeil tradition corresponds to that of the Makari tradition insofar as the reference to the beginning of the history of the people in Israel is inconceivable without some kind of Israelite leadership, at least at the intellectual level, during the process of founding the Sao-Kotoko city-state south of Lake Chad. With respect to the second founding in the Gulfeil tradition, we have to consider the meaning of the story with respect to ancient Near Eastern history, as well as the local implications.

Contrary to Musakala in Makari, Baggari, the leader of the second putative group of migrants to Gulfeil, is not mentioned in the context of local but of ancient Near Eastern history. In view of the similarity of the names, we realize that he is the same as Baggari-the-Great, the tenth king in the GKL.[61] Following the three Israelite kings, Mashi (7)/Menahem (749-738), Semade (8)/Pekah (736-732) and Ahmad (9)/Hoshea (732-722), he most likely corresponds to a composite figure including Tiglath-pileser III (744-727) and Sargon II (722-705). While the first of these Assyrian kings was the great organizer of the Assyrian Empire who conquered the northern part of Israel in 734-3 and subjected the Israelite rump state of Ephraim and Menasseh to severe control, the second

---

[59] Griaule / Lebeuf 1948 : 6; Lebeuf 1969 : 68-69.

[60] Lebeuf / Detourbet 1950 : 48-49; Lebeuf 1969 : 61-64.

[61] Lebeuf / Rodinson 1948 : 36; Lebeuf 1969 : 78.

conquered Samaria in 722 and organized the deportation of numerous Israel-ites.[62] The name Baggari itself seems to be derived from that of Tiglath-pileser/*Tukulti-apil-Eyarra* by the following phonetic transformations: *(Tu)k(ulti)* > (Ba)gga-, and by inversion of the vowels *i and a (ap)i(l-Eya)rr(a)* > -ri. Mentioned in the Hebrew Bible as Tiglath-pileser and as Ful, the great As-syrian conqueror also figures prominently in the Makari King List, the *D3wın* and the Kebbi King List, and above all his name is probably reflected in the autonym of the Pullo;Fulani.[63] Therefore the second peopling of the town of Gulfeil under the leadership of Baggari, *i.e.* of Baggari-the-Great, most likely not only dramatizes the Assyrian conquest of Israel by Tiglath-pileser III and Sar-gon II, but also reflects the mass deportation of Israelites and the resettlement of Israel by other deported people.[64] By apparently commemorating this prac-tice of the Assyrian rulers, the Gulfeil tradition insists on the most oppressive aspect of interactions between the Assyrians and other people of the Ancient Near East.

With respect to the three male heroes in the second group of migrants men-tioned by the narrative tradition, Madao, Alanga and Aldjauna, we note that most likely they were Assyrian kings who succeeded to Tiglath-pileser III (744-727). On account of his association with the 'conversion to Islam', Madao should be considered as the youngest of the four heroes, in spite of his position after Baggari.[65] By reason of their being brothers, it is tempting to identify Alanga and Aldjauna with Shalmaneser V (726-722) and Sargon II (722-705), two kings who were the only brothers to reign in Assyria in the second half of the eighth and the first half of the seventh century BCE.[66] As the youngest of the four heroes, Madao seems to be equivalent to Esarhaddon (680-669), the organizer of the first Assyrian campaign against Egypt.[67] Though not entirely satisfying, the phonological resemblances between the names of the compan-ions of Baggari and those of the successors to Tiglath-pileser III can be visual-

---

[62] Bright 2000: 274-5; Liverani 2005: 144-5.

[63] 2 Kgs 15:19, 29; 16:7-10; Lebeuf 1969: 77; Lange, 2011 : 13; 2009: 374.

[64] Na'aman 1993 : 104-112; Younger 1998 : 201-227.

[65] Lebeuf/ Detourbet 1950 : 46-47; Griaule / Lebeuf 1948 : 5-6; Lebeuf 1969 : 68-69.

[66] Van de Mieroop 2007: 316-7; Saggs 1984: 92.

[67] Roux 1992: 327-8; Saggs 1984: 108.

ized as follows: Alanga equals Shalmaneser / *Yulmɪnu-ayar2d* V *(Yu)l(m)ɪn(u)-a(yar2d)* > (A)lan(g)a), Aldjauna corresponds to Sargon/*Yarru-k3n* II *(Y)arru-(k3)n* > Al(dja)un(a), and Madao is the same as Esarhaddon/*Ayyur-axa-iddina(Ayyur-ax)a-idd(in)a* > (M)da(o).[68] Some support for these equivalences is provided by further elements in the tradition of Gulfeil. The leading role of Baggari himself is emphasized by his installation of Aldjauna in his function as the first Me Galgué, the military leader of the town, and by his building of the palace.[69] At the same time it should be noted that Aldjauna's incarnation in the military leader of Gulfeil concords well with the warlike character of Sargon II.[70] The brotherhood of Alanga/Shalmaneser V and Aldjauna/Sargon II is materialized by the common residence of their putative descendants in the southern quarters of the town.[71] The insertion into the list of the relatively unimportant Alanga/Shalmaneser V (726-722), who is mainly known for his siege of Samaria, may be explained by the overall Israelite focus of the narrative tradition of Gulfeil.[72] By mentioning Aldjauna in addition to Baggari, the narrative tradition distinguishes at this point between Tiglath-pileser III and Sargon II and thus restricts the name of Baggari to Tiglath-pileser III. By having Baggari-the-Great (10) follow Semade (8)/Pekah (736-732) and Ahmad (9)/Hoshea (732-722), the dynastic tradition, by contrast, connects the figure with Sargon II (722-705). Therefore the two complementary traditions of Gulfeil confirm the double nature of Baggari/Baggari-the-Great. Both offer an Israelite perspective on the Assyrian expansion, but the dynastic tradition does this more clearly than the narrative tradition.

Information on Magira Dama, the important female figure in the group led by Baggari, seems to reflect a tradition of a more legendary nature. Though the second great heroine of Gulfeil tradition may have been the otherwise unknown mother of Tiglath-pileser III, she more likely corresponds to a figure of oral traditions. It is tempting to think of the wife of the famous substitute king,

---

[68] With respect to the derivation Esarhaddon/*Ayyur-axa-iddina* *(Ayyur-ax) a-idd (in) a* > (M) da (o) we note that the Hebrew form ('2*sar-hadd4n*) is closer to the Gulfeil form that the Akkadian form (A. Kirk Grayson, 'Esarhaddon', ABD, II, 574).

[69] Griaule/Lebeuf 1948 : 6, 7, 14; Lebeuf 1969 : 187.

[70] Roux 1992 : 310-7; Grayson 1975 : 86-102.

[71] Griaule / Lebeuf 1948 : 6, 13; Lebeuf 1969 : 69.

[72]; Roux 1992 : 310.

Damqi, from about 670 BCE, mentioned in the dynastic tradition of Makari, as Amsa Makiya, the wife of Damda Keberaka (*Dam(qi)* > Dam(da).[73] Having projected the founding gesture of Sao Udme Mashilanga onto the period of the first group of immigrants, and under the influence of the legend of Semiramis, the early traditionalists of the town might have felt that is was necessary to attribute a powerful female founding figureto the second group of immigrants. Moreover, since both the formerly sacrificed mother of the enthroned king and Magira Dama were Queen Mothers, it is tempting to believe that both accounts are connected. The transmission of the tradition was certainly encouraged by the existence in Gulfeil, as in all other city-states of the Kotoko, of the office of Queen Mother, involving a strictly applied avoidance relationship between the king and his mother, the Magira.[74] The existence of this office probably supported the continued commemoration of the two ancient queens.

Turning our attention to the non-narrative dynastic tradition, we note that according to the counting of the GKL, the 'Sao period' extends over twenty-two rulers, including five forgotten kings.[75] Following Baggari-the-Great (10), it has Mater (11), Bulal-the-Great (12), Ahmad-the-Great (13) and at the end Busi (22). Judging from the identification of Baggari-the-Great (10), these remaining names probably designate Assyrian and other non-Israelite kings of the Fertile Crescent. Like the later kings Matar (40), Mader (41), Mader (45) and Matar (46) of the 'Muslim period', Mater (11) seems to correspond to one of the numerous royal Babylonian names containing the name of the Babylonian state god Marduk. By inversion of the consonants *r* and *d* and elision of the final syllable –*uk*, the name seems to have resulted from the following phonetic shift: *Ma(r)d(uk)* > Ma-de-r.[76] On account of their position after Baggari-the-Great (Tiglath-pileser III – Sargon II), it is tempting to recognize in Mater (11) Marduk-apla-iddina (721-703), in Bulal-the-Great (12) Nabopolassar (626-605) and in Ahmad-the-Great (13) Nebuchadnezzar II (605-562). If these identifications are correct, the author of the list would have proceeded from the beginning of

---

[73] Olmstead 1923 : 175-181; Roux 1992 : 305-310; Lebeuf 1969 : 77.

[74] Lebeuf 1969 : 70, 130-1, 340; Lange 1990 : 142.

[75] Lebeuf / Rodinson 1948 : 38 n. 14; Lebeuf / Masson 1950 : 183.

[76] For the period concerned M. Van De Mieroop (van Mieroop 2007: 313) lists four Marduk names. (Lebeuf / Rodinson seem to imply that an alternative list omits Mater (11) and has Baggari-the-Great instead (Lebeuf / Rodinson 1948:36).

the Neo-Assyrian period straight to the end. Disregarding for a moment the whole list of Assyrian kings of that period available to him probably led astray by oral traditions he seems to have first mentioned the two Babylonian kings from the end of the Near Eastern history of the migrants before continuing to record the other onomastic elements available to him concerning the Assyrian domination of the Fertile Crescent.

On the assumption that Bulal-the-Great (12) and Ahmad-the-Great (13) were later and falsely inserted into an original list of chronologically correctly ordered Assyrian kings, we may tentatively identify all the nine remaining kings of the 'Sao period' with precise ancient Near Eastern prototypes: Sale (14) with Shalmaneser V (726-722) $(Y(u)l(mınu-ayar)2(d)$ > S-l-e by an inversion of the vowels u and a), Bulal-the-Small (15) with Nabu-nasir (747-734) by comparison with the later 'great' Babylonian king, Nabopolassar, Baggari-the-Small (16) with Sargon II (722-705) by extension of the figure of Baggari-the-Great to include the conqueror of Samaria, Ter (17) with Sennacherib (704-681) $(S(în-axx2)-er(3ba)$ > T-er by the shift of the consonant s to t), Abali (18) with Assurbanipal (668-627) $(A(yyur)-bı(ni-apli)$ > A-ba-li), Zeribe (19) with Assur-etililani (627-623) $(A)yyur-(et)elli(-ilıni)$ > Zer-ibe by the shifts of yy to z and of l to b and by the inversion of the vowels e and i), Dahala (20) with Sin-shar-ishkun (623-612) $(Sîn-yarra-iykun: Sîn$ > Da-, $yarra$ > -hala by the shift of the consonants y to h and r to l and the omission of $-iykun$), Ahmad (21) with Nabopolassar (626-605) and Busi (22) with Assur-uballit II (612-609) $(Ayyur-u)balli(t)$ > Bu-si by the inversion of the vowels u and a and the shift of the consonant l to s). If these equivalences are correct, eight of the twenty-two kings of the 'Sao period' were Assyrian rulers. Apart from the later superimposed Bulal-the-Great (12)/Nabopolassar and Ahmad-the-Great (13) / Nebuchadnezzar II, they are listed in exact chronological order. Moreover, with the exception of the subsequently mentioned Madu (24)/Esarhaddon, they include all the Assyrian kings from Baggari-the-Great (10)/Tiglath-pileser III (744-727) to Busi (22)/Assuruballit II (612-609). However, this surprising completeness of the list of Neo-Assyrian kings presupposes several errors committed by the author of the king list: he mentions Bulal-the-Great (12)/Nabopolassar and Ahmad-the-Great (13) / Nebuchadnezzar II (605-562) far too early, he refers to the Babylonian king Nabu-nasir (747-734) anachronistically as Bulal-the-Small (15) by comparison with Bulal-the-Great (12) / Nabopolassar, he refers to Nabopolassar twice, once as Bulal-the-Great (12) and once as Ahmad (21), and he calls the Israelite king and the Neo-Babylonian ruler by the name Ahmad. Though Ahmad is an Arabic name which in this form must have resulted from the Arabic translation of an earlier written or oral list, the previous version of the list most likely had a similar name. Hence, the incongruities pointed out appear to go back to the differences between the literary and oral heritages of the immigrant state builders.

Having access apparently to both an Israelite king list and a distinct Assyrian-Babylonian synchronistic king list, the early chronicler of Gulfeil seems to have tried to combine the available oral traditions on the conquest of Israel and the conquest of Assyria by onomastic means with his list material.[77] By doing so he introduced two major distortions into his list material, one concerning the figure of Baggari-the-Great (10) and the other the two successive figures of Bulal-the-Great (12); Nabopolassar and Ahmad-the-Great (13), Nebuchadnezzar. The first consisted in the insertion of the figure of Baggari, Tiglath-pileser III into the position of the Assyrian conquest of Israel and thus its extension to include Sargon II, the conqueror of Samaria. Realizing the duplication of this figure, he distinguished between the conqueror of Samaria and the predecessor of Ter (17) Sennacherib by calling the first Baggari-the-Great (10) and the second Baggari-the-Small (16). The second distortion of the list material concerns the Babylonian conquest of Assyria and consists in the erroneous insertion of the names of the two Babylonian conquerors, Bulal (12) Nabopolassar (626-605) and Ahmad (13) Nebuchadnezzar (605-562), before Sale (14) Shalmaneser V (726-722) and other Assyrian kings. Probably this error resulted from the orally transmitted connection between Madao, the youngest companion of Baggari (10), and the incursion of Bulal (12), producing the dispersal of the Sao under Madao. In order to bring these names as close to each other as possible, Bulal-the-Great (12) and Ahmad-the-Great (13) the author has made them appear as close successors of Baggari/Baggari-the-Great (10). Though Ahmad (21) and Duna (23) are still recognizable as the two conquerors of Assyria, their originally oral counterparts, Bulal-the-Great (12) and Ahmad-the-Great (13), are given a much higher position, by the epithet 'the Great', than the two figures in the GKL as it stands who epitomize the fall of the Assyrian Empire in 605 BCE.

According to the above analysis, the 'Sao period' does not cover all the 'pre-Islamic

---

[77] For similar written documents available to the early chroniclers of Kanem-Bornu and Kebbi, see Lange 2011: 11-18, and Lange 2009: 369-379.

kings' of the Gulfeil dynastic tradition, as generally assumed, but only a certain number of them.[78] As putative descendants of Sao Udme Machilanga/Semiramis, the Sao are clearly distinguished from the descendants of Mamba/Eber and those of Abra-Shemshem/Abraham and Sodo/Sarah. The Sao therefore seem to be only those ancestors of the Gulfeil people who claim descent from Sao Udme Mashilanga and not those claiming descent from Mamba/Eber, Abrabimo and Abra-Shemshem, Abraham. Insofar as the end of the Sao is clearly related to the flight of the majority of the people under the leadership of Madao (24)/Esarhaddon (for: Assur-uballit II), their arrival in the town must have coincided with the immigration of Baggari (10)/Tiglath-pileser III (744-727) and hence with the Assyrian conquest of Israel. It would therefore appear that the Sao were Assyrians and hence agents of the limited period of Assyrian hegemony over the Fertile Crescent extending from the conquests of Tiglath-pileser III to the fall of the Assyrian Empire in 605 BCE. They were thus distinct from the Israelites, Babylonians and other peoples submitted by the Assyrians.

Though according to the Gulfeil tradition the Assyrian ancestry of the Sao is linked to the eponymous ancestress Sao Udme Mashilanga/Semiramis, their name is possibly derived from that of Shu-Ninua ('the one of Nineveh'), the 54th king of the Assyrian King List. Shu-Ninua, who ruled around 1550 BCE, was probably the real founder of the Assyrian dynasty, rather than the more often mentioned Adasi (47) and Belu-bani (48).[79] Just as the Assyrian legends recorded by Greek authors credit Ninos, *i.e.* Ninua, with the founding of the Assyrian dynasty, legends of the region of Lake Chad seem to attribute the same feat to a male or female figure called Sao/Saw.[80] Whether male or female, and whether really related to Shu-Ninua (54) or not, according to the narrative tradition of Gulfeil the great ancestral figure of the legendary Sao was clearly connected with the Assyrian domination of the Fertile Crescent. On account of the demise of the Sao after the rule of Busi (22)Assur-uballit II (612-609), strictly speaking the rule of the Sao was limited to the ancient Near East. It is only insofar as all traditions transposed from the Fertile Crescent to the Central

---

[78] Contrary to Lebeuf 1969: 77-79, Lebeuf / Rodinson and Lebeuf / Detourbet are hesitant to use the notion of a 'Sao period' in the history of the Kotoko city-states (Lebeuf / Rodinson 1948:33-46; Lebeuf / Detourbet 1950: 183-5).

[79] Pettinato 1991: 67; Olmstead 1923: 30.

[80] Ctesias *in* Diodorus II: 1-7; Pettinato 1991: 58-82.

Sudan also have implications for the local history of the Kotoko that we may extend the notion of the civilization of the Sao to the region of Lake Chad.

In spite of the emphasis on the SaoAssyrians, the narrative tradition of Gulfeil offers a sagacious combination of Israelite and Assyrian perspectives. The sacrifice of Oumar's/Omri's three daughters, GaraShawe, Garé and WindéMassassaryo, and the beneficial effect of their immurement in the city walls appear to represent a special blessing related to Israel.[81] As for WindéMassassaryo, whose identity as an Assyrian legendary founding figure sets her apart from her Israelite companions, her presence among the three sacrificed girls can perhaps be explained by the general attempt to expand the beginning of Israelite history by an Assyrian factor and by the wish to balance Israelite and Assyrian influences. Similarly the regalia, including a spear and a Quran in a leather bag believed to have been brought originally by the first group of migrants, identified as Israelites, should be considered as symbols of specifically Israelite authority.[82] Since the 'Quran' is said to have been covered in earlier timesby the skin of the king's sacrificed mother and by the skin of a bull at the enthronement of a new king, the custom is attributed to the time of the Israelite kings before Mashi / Menahem, who is said to have abolished the custom of sacrificing the queen mother.[83] The notion of an ancient 'Quran' should thus be considered as the substitution of a modern term to designate a rudimentary Israelite not Judahite Torah.[84]

Fourth episode: The 'Muslim period' Fall of Assyria and the trans-Saharan exodus

The last event mentioned in the narrative tradition of Gulfeil is a great disruption resulting in the flight of Madao and the majority of the townspeople. The turmoil

---

[81] Griaule/Lebeuf 1948 : 6, 6 n. 4; Lebeuf 1969 : 69.

[82] Griaule and Lebeuf note two different versions; according to the first the objects were found on the mound and according to the second they were brought by the migrants ('Fouilles (III)', 5, 5 n. 6).

[83] Lebeuf 1969 : 273-4; Boulnois 1943 : 105-8, Pl. I, Pl. II.

[84] Cf. R. E. Friedman, 'Torah', ABD, VI, 612-4.

is said to have been produced by the introduction of Islam by foreign invaders led by a Bulala family. Strangely enough it is the father of Madao who is believed to have told his son to get himself circumcised, although he himself appears not to have been a Muslim. Madao refused to obey this injunction and when he was attacked by the invaders he opened a breach in the eastern wall of the town which formerly had no gate. Followed by those townspeople who likewise refused to be converted to Islam, he escaped through it to the region of Mora, situated 170 km south of Gulfeil, and settled on a nearby mountain, called ever since the 'rock of Gulfeil'.[85] Even more disturbing than the prime cause of the foreign invasion is the account according to which the majority of the people fled from the town in order to avoid circumcision and conversion to Islam.

Doubtlessly the conversion story is a recent tale reflecting the preoccupation of the people with the spread of Islam, and the resulting conflicts with respect to traditional practices. In particular, it is hard to believe that a non-Muslim father would instruct his son to get circumcised for the sake of Islam, or that the threat of enforced conversion to Islam led to a great exodus. We should therefore consider the conversion story as an Islamic reinterpretation of an older form of the tradition. Parallel to the story of Mashi's/Menahem's refusal to sacrifice his mother, the original tradition probably provided an explanation for the conquest of the town by foreign people, by attributing the cause of this disaster to disrespect of an important ritual practice. Having disobeyed the injunction of his father, Baggari/Tiglath-pileser III, to comply with a religiously sanctioned convention such as circumcision, the culprit, Madao/Assur-uballit II, was punished by a foreign invasion. If this assumption is correct, the Bulala, and hence the Babylonians, were thought to have been the instruments of divine wrath, subsequent to the transgression of important injunctions.[86]

A comparison with the information derived from oral traditions shows that the written GKL reflects the devastating event of the Babylonian conquest of Assyria twice. The first reflex consists in the insertion into the list of the name of Bulal-the-Great (12), derived from that of Nabopolassar (625-605), designating the Babylonian conqueror of Assyria, and the name of Ahmad-the-Great (13) Nebuchadnezzar II,

---

[85] Lebeuf / Detourbet 1950 : 46-47; Griaule / Lebeuf 1948 : 6; Lebeuf 1969 : 69.

[86]For a similar motif in the Oyo-Yoruba narrative tradition, see Johnson 1921: 160; Hess 1898: 157.

designating his son and successor. Applied originally to Nabopolassar, the name of Bulal also refers to the eponymous ancestor of the Bulala of Lake Fitri, whom the Bulala themselves remember in the slightly truncated form of Bulu: *(Na)bû-(ap)la-(usu)r* > Bu-la-r/l.[87] The neighbouring Kanembu-Kanuri and the Buduma know the Babylonian conqueror of Assyria likewise by the name of Bulu.[88] It is obviously also on account of this great conqueror that the narrative tradition of Gulfeil refers to the foreign invader who 'converted the people to Islam' as an unnamed hero who had married a Bulala woman, who led a family of the Bulala or who was himself a Bulala.[89] By combining the available information, we realize that the association of Bulal-the-Great (12) Nabopolassar with the Bulala sheds important light on the ancient Near Eastern antecedents of the Bulala ethnogenesis.[90] Though the two names, Bulal-the Great (12) Nabopolassar and Ahmad-the-Great (13)/Nebuchadnezzar II, are wrongly inserted into the middle of the section of the GKL devoted to the Sao/Assyrian period, they allow us, by the association of Bulal-the-Great with the Bulala, to date the beginning of the 'Muslim period' of Gulfeil history to the Babylonian conquest of Assyria in 605 BCE.

The second reflection of the Babylonian conquest of Assyria in the GKL is at the beginning of the section devoted to the 'Muslim period'. By indicating that the reign of Busi (22)Assur-uballit II was followed by the period of 'Muslim kings', the dynastic tradition of Gulfeil clearly makes the beginning of the new era coincide with the fall of the Assyrian Empire. With respect to Duna (23), the first 'Muslim king', some traditionists of Gulfeil claim that he arrived from the east after having married a Bulala woman from the region of Lake Fitri, while others affirm that he was a Bulala himself. This explanation of the origin of Duna (23) closely resembles the association of Bulal-the-Great (12) by some traditionalists with 'the Blala, Bulala, Bilala people who live close to the Lake Fitri'.These comments concerning two different figures in the GKL seem to echo the oral tradition dealing with the flight of Madao, according to which invaders of the town, led by a family of Bulala, imposed Islam on the inhabi-

---

[87] Carbou 1912 : 263, Palmer 1928 : 99, and Le Rouvreur 1962 : 78, 116.

[88] Lange 2009: 27, 67; 2011: 13-14, 17; Palmer 1928: 30 (Magumi Bulwa); Landeroin 1911: 311-2.

[89] Lebeuf / Rodinson 1948 : 38; Griaule / Lebeuf 1948 : 6.

[90] The onomastic evidence confirms the Duguwa/Babylonian identity of the Bulala (Lange 2008: 144-5, 248).

tants of Gulfeil.[91] In all likelihood the three different episodes associated with the Bulala reflect in fact one and the same great disruption in ancient Near Eastern history. This event can best be situated in time by considering the onomastic evidence provided by the GKL with respect to the end of the 'Sao period'. After Baggari-the-Small (16) Sargon II (722-705), the list enumerates a series of six Assyrian rulers following each other from Ter (17)Sennacherib (704-681) to Busi (22) Assur-uballit II (612-609), in correct order, with the sole exception of the penultimate king, Ahmad (21), corresponding to the Babylonian conqueror Nabopolassar (626-605).[92] It is quite remarkable that the last two figures in this series, Ahmad (21)/Nabopolassar and Busi (22)/Assur-uballit II, follow each other in the same onomastic sequence as the corresponding kings in the dynastic records of Kanem-Bornu and Kebbi – Bulu (8) and Arku (9) in Kanem and Maru-Tamau (32) and Maru-Kanta (33) in Kebbi.[93] Thus, in all likelihood the GKL deals at this point with the destruction of the Assyrian Empire by the Babylonian insurgents in 605 BCE, like the two other dynastic records. It should further be noted that the date for the end of the reign of Busi (22) Assur-uballit II (612-609) apparently does not coincide with that for the beginning of the rule of Duna (23) Nebuchadnezzar (605-562), but this discrepancy is not significant since the final date of Assur-uballit's reign, 609 BCE, is purely conventional because it is based on the last documented reference to the king.[94] We may assume that in fact the last Assyrian king lived at least until 605, when the troops of the Egyptian-Assyrian alliance were finally defeated by Nebuchadnezzar in the battle of Carchemish.[95] As suggested by the list, Busi (22) Assur-uballit II and Duna (23) Nebuchadnezzar may therefore indeed be considered as two successive kings, but while the first stands for the fall of Assyria, the second represents the rise of Babylonia and the exodus of the refugees.

The inauguration of the 'Muslim period' by Duna (23) Nebuchadnezzar II (605-562) can be understood as a tribute to the Babylonian conqueror. By designating him as the first 'Muslim ruler', the GKL or the associated oral tradition

---

[91] Lebeuf / Rodinson 1948; Griaule / Lebeuf 1948 : 6.

[92] Lebeuf / Rodinson 1948 : 36-38; Lebeuf, 1969 : 78.

[93] Lange 2011: 14; Lange 2009: 370.

[94] Grayson 1975: 96; Glassner 2004: 223.

[95] Labat 1967: 97-98; Oates 1991: 182-3.

qualifies the liberator of numerous national communities deported by the Assyrians to Syria-Palestine as the saviour of those people who fled to the region
of Lake Chad subsequent to the breakdown of public order resulting from the
defeat of the Assyrian-Egyptian alliance.[96] In Makari, the role of the first 'Muslim ruler' is attributed to the Israelite 'father of nations', Biri (25) Abraham, and
among the Oyo-Yoruba the similar role of the national redeemer is bestowed
on Abiodun, the Babylonian conqueror Nabopolassar and father of Nebuchadnezzar II.[97] Like the well-known royal name Dunama of the neighbouring Se-
fuwa, the name of Duna is derived from Akkadian *dannu* 'strong', which figures
in a frequently employed Babylonian-Assyrian royal epithet as *yarru dannu*'
strong king', and its non-royal meaning is preserved in Kotoko as *dón4*.[98] The
proper name attested in the GKL therefore points to the strength of Nebuchadnezzar II as the conqueror of the Assyrian Empire.

The second 'Muslim king' is Madu (24) who is also known by the variant names of
Bulal and Muhammad.[99] He seems to be the same as Madao, the youngest companion of Baggari (10)/Tiglath-pileser III whom we have previously identified with
Esarhaddon (680-669), the Assyrian conqueror of Egypt. According to the narrative tradition of Gulfeil, Madao/Esarhaddon was the leader of the numerous migrants from the town who fled from the threat of circumcision and conversion to
Islam.[100] Owing to the continued Assyrian pre-eminence during the final period of
turmoil of the Assyrian Empire for the communities of deportees, most of the
Central Sudanic state founders seem to have attributed the leadership of the trans-
Saharan exodus to an Assyrian king. Thus the non-narrative traditions of Makari
and of Kebbi designate the last Assyrian king, Assur-uballit II, as the final leading
figure of the people after the Babylonian conquest.[101] The narrative traditions of
Kebbi and of Oyo refer more vaguely to previous Assyrian rulers or to an abstract
Assyrian figure as leaders of the trans-Saharan exodus.[102] The great Hausa tradition

---

[96] Oates 1991: 182-3; Lange 2012: 150-4.

[97] Lebeuf 1969: 77; Lange 2011: 590-1.

[98] CAD III, 92-100; Seux, *Épithètes*, 68-70; Lange 2009: 13.

[99] Lebeuf / Rodinson 1948 : 36.

[100] Lebeuf/ Detourbet 1950 : 46-47; Griaule / Lebeuf 1948 : 6; Lebeuf 1969 : 69.

[101] Ligonge Aladinga (24) and Maru Kanta (33) (Lebeuf 1969 : 77; Lange 2009 : 374-5).

[102] Kotai/Sargon II, Ganbi/Assurbanipal and Sakai/Sin-shar-ishkun (Lange 2009 : 365, 374-5)

of Daura offers a combination of both possibilities; while the dynastic aspect of the tradition emphasizes the role of the last Assyrian king, Bayajidda Assur-uballit II, the non-dynastic aspect highlights the role of the Mesopotamian epoch ruler, NimrodSargon of Akkad.[103] It is therefore not surprising that the GKL inaugurates the post-imperial history with Duna (23) Nebuchadnezzar II, the final destroyer of the Assyrian Empire, and continues it with Madu (24) Esarhaddon, the figure which the narrative tradition of the town describes as the great leader of the exodus. Like Bulal-the-Great (12) Nabopolassar and Ahmad-the-Great (13) / Nebuchadnezzar II, Madu (24) Esarhaddon is inserted into the GKL by the influence of oral traditions, while in fact he stands in this position for Assur-uballit II or, if he existed, his Assyrian successor. Although in fact Madu (24)/Esarhaddon was a figure of the Assyrian regime, he is positively connoted as a 'Muslim' ruler, in spite of the overthrow of Busi (22)/Assur-uballit II and the end of the repressive Assyrian hegemony. In this respect the alternative names, Bulal/Nabopolassar and Muhammad, give the impression that some ancient chronicler attempted to eliminate from the 'Muslim period' a figure inadequately inserted into the post-Assyrian portion of the king list and to replace it by a second Babylonian figure.[104]

Following Madu (24)/Esarhaddon (or Assur-uballit II) we find up to Bulal (54) a list of thirty rulers whose names point to strong connections with the previous 'Sao period', i.e. the period of Assyrian hegemony in the ancient Near East. Most significant in this respect are the names Mater (40), Madér (41), Madir (45) and Matar (46). Like Mater (11)/Marduk-apla-iddina II (721-703) from the 'Sao period', on account of their phonetic closeness to the name of the Babylonian state god, Marduk, they seem to originate from Babylonian theophoric royal names containing that of the state god.[105] Babylonian influence also appears to be indicated by the names Bulal-the-Small (49) and Bulal (54), which as in the previous section could designate either the dynastic founder Nabopolassar himself or, perhaps more likely, kings equated with him. Israelite influence seems to be attested by the names Buri (25), Buri Maranga (28), Biri (44), Birimi (48), which are apparently based on Abram Abraham, and perhaps by

---

and Oduduwa (Lange 2011 : 583-4).

[103] The army leader Bayajidda and Najib / Nimrod, the leader of the people (Lange 2012: 150-4).

[104] Lebeuf/Rodinson 1948: 38.

[105] Grayson (1975) lists 15 names beginning with Marduk ('Königlisten und Chroniken', RLA, VI, 130-131).

Masi (35) possibly derived from Moshe/Moses. Assyrian influence might be inferred from the names Sale (47) and Sale (51) which, like the previous Sale (14), in all likelihood reflect the exclusively Assyrian royal name Shalmane-ser/$Yulmınu$-$ayar2d$applied to five kings, reaching from Shalmaneser I (1273-1244) to Shalmaneser V (726-722).[106] Mentioned in the GKL after the onomastic pointers to the fall of the Assyrian Empire, the series of thirty ancient Near Eastern royal names correspond to a confused assemblage of kings from different nations. It can be compared to the non-chronological sections of several other Central Sudanic king lists. Figuring either at the beginning of the king lists or in the Neo-Assyrian section, the names usually designate specific communities of deportees from the Assyrian Empire.[107] Inserted at the end of the GKL, these names indicating the presence of Babylonians, Israelites and Assyrians seem to refer more particularly to the group of refugees which participated in the trans-Saharan exodus after the collapse of the Assyrian Empire.

According to this analysis the great majority of the kings listed in the GKL were rulers of the ancient Near East. Real conversion to Islam appears to have begun at the dynastic level only with #Al3 Basan (55) who was the sixth predecessor of #Umar (61), a contemporary of the Bornuan ruler Rıbih (1893-1900).[108] If the list is complete at this point, #Al3 Basan (55) cannot possibly have ruled before the eighteenth century. With the exception of Kanem the relatively recent spread of Islam in the Central Sudanic kingdoms is ironically more difficult to ascertain than the date and the circumstances of departure of the pioneer state founders from the ancient Near East about 605 BCE.

Conclusion

Parallel to the dynastic tradition of Oyo, we note in the narrative traditions of Gulfeil a combination of Israelite and Assyrian perspectives. Beginning with the ancestry of Mamba/Eber and the presence of a pre-exilic version of the Torah,

---

[106] A. K. Grayson, 'Königslisten und Chroniken', RLA, VI, 134.

[107] In Makari, Kanem-Bornu, Kebbi and Oyo (Lebeuf 1969: 77; Lange 2011: 16-17; Lange 2009: 369-373.

[108] Lebeuf / Rodinson 1969 : 38; Brenner 1973 : 123-130.

Gulfeil traditions cherish the refusal of Mashi/Mehahem to sacrifice his mother by calling him Mamat Muhammad, they attribute the sacrifice of Oumar's Omri's daughters to the Assyrian conqueror Baggari Tiglath-pileser III, and they praise the last Israelite king, Hoshea, by calling him Ahmad Muhammad. This insistence on Israelite achievements is to some extent counterbalanced again similar to Oyo tradition by the positive depiction of the Assyrian contribution to the ancient history of the people of Gulfeil: the important role attributed to Sao Udme Mashilanga Semiramis, the appreciative reference to the early Assyrian connections with Israel, the favourable mentioning of Baggari'sTiglath-pileser's conquest of Israel, and the designation of Madao Shalmaneser II as the leader of the trans-Saharan exodus. In the attribution of the ancestry of the six quarters of Gulfeil to Israelite and Assyrian figures the north to AbrabimoAbraham and Magira Dama, the centre to Mamba/Eber and Madao Shalmaneser V, and the south to Teri Terah and Alanga Sargon II and Aldjauna Esarhaddon we find a remarkable attempt to balance Israelite against Assyrian interests.

In connection with the GKL, the narrative traditions of the town can be interpreted as reflecting very specifically the Near Eastern history of the migrants from Syria-Palestine in consequence of the fall of Assyria at the end of the seventh century BCE. The onomastic details of the king list confirm the validity of the above analysis according to which the successive local founding events under Mamba, Abra-Shemshem and Baggari, and the flight of the majority of the inhabitants of the town to the south under Madao, correspond to local reflexes of three major disruptions in the history of Israel: the arrival of a small group of immigrants under the leadership of Eber or Abraham in Israel, the Assyrian conquests of Israel by Tiglath-pileser III in 734-733 and Sargon II in 722, and the exodus of numerous resettled deportees from Syria-Palestine to the region of Lake Chad after the collapse of Assyria in 605 BCE. The projection of the key events of Israelite-Assyrian history onto the local conditions of Gulfeil can therefore be understood as the natural attempt of the newcomers to preserve their ancient history in their new environment in spite of their long-distance migration.

The Gulfeil narrative traditions are particularly instructive with respect to the identity of the Sao. Like the Near Eastern prototype Sammuramat/Semiramis (810-806), they largely dissociate the eponymous ancestress of the Sao, Sao

Udme Machilanga, from her original historical context and project her onto the beginning of Sao/Assyrian history. Though neither the narrative nor the dynastic traditions of Gulfeil offer any precision with respect to the real interference of the Sao in Israelite history, it appears that this important event did not take place before the demise of the last Israelite king, Ahmad (9)/Hoshea, in 722 BCE. At that time Baggari (10), who incorporates the two Assyrian conquerors, Tiglath-pileser III (744-727) and Sargon II (722-705), took possession of the 'abandoned city'. In fact, the takeover of the city of Gulfeil, abandoned by its inhabitants, most likely reflects the resettlement of the country of Israel after the departure of its original Israelite inhabitants in consequence of the Assyrian praxis of mass deportations.

The onomastic evidence of the GKL offers a number of details which supplement the settlement aspect of Israelite-Assyrian history by the political dimension. Also beginning with Mamba (1) Eber, the eponymous ancestor of the Hebrews, the king list continues with the Israelite kings Mashi (7) Menahem (749-738), Semade (8) Pekah (736-732) and Ahmad (9)Hoshea (732-722). Probably referring by the five 'forgotten names' to the legendary patriarchs Teri Terah and AbrabimoAbraham, and to the historical figure of Oumar Omri (884-873), the list contains onomastic information which in immensely reduced fashion parallels Israelite legendary and factual history transmitted by the Hebrew Bible. Contrary to other Central Sudanic dynastic traditions, it highlights the importance of Tiglath-pileser III (744-727) and Sargon II (722-705) by including the two Assyrian conquerors in the single figure of Baggari (10) qualified as 'the Great'. As the organizer of the resettlement of the town, Baggari-the-Great (10)/Tiglath-pileser-Sargon was probably the most powerful figure in Gulfeil history.

Most likely the history of the Sao / Assyrians begins with the incursions of Baggari-the-Great (10)/Tiglath-pileser-Sargon in Israel and elsewhere in the Fertile Crescent. Mentioned by the GKL, the Babylonian kings Mater (11)Marduk-apla-iddina II, Bulal-the-Great (12)Nabopolassar and Ahmad-the-Great (13) Nebuchadnezzar only bear witness of the great influence of Babylonia on Assyrian history, and in particular on its end. All the other figures of the 'Sao period', from Sale (14) Shalmaneser V (726-722) to Dahala (20) Sin-shar-ishkun (623-612), are Assyrian kings of the imperial period listed in correct chronological order. Extending from Baggari-the-Great (10) Tiglath-pileser-Sargon to Dahala

(20)/Sin-shar-ishkun, the period of Sao/Assyrian hegemony saw the greatest extent of imperial power in ancient Near Eastern history, and its reflex on the narrative traditions of Gulfeil bears witness to the great impression it made on the immigrant town builders.

The end of Sao/Assyrian domination was inaugurated by Ahmad (21) Nabopolassar (626-605), the penultimate ruler of the 'Sao period'. According to the GKL, the Babylonian king Ahmad (21)Nabopolassar was followed by the Assyrian king Busi (22) Assur-uballit II (612-605), who in turn was followed by Duna (23)Nebuchadnezzar (606-562). This succession of rulers during the great disruption between the rule of the Sao Assyrians and that of the post-Sao post-Assyrian refugees mirrors exactly the events between the rise of the Chaldean insurgent Nabopolassar and the final collapse of the Assyrian Empire. Having been ousted from their Assyrian mainland by Ahmad (21)Nabopolassar in 612 BCE, the SaoAssyrians retreated to Syria, where, assisted by the Egyptian army, they continued their desperate struggle for survival. This fight was lost in 605 BCE, when Duna (23)/Nebuchadnezzar, the son and successor of Ahmad (21) Nabopolassar, defeated the Egyptian-Assyrian alliance in the battle of Hamath. After that battle, any organized resistance to the new Babylonian-Median order in the Fertile Crescent ceased to exist.[109] Interpreted in the light of reconstructed ancient Near Eastern history, the onomastic evidence provided by the GKL therefore exactly reflects the fall of the Assyrian Empire and the rise of the Babylonian-Median hegemony in the Fertile Crescent between 612 and 605 BCE.

The kings of the post-Sao period are indicative of the situation prevailing between the collapse of the Assyrian Empire and the emergence of the city-state of Gulfeil south of Lake Chad. The onomastic pointers of the post-Sao period suggest that the apparent continuity of the Sao regime concerns the trans-Saharan exodus of the refugees following the Assyrian defeat and the beginning of the state-building process south of Lake Chad. Though the Sao Assyrians were certainly not the new rulers of the polity, the solidarity felt by the communities of deportees towards the former Assyrian rulers during the final onslaught of the local enemies, and the common experience of the long and

---

[109] Grayson 1975: 99; Labat, 1967: 97-98; Redford 1992: 452-5.

painful exodus, constituted the basis of a new evaluation of the past. Thus, the ambiguity felt by the people of Gulfeil and other Sao-Kotoko city-states towards the Sao Assyrians is most likely the result of the largely oppressive, but in the end also supportive, role played by the Sao Assyrians in the ancient Near East. Whatever the case, due to their overthrow in the Fertile Crescent and the un-forgotten cruelty of their hegemony in the Near East, the Sao Assyrians played only a commemorative and not an effective role in the region of Lake Chad.

References cited

Boulnois, J., 1943, 'La migration des Sao au Tchad', *Bulletin de l'IFAN*, 5 80-120.

Brenner, L., 1973, *The Shehus of Kukawa*, Oxford.

Bright, J., 2000, *History of Israel*, 4th ed., Louisville.

Cancik, H., & H. Schneider, eds., 1996-2003, *Der Neue Pauly. Enzyklopädie der Antike*, 15 vols. Stuttgart.

Carbou, H., 1912, *Le région du Tchad et du Ouaddaï*, vol. I, Paris.

Ebeling, E., & B. Meissner, eds., 1932-2016, *Reallexikon der Assyriologie*, 13 vols.

Freedman, D.N., ed., 1992, Anchor *Bible Dictionary*, 6 vols. New York.

Garbini, G., 1988, *History and Ideology*, London.

Gelb, J., *et al.*, eds., 1956-2006, *The Assyrian Dictionary of the University of Chicago*, 21 vols.

Gesenius, W., 1842, *Hebräische Grammatik*, 13th ed., Leipzig.

Glassner, J.-J., 2004, *Mesopotamian Chronicles*, Atlanta.

Grayson, A. K., 1975, *Assyrian and Babylonian Chronicles*, New York.

Griaule, M., & J.-P. Lebeuf, 1948, 1950, 1951, Lebeuf, 'Fouilles de la région du Tchad, (I), (II), (III)', *Journal de la Société des Africanistes* 18, 1 (1948), 1-116; 20,1 (1950), 1-151; 21, 1 (1951), 1-95.

Griaule, M., 1943, *Les Saô legendaires*, Paris.

Hess, J., 1898, *L'Âme nègre*, Paris.

Johnson, S., 1921, *The History of the Yorubas*, London.

Labat, R., 1967, 'Assyrien und seine Nachbarländer von 1000 bis 539 v. Chr. ', *in:* E. Cassin *et al.* (eds.), *Fischer Weltgeschichte, III*, Frankfurt/M, 9-111.

Landeroin, M.-A., 1911, 'Du Tchad au Niger – Notice historique', *in:* J. Tilho, *Documents scienti-fiques de la Mission Tilho (1906-1909)*, vol. II, Paris, 309-537.

Lebeuf, A.-M., 1969, *Les principautés Kotoko*, Paris.

Lebeuf, J.-P. & A. Masson Detourbet, 1950, La *civilisation du Tchad*, Paris.

Lebeuf, J.-P.& M. Rodinson, 1948 , 'Généalogies royales des villes kotoko (Goulfeil, Kousseri, Makari)', *Études Camerounaises* 23-24, 31-46.

Lange, D., 1977, *Le Dзwın des sultans du Kınem-Born५: Chronologie et histoire d'un royaume africain*, Wiesbaden.

Lange, D., 1990, 'Die Königinmutter im Tschadseegebiet: historische Betrachtungen', *Paideuma* 36, 139-156.

Lange, D., 2008, *Ancient Kingdoms of West Africa: Africa-Centred and Canaanite-Israelite Perspectives*, Dettelbach 2004. 'Islamic feedback or ancient Near Eastern survivals?', *Paideuma* 54, 253-264.

Lange, D., 2009a, 'Biblical patriarchs from a pre-canonical source mentioned in the *D3wɪn* of Kanem-Bornu (Lake Chad region)', *Zeitschrift für Alttestamentliche Wissenschaft* 121, 4, 588-598.

Lange, D., 2009b, 'An Assyrian successor state in West Africa: the ancestral kings of Kebbi as ancient Near Eastern rulers', *Anthropos* 104, 359-382.

Lange, D., 2011, 'Origin of the Yoruba and the 'Lost Tribes of Israel', *Anthropos* 106, 579-595.

Lange, D., 2012, 'The Bayajidda legend and Hausa history', E. Bruder and T. Parfitt (eds.), *Studies in Black Judaism*, Cambridge, 138-174.

Leick, G., 1991, *A Dictionary of Ancient Near Eastern Mythology*, London.

Le Rouvreur, A., 1962, *Saheliens et Sahariens du Tchad*, Paris.

Liverani, M., 2005, Israel's *History and the History of Israel*, London.

Luckenbill, D. D., 1926-7, Ancient *Records of Assyria*, vols. I, II, Chicago.

Lukas, J., 1936, *Die Logone-Sprache im Zentralen Sudan*, Leipzig.

Na'aman, N., 1993, 'Population changes in Palestine following Assyrian deportations', *Tel Aviv* 20, 104-124.

Oates, J., 1991, 'The fall of Assyria (635-609 B.C.)', *in:* J. Boardman *et al.* (eds.), *The Cambridge Ancient History*, 2nd ed.,vol. III, Cambridge,162-193.

Olmstead, A., 1923, *History of Assyria*, New York.

Palmer, H. R., 1928,   *Sudanese Memoirs*, 3 vols. Lagos.

Palmer, H. R., 1936, *The Bornu Sahara and Sudan*, London.

Pettinato, G., 1991, *Semiramis: Herrin über Assur und Babylon*, Munich.

Pritchard, J. B., 1969, *Ancient Near Eastern Texts*, 2nd ed., Princeton.

Redford, D. B., 1992, *Egypt, Canaan and Israel in Ancient Times*, Princeton.

Roux, G., 1992, *Ancient Iraq*, 3rd ed., London.

Saggs, H., 1984, *The Might that was Assyria*, London.

Seux, M.-J., 1967, *Épithètes royales akkadiennes et sumériennes*, Paris.

Soden, W. von, 1952, Grundriss der Akkadischen Grammatik, Rom.

Van de Mieroop, M., 2007, *A History of the Ancient Near East*, 2nd ed., Oxford.

Wehr, H., 1979, *Dictionary of Modern Written Arabic*, 4th ed., Wiesbaden.

Younger, K. L., 1998, 'The deportations of the Israelites', *Journal of Biblical Literature* 117, 2, 201-227.

Sacred deer in Japan

by Kazuo Matsumura

Abstract: Deer has been probably the most popular hunting game across Eurasia. To the author, the most impressive deer images are presented by the golden artefacts that have been excavated from the tombs of nomadic Scythians. Deer images and motifs however are also found in other areas too, almost all over Eurasia: in Japan, China, Greece, among the Mongolians, and among the Celts. In Japan, deer is the messenger of Shinto god in ancient capital city of Nara and thus deer is still regarded as sacred just like a cow in India or a white elephant in Thailand. In China, deer horn is attached to the image of dragon, Taotie (animal motif of ritual bronze vessel), and qulin, a legendary animal. In Greece, hind (with unusual horns) is sacred to Artemis. Among the Mongolians, the legend tells that the line of Chingis han come from the union of a blue wolf and white hind. Lastly, among the Celts, we know god Cernonnus, Gunestrup horned-figures surrounded by deer, and Finn and his finneans who run woods like deer. All such images and stories about deer are just a coincidence, since deer is familiar to all people of Eurasia? I think our ancestors across Eurasia have been attracted by deer because they found deep religious, spiritual, and symbolical meaning to this animal.

Keywords: Deer, Nara, Japan, Scythians, Eurasian cultures

Introduction

Mankind has been in the habit of expressing their ideas of the environment and life by employing various familiar animals as agents in their myths. Across the Eurasian continent, such animals as deer, wolves, horses, and birds abound in myths. Wim and I had many good discussions as members of IACM from 2006 to the present in various places such as China, the Netherlands, the USA, and Japan. We share a preference for comparisons over wider scales, in the case of Wim starting from Africa (van Binsbergen 2012) and in my case starting from

Japan (Matsumura 2010; 2015). This essay is written in his honor. [1]

On this occasion, therefore, I would like to focus on the phenomenon of the wide-spread worship of deer. The investigation starts from the reason why the deer in Nara Park in contemporary Japan are given the status of a national treasure and the search for that reason leads to the traditional belief about deer in ancient times. The author hopes to expand the themes of the religious significance of deer across the Eurasian continent, touching upon the cases of the Chinese, Mongolians, and Scythians, eventually reaching Ireland, the other end of the continent in the future, although it may take a long time to get there. Comparison with the cases of other animals (wolves, birds) is also in view.

Nara and deer

I lived in Nara, the ancient capital city of Japan (710-793), for ten years (1989-1999), when I was working as a professor there. There is a public park of about 502 hectare called Nara Park, established in 1880, at the foot of Mount Waka-kusa. It is like Central Park in New York or Hyde Park in London, and people walk around and enjoy the open air. But there is one difference. Nara Park is full of wild deer.

Around the park freely roam over 1,200 deer that are classified as a natural treasure. The park is surrounded by old and famous temples and shrines such as the Tōdai-ji Temple (which houses a gigantic Vailocana Buddha statue inside the largest wooden building in the world), Kōfuku-ji Temple, and Kasuga Shrine. Deer are also free to roam around these places, so their territorial scope could be as large as 660 ha. According to local folklore, deer from this area were considered sacred due to a visit from Takemikazuchi, one of the four gods of Kasuga Shrine. He was said to have been invited there from Kashima, Ibaraki, and appeared on Mt. Mikasa riding a white deer. From that point, the deer were

---

[1] Many thanks go to my friend Anthony Boys for not only correcting and improving this paper but also giving me valuable information on deer in Japan. Many thanks also go to another friend Attila Mátéffy for supplying me with important bibliographical information and encouraging me for the completion of this paper. This work was supported by JSPS KAKENHI Grant Number 16K02185.

considered divine and sacred by both Kasuga Shrine and Kōfuku-ji. Killing one of these sacred deer was a capital offense punishable by death up until 1637, the last recorded date of a breach of that law. After World War II, the deer were officially stripped of their sacred / divine status, and were instead designated as national treasures and are now protected as such (Wikipedia 'Nara Park').

Archaeological data

According to a Japanese archaeologist Hideji Harunari, many bronze bells had their magical effectiveness reinforced by thread relief figures. He says that some forty-one bells have pictorial decorations, on which he counts 129 deer, 58 humans, 31 fish, 27 cranes, and 18 wild boars. Deer appear on 63.4 percent of the decorated bells, showing their importance for the people of Yayoi agricultural societies (Kidder 2007: 144).

One bronze bell, said to be from Kagawa Prefecture on Shikoku Island (Ht. 30.75 cm), has the largest numbers of decorated panels. The themes are of food acquisition, storage, and preparation, accompanied by symbols of the harvest season. There are pictures of dragonflies, salamanders, egrets with fish in their beaks, tortoises, people pounding rice, a shaman in an ecstatic leap, a deer hunt, a wild boar hunt with a pack of five dogs, a granary, a praying mantis and a spider (Kidder 2007: 144-145). Judging from these panel pictures on the bronze bell, deer seems to be associated with the autumn harvest.

Pieces of deer and wild boar shoulder blades from the Yayoi period (BC 300-250 AD) through the Kofun period (250-538), the Aska period (538-710), into the Nara (710-794) and Heian (794-1185) periods have been discovered in various parts of Japan. They are remnants of oracle bones. Over a hundred pieces are known today from more than thirty sites across Japan from the Island of Tsushima and Iki in proximity to the Korean peninsula, to the central part of the main island of Honshū. 70 percent were dated to Yayoi, 16 percent to Kofun, and 14 percent to the Nara period (Kidder 2007: 154-156).

Kidder says that deer were religiously more important than wild boar: 'Unlike boar bones in Yayoi sites, which are rather plentiful, deer bones are relatively rare, so it is likely that deer were hunted specially for this purpose with all due ceremony, and an established business existed in supplying shoulder blades for diviners. The deer hunts pictured on the bronze bell may be illustrative of these occasions' (Kidder 2007: 156).

Deer in Japanese myth

The *Kojiki*, Records of Ancient Matters, the oldest surviving Japanese book, was compiled for the imperial family in 712. In the myth that gods in heaven are trying to draw the sun goddess Amaterasu out of a cave in which she had locked herself up, one of the magical means is the usage of deer bones. Omoi-kane, the god of wisdom, orders two priestly gods Ameno-koyane and Futo-tama to catch a stag from the Heavenly Mount Kagu. Two gods catch and kill the stag and extract the bones. At the burning sacred tree of Heavenly Mount Kagu, the two gods divine the future by examining cracks in the stag's shoulder blade. (vol. I, sec. XVI 56, Chamberlain 1982: 64)

In the *Nihongi* (or *Nihon-shoki*), the first official chronicle of Japan, starting from the mythological age down to the historical period of 697 and compiled in 720, another episode about stag is described. In this story, Ishi-kori-dome, an artisan god, stripes off in one piece the hide of a stag and makes Heavenly bel-lows out of it. (I. 42, Aston 1972: vol.1, 47)

It is notable that a stag's shoulder blade and a stag's hide are mentioned as magical objects in 'the Hidden Sun' episode, the most important section of Japanese mythology.

Deer in Japanese *Ancient Records*

In the *Nihongi*, we find a story that shows a close tie between kingship and a deer's cry. The section on the 16th legendary emperor Nintoku in the autumn, the 7th month of his 38th year of reign (350 AD in the *Nihongi*'s calculation) says that the emperor and empress were enjoying the cry of a stag from a nearby moor every night. One night the cry stopped. The next morning, when the em-peror was worrying about the stag, a servant belonging to the *SahekiBe* group brought venison of the stag he had killed. The meat was of the stag of the moor.

The emperor said to the empress: 'It is true that that man was not aware of our feeling of affection, and that it was by chance that he came to take it. We nevertheless cannot resist a feeling of resentment. It is therefore our wish that the *Sakaki Be* shall not approach the Imperial Palace.' The emperor made that group move to other place (XI. 21-22, Aston 1972: vol.1, 289-290).

The oldest existing collection of Japanese poetry, the *Man'yōshū* (literally 'Collection of Ten Thousand Leaves') was compiled under imperial order sometime after 759, about 40 years after the *Nihongi*. In this poetic collection, two poems by historically attested emperors about the cry of deer are recorded. One is said to have been sung by an emperor called Okamoto named after the location of the palace. This could be either Emperor Jomei (34th) or Empress Saimei (37th) whose palaces were there. It is in Book 8, the first poem of the autumn season (no.1511): 'In the evening, I usually hear a cry of a deer of the Ogura Mountain, but not today. I wonder if the deer is already in sleep.' A practically identical poem is recorded at the top of Book 9 (no. 1664), said to have been sung by Emperor Yūryaku (21st). The only difference is that in the former it is a 'crying deer' (*nakushika*) whereas in the latter, it is a 'lying deer' (*fusushika*). As a poem, the latter is regarded better by specialists and, furthermore, the former piece lacks a clear name of the emperor (or empress) indicating only his (or her) palace's location *Okamoto*. Emperor Yūryaku is a very important figure in the ancient history of Japan. He seems to have been a powerful lord who succeeded in unifying the state during the Kofun period (Matsumura 2015: 33-34).

No matter who was the true composer of this poem (often traditional poems were ascribed to emperors), it is important that in both an official chronicle and in an official poetic anthology the emperors were said to have had a particular interest in the cry of deer (in autumn).

In the *Man'yōshū*, deer are sung about in over fifty poems and almost all are about a stag's cry. In the *Fudoki* (Local Gazette), the local records whose compilation began in 713 and were completed in 20-year's time, there are also many allusions to deer.

Many wild animals inhabit in Japan, such as wolves (now extinct), monkeys, boars, bears, deer, and mountain deer (*kamoshika*). Of these wild animals, deer are by far the most mentioned in the *Nihongi*, the *Man'yōshū*, and the *Fudoki*. The reason must be that deer were regarded special, not only as food, but more as a symbol of sacredness. That is why the deer in the contemporary Nara Park are a national treasure.

Deer in Ancient Japanese rituals

According to the *Nihongi*, Buddhism was officially introduced into Japan in 552 from Baekje, Korea by Buddhist monks.(13th year of Emperor Kimmei, XIX 33, Aston vol.2: 65) Since then emperors have accepted Buddhism and have built many state temples. The Buddhistic aversion to meat led Emperor Temmu in April of the 4[th]year of his reign (675) to issue an order saying,'Let no one eat the flesh of kine, horses, dogs, monkeys, or barn-door fowls.' (XXIX 10, Aston vol.2: 329) Note that deer and boar are exempted. Two interpretations seem possible: deer and boar were harming crops and hunting them was necessary for farmers; in spite of Buddhist teaching, people kept eating meat so that only animals that were usually not eaten are listed in the prohibition order. I think both are possible and they are not incompatible. As to deer, one more explanation is possible. The story of Emperor Nintoku and the killed stag, mentioned above, indicates that there was a belief in the magical power of a stag's cry that would affirm a rich harvest.

White Deer (*hakushika*) is the name of a popular sake wine brand in present day Japan. In Book 21 of the *Engi-shiki* (Procedures of the Engi Era), a court ritual book completed in 927, various auspicious animals are mentioned. The highest ranked animals are the dragon, phoenix, six-legged animals, white elephants and white deer. Second ranked are white pigeons, white birds, white pheasants, and white rabbits (Tarao 1993: 3-4). The *Nihon Sandai Jitsuroku* 'The True History of Three Reigns of Japan' is one of the official history books completed in 901, covering the period of 858-887. In that book, two entries mention white deer:

> 1. 4th year of the Jōgan era (862): 'On September 27th, a white deer was presented to the emperor from Mimasaku (present day Okayama Prefecture).' (Takeda and Satō 1986: 171)

> 2. 1st year of the Genkei era (876): 'A white deer was presented to the emperor from Bingo (present day Hiroshima Prefecture). Lovely in snow white color. After seen by the emperor, the deer was set free at the Shinsen-en (imperial pleasure garden).'(Takeda and Satō 1986: 736)

Belief in the sacredness of deer by ancient Japanese people can also be confirmed from these historical records.

Deer described by the nobility of the Middle Ages

From diaries written by high nobilities of the Heian period (794-1191), continuous reverence of deer is confirmed (quotations in English are from Ogata Noboru: 'Messengers of the Gods-Deer of Nara', slightly abridged and modified).

Fujiwara Yukinari (972-1027) was a court official belonging to the powerful Fujiwara clan and left a diary named *Gonki* (which exists in full form from 991 to1011, with fragments up to 1026). The following entry mentions his encounter with a deer in Nara (the Japanese text is in Fujiwara no Nariyuki 2012: 442-443).

> 15th day [1st month, 3rd year of the Kankō era] (1006); At 10 o'clock, I departed for Kasuga [Shrine]..... 16th day; I purified myself in the early morning. .... At my entrance of the Shrine, a pheasant chirped. During my prayer, a crow perched on the third main shrine. Then at my exit from the Shrine, I encountered deer. These are all good omens.

The next quotation is from Fujiwara no Munetada's diary *Chūyūki*. Munetada (1062-1141) is also a court official belonging to the Fujiwara clan. (the Japanese text is in Fujiwara no Munetada 1965: 169)

> 16th day [6th month, 3rd year of the Ten-ei era] (1112); I traveled to Nara to examine the location of the pagoda of Kasuga Shrine which the Regent [Fujiwara no Tadazane] is planning to build. .... 17th day; .... I examined the location of the planned pagoda within the forest of the Shrine. .... During our activity, 40 or 50 deer appeared from the forest and from the direction of the Shrine, and roamed accompanying with us. Saying that this was really a very good omen and a sign of the gods, monks of the Temple showed their delights.

The third and fourth quotations are both from KujōKanezane's diary *Gyokuyō*. (KujōKanezane 1998: 36; KujōKanezane 2011: 102-103) KujōKanezane (alias Fujiwara no Kanezane, 1149-1207) was of the Fujiwara clan and held the highest positions of regent and Chief Minister in the court. His diary *Gyokuyō* covers the period from 1164 to 1200.

> 26th day [2nd month, 3rd year of the Angen era] (1177); ..... During our visit to [Kasuga] Shrine, many deer appeared in the morning darkness. These are all signs from the gods and good omens. People say that when one encounters deer, he or she should get out of the carriage and bow to the first one. Accordingly, the boy [probably Kanezane's son Yoshitsune] got out of the carriage and bowed [to the deer] ....

22nd day [8th month, 5th year of the Bunji era] (1189);Clear weather. On this day, I travelled to Nara to worship the Bodhisattva of Nan'en-do, and to inspect building of the Temple. According to the precedent, the head of the clan should visit the Temple soon after the framework-raising .... After entering through the western gate, I walked around passing the northeastern temporary building and the site of the middle gate. Then I visited the middle hall. At that moment, a deer appeared from inside the hall and ran westward. It was so miraculous that I cannot describe it in letters. I was momentarily bewildered, then joined my hands and bowed to the deer. Everyone attending the ritual was moved to tears. Anyone who travels to Nara to visit the Shrine considers encountering deer as a good omen. Anyone who meets with a happy event sees this sign. Generally encountering deer is confined to hills and wilderness because they tend to avoid humans. Although the precincts of the Temple were crowded by the people who prepared the ritual since yesterday, even in the crowd, even appearing from the hall, a messenger of the gods sent a lucky signal to the head of the clan. Embarrassed and delighted at the same time, I shed tears. In spite of my incompetence, I have always worshipped the Temple and the Shrine. Is it the reason why I met with this miracle?

Concluding question: Is the sacred deer in the Japanese tradition related to the sacred deer in the Eurasian tradition?

Archaeological and scriptural evidences from the Yayoi period down to the Heian period consistently show the sacredness of deer in Japanese culture. While the notion of the sacredness of deer in Japan must have originated on the Eurasian continent, there appears to have been a special reverence for deer in Japan since ancient times. Present-day Japanese have not forgotten the 'magic' qualities of deer, but the reason why deer occupy this place in their hearts seems to have been lost in the mists of time and has not been fully discovered.

Sacred deer have also been investigated by specialists of various areas of Eurasia from different academic angles. For instance, in the case of China, by Salmony 1954, in the case of Scythians, by Tchlenova 1963, in the case of Ancient Caledonians, by McKay 1932, in the case of Celts, by Fickett-Wilbar 2003; Hemming 1998, in the case of Hungrians, by Mátéffy 2012, in the case of Siberians, by Jacobson 1983 and 1993, and from the point of art history, by Cammann 1956; Doan 1983; Hančar 1952. It is my hope to seek connection between the image of sacred deer in Japanese culture and the images of sacred deer in other Eurasian cultures. Japan certainly received continuous cultural influence from China: in the case of the sacred deer, scapulimancy (the practice of divination by use of scapulae (shoulder blades), is mentioned above. In China, both scapulae and the plastrons of turtle were used) and auspicious animal images (dragons, phoenixes, tigers *etc.*) are clearly discernable. Still, there are possible candidates that also gave impetus to the formation of the sacred image of deer in Japan: the Scythians, the Iranians, the Sogdians, and the Mongolians.

In order to reconstruct the sacred animal images of Eurasian cultures, other animals such as birds (woodpeckers, eagles) and wolves must also be taken into consideration because an animal is not isolated in myths but forms a structure with other animal species.

References Cited

Non-Japanese texts

Amandry, Pierre 1965: 'Un motif 'scythe'en Iran et enGrèce', *Journal of Near Eastern Studies* 24, 149-160.

Aston, W. G. trans. 1972: *Nihongi*, Charles E. Tuttle.

Cammann, Schuyler 1956: 'The Animal-Style Art of Eurasia', *The Journal of Asian Studies* 17, 232-239.

Chamberlain, Basil Hall trans. 1982: *The Kojiki*, Charles E. Tuttle.

Doan, James E. 1983: 'The Animal Style in Celtic and Thracian Art', *Proceedings of the Harvard Celtic Colloquium* 3, 149-167.

Fickett-Wilbar, David 2003: 'Cernunnos: Looking a Different Way', *Proceedings of the Harvard Celtic Colloquium* 23, 80-111.

Hančar, Franz 1952: 'The Eurasian Animal Style and the Altai Complex', *ArtibusAsiae* 15, 171-194.

Hemming, Jessica 1998: 'Reflections on Rhiannon and the Horse Episode in 'Pwyll'', *Western Folklore* 57, 19-40.

Jacobson, Esther 1983:'Siberian Roots of the Scythian Stag Image', *Journal of Asian History* 17, 68-120.

Jacobson, Esther 1993: The Deer Goddess of Ancient Siberia: A Study in the Ecology of Belief, Leiden, New York, Köln: E. J. Brill.

Kidder, Jr., J. Edward 2007: *Himiko and Japane's Elusive Chiefdom of Yamatai*, Honolulu: University of Hawai'i Press.

Manyōshū: one thousand poems 1940:Tokyo : Published for the Nippon GakujutsuShinkōkai by the Iwanami Shoten.

McKay, J. G. 1932: 'The Deer-Cult and the Deer-Goddess of the Ancient Caledonians', *Folklore* 43-2, 144-174.

Mátéffy, Attila 2012: 'The Hind as the Ancestress, Ergo Virgin Mary – Comparative Study about the common origin myth of the Hun and Hungarian People', *Sociology Study* 2, 941-962.

Matsumura, Kazuo 2010: 'Can Japanese mythology contribute to comparative Eurasian mythology?', in Wim M. J. van Binsbergen& Eric Venbrux eds., *New Perspectives on Myth*, PIP-TraCS, 253-264.

Matsumura, Kazuo 2015: 'Heroic Sword God: A Possible Eurasian Origin of a Japanese Mythological Motif', *Cosmos* 31, 27-49.

Ogata Noboru: 'Messengers of the Gods – Deer of Nara': http://www.hgeo.h.kyoto-u.ac.jp/soramitsu/NaraDeer.html ; retrieved2016/09/29

Salmony, Alfred 1954: 'Antler and Tongue: An Essay on Ancient Chinese Symbolism and Its Implications', *ArtibusAsiae. Supplementum*, Vol. 13, Antler and Tongue: An Essay on Ancient Chinese Symbolism and Its Implications, 1+3-57

Tchlenova, N. L. 1963: 'Le cerf scythe', *Artibus Asiae* 26, 27-70.

van Binsbergen, Wim M. J., 2009-2010, 'Before the Presocratics. Cyclicity, transformation, and element cosmology: The case of transcontinental pre- or protohistoriccosmological substrates linking Africa, Eurasia and North America', *Quest: An African Journal of Philosophy / Revue Africaine de Philosophie* vols. 23-24, nos 1-2.

Wikipedia 'Nara Park': https://en.wikipedia.org/wiki/Nara_Park; retrieved 20160923

Japanese texts

Fujiwara no Munetada 1965: *Chūu-ki*, vol.4 (ZōhoShiryō Taisei Kankō-kai ed., ZōhoShiryō Taisei 12), RinkawaShoten.

Fujiwara no Nariyuki 2012: *Gonki* (trans. Kuramoto Kazuhiro), vol.2, Kōdanshā (KōdanshāGakujutu Bunko 2085)

KujōKanezane 1998: *Gyokuyō* (ToshoryōsōkanKujōkehonGyokuyō 5), KunaichōShoryōbu.

KujōKanezane 2011: *Gyokuyō* (ToshoryōsōkanKujōkehonGyokuyō 12), KunaichōShoryōbu.

Takeda Yūkichi and SatōKenzō eds. 1986: *Kundoku Nihon SandaiJitsuroku*, RinkawaShoten.

Torao Toshiya ed. 1993: *Shinto TaikeiKoten-hen 12, Engi-shiki, ge*, ShintōTaikeiHensan-kai.

Urban female initiation rites in globalised Christian Zambia

An example of virtuality?

by Thera Rasing

Abstract: In this chapter, I will discuss some ideas on globalisation, ethnicity and virtuality, as they have been analysed and used by van Binsbergen. The concept of virtuality has been used by van Binsbergen to explain female initiation rites in urban Zambia. Therefore, the article will particularly focus on female initiation rites in urban Zambia, and examine if and how van Binsbergen's ideas on globalisation, ethnicity and virtuality can be used to understand female initiation rites in contemporary urban Zambia. It will show the benefits of these concepts, as well as express some critiques on them, and explore other concepts to analyse initiation rites in current urban Zambia. It will argue that, while virtuality can be used to analyse them, this is too narrow a concept. These rites should also be seen as the construction of gender and adulthood, and the creation of a female domain, which is neither urban nor rural.

Keywords: female initiation rites, Zambia, virtuality, urban, globalisation

Introduction

Since the 1990s globalisation has been a topic of discussion in several sciences, including anthropological studies. In line with this, the Netherlands Foundation for Advanced Tropical Research (WOTRO) established a research programme on 'Globalization and the Construction of Communal Identities' (WOTRO, 1995-1999).[1] In this programme, discussions took place on the rapidly growing social-science literature on globalization, as well as on presentations of the newly conducted research by the members of that programme. During this programme, it was envisaged that there was need for further conceptual development in anthropological globalization studies. My contribution to this programme was a Ph.D. study (funded by the Programme) on female initiation rites in today's urban Zambia, in the context of the Roman Catholic Church as a globalising factor.

In the same research programme, Wim van Binsbergen applied the concept of virtuality, which he considered a key concept for a characterization and understanding of the forms of globalization in Africa. In his publications on the concept of virtuality (1997, 1998a, 2015), van Binsbergen uses several empirical case studies of contemporary Africa to apply the theory on virtuality, thereby examining the problem of meaning in the African urban environment. Apart from using his earlier study on the Kazanga festival in rural Western Zambia, he also uses Rasing's (1995) case study on female initiation rites in urban Zambia, claiming that they illustrate particular forms of virtuality as part of the globalization process. As these initiation rites were the topic of my long-term study (both for my Master's, my Ph.D. and thereafter), this chapter will mainly focus on these initiation rites.[2] The chapter explores whether the concept of virtuality is helpful to understand and analyse female initiation rites in urban settings in today's Zambia, and I will examine other ways of analysing and interpreting these initiation rites.

The chapter starts with some definitions and explanations of the concepts of globalisation, ethnicity and virtuality respectively. It will pay attention to localisation and meaning in an African urban context. After this, female initiation

---

[1] At the initiative, and initially under the directorship of, Peter Geschiere and Wim van Binsbergen, subsequently joined by Bonno Thoden van Velzen and Peter van der Veer .

[2] See Rasing 1995, 2001a, 2001b, 2003a, 2003b, 2003c, 2004a, 2004b, 2004c, 2007a, 2010, 2014a, 2014b, 2015a, 2015b. Both my MA research and my PhD research were conducted at the initiative, and under the supervision, of Wim van Binsbergen (inspired by his own earlier research on Nkoya female puberty rites in Western Zambia); he also secured the funding for my PhD research in the context of the WOTRO national globalisation programme. I here register my considerable indebtedness to him.

rites in urban Zambia will be scrutinised. It will examine whether the concept of virtuality can be applied to analyse these urban rites, as well as explore other ways to analyse them. It will argue that, while virtuality can be used to analyse them, this is too narrow a concept. These rites should also be seen in terms of gender and adulthood construction, as well as the construction of a female domain, which is neither urban nor rural.

The concept of globalisation

Since the 1990's, social studies, following economic studies, have paid attention to globalization. Globalization revolves the world-wide flows of goods and ideas, not only the moving of these goods, but also the way these goods and ideas organize people, objects and individual experience, thereby creating new ideas, new groups, members' interaction and particularism, and social localization, within the global flow of goods. In this process of globalisation, the receiving society changes the ideas attached to certain goods, and subsequently the society is changed by creating new ideas, new groups, new sense of belonging and individualism.

In this sense, globalisation is usually seen as the spread of western ideas to non-western cultures. The process of globlisation, however, also involves creative reactions from 'the South', and hence the South also influences western societies. For instance, since a few decades societies have been influenced by the Internet, while it is easy to travel all over the world, recently also for people living in the South. This means that ideas about time and space change, whereby cultures come close to each other, or come together, especially in the case of migration.

Within globalisation, we usually refer to electronic products, such as television, computers, Internet, means of transport, *etc.*, that reduces time and space. These electronic elements of globalisation affect daily life and societies, and subsequently cultures. In addition to the world wide flow of goods and ideas, globalization is related to the reduction of time and space: It is 'the compressing of time and of time costs in relation to spatial displacement, as well as the meaning and the effects of such displacement' as van Binsbergen (1998a : 875) put it. Therefore,

'Globalization is not about the absence or dissolution of boundaries, but about the dramatically reduced fee imposed by time and space, and thus the opening up of new spaces and new times within new boundaries that were hitherto inconceivable' (van Binsbergen 1998a; 875).

Human cultures could perhaps be said to have always been subject to globalizing tendencies, which van Binsbergen calls proto-globalisation (van Binsbergen 1997, 1998a, 1999b). In the distant past, when the management of time and space was so limited that the social and cultural life-world was bounded by geographical limits, people, ideas, and goods did travel, often across wide distances, by the then common means of transport such as certain animals and sailing boats. Where no such conditions prevailed, movement meant dissociating from the social setting of origin, and establishing a new local world elsewhere, dislocated and disconnected from the world of origin.

Today, most people live in a world that is in contact with other parts of the world, and therefore deal with objects and ideas of other partsof the world.

Social aspects of globalization

Globalization has profoundly changed societies and cultures, and has far reaching consequences (see Featherstone 1990, Giddens 1990, 1991, Friedman 1995, van Binsbergen 1999b). Changes in time and space, as have been provoked by the globalization process, such as the reduction of time to reach a certain place or person – either physically, or virtually by the Internet, telephone, *etc.*, and changes in place, *e.g.* travelling between geographical spaces goes rather fast, by road or air; and takes a few hours instead of days or weeks. So, rather fast travelling between an African village and town has influenced village life in many parts of Africa. Even remote villages can be reached by telephone, while other parts of the world enter the village through television.

This means that Western and Southern worlds meet; (individuals of) both worlds see and influence each other. In this process of taking over aspects from other cultures, these aspects are changed. Obviously, this leads to the construction of new identities, as especially people in the South wish to take over certain objects and aspects from the west. In this process, in line with the creation of new identities, new rituals, too, are constructed. For instance, Zambian kitchen parties (that are huge women-only parties in which a bride receives her kitchen utensils, and 'white weddings' have been constructed as new rituals (Rasing 1997, 1999 and 2001, see also Tranberg-Hansen 1996).

So, with globalization, we are dealing with social aspects of what were initially technical issues. Therefore, globalization is a deeply and profound transformation of today's experience, especially in the South, but also in the Western world. It deals with the process of the appropriation of globally available objects, images and ideas in local contexts, which may constitute itself in that-process of appropriation (van Binsbergen 1998a: 880).

In line with this, globalisation studies have paid attention to the global – local debate. The local is not a given, let alone a solution (Appadurai 1990, 1995). As Appadurai (1995) put it, locality can not only be seen as social space but also as problematic, it is to be actively constructed in the face of a situation of non-locality.

Under today's conditions of communication technology and the social construction of self-organization for identity, the socially local is no longer necessarily the geographically near. However, in the distinction between social space and geographical space the material technologies of geographical space are still relevant in the face of locality construction.

Both Appadurai (1995) and Hannerz (1987, 1990, 1992) stressed the paradoxical phenomenon of the globalization theory that the increasing unification of the world in political, economic, cultural and communication terms does not lead to increasing uniformity but, on the contrary, goes hand in hand with a proliferation of local differences.

At present, a group of (initially locally) connected people may no longer be seen as a localized group or set of people who are connected by enduring social relations and forms of organization. The individuals who constitute that group do not necessarily construct their social experience as an ideal community with a name, an identity, moral codes and values. Their cultural and organizational identity and experiences are fragmented, heterogeneous, and might result in alienation.[3]

Thus, next to being global citizens, the construction of cultural and ethnic identities is an important aspect, as people feel the need to have their own identity. So, while cultures mix, and some individuals (seek to) have multiple ethnic identities, there is an increasing search for ethnic and cultural identity.[4] For instance, both in the West and in the South, there is increasingly more attention for cultural issues such as folklore and local languages or dialect, and increasing claims on local culture, rituals and history as part of one's own cultural identity. Examples of such new rituals are among others, cultural festivals such as the Kazanga (van Binsbergen 1992a and 1994b) and the Mwenya ceremony that has been installed in 1997 as a yearly ceremony for the Bemba, at chief Chitimukulu's palace in Northern Zambia. Perhaps also the claims to re-install the rapidly disappearing female (and male) initiation rites as they are severely curtailed and changed today (Rasing 2001, 2004, 2014b), can be seen in this context. As these female initiation rites are since 1995, considered 'gender based violence' by the UN and subsequently by Western NGO's (see Rasing 2013 and 2014b) this has led to a silent protest by elderly Zambians, both men and women, who claim that their children should know their culture, background and history, and to confusion among youngsters, who lack such knowledge. Before we turn to the section of initiation of urban rites and their meaning, we will now focus more on the concept of ethnicity.

The concept of ethnicity in (urban) Zambia

Ethnicity is another topic that van Binsbergen has extensively published about.[5] Van Binsbergen has shown how ethnic identity has changed and how ithas been constructed, particularly in the urban settings, where the population is comprised of multi-ethnic groups from all over Zambia. He has illustrated how ethnic identity is not (only) based on ethnic lines, whereas ethnic groups rarely exist of one purely ethnic line, but how this is being created in certain groups,

---

[3] See van Binsbergen 1998a: 886; see also Ferguson 1999.

[4] See van Binsbergen 1988a, 1999a, 1999b; Appadurai 1990, 1995; Hannerz 1986, 1987, 1990, 1992.

[5] See van Binsbergen 1992a, 1992b, 1994a, 1994b, 1994c, 1996, 1997a.

such as church lay groups and political parties. Hence ethnicity has to do with inclusion and exclusion of certain religious, political and other groups.

The classic example of how ethnicity is constructed is Mitchell's (1956) famous publication *The Kalela Dance*, which shows how, on the Zambian Copperbelt, through lines and boundaries of urban settings, songs, dance, and language ethnicity is created.

The construction of ethnicity is related to the concept of 'culture', which has become increasingly problematic. There are over two hundred definitions of culture, which means that the concept of culture is too complex to be grasped in a few words. Even norms and values in a given culture are not the same for everyone in that particular culture. This is related to the position, roles, age and gender of that person. Hence culture has to do with human behaviour, which may be partly prescribed by that culture, but which is also, at least partly, a choice of that person.

Yet, generally, anthropologists agree about certain characteristics of culture. The most frequently used definition is that culture is 'the whole set of learned customs and habits, attitude, symbols, images, values and norms of a particular group' (Kloos 1991: 15). In addition, as particularly globalization studies have made clear, this whole set of learned customs, attitudes, images and values change over time and according to its context and situation. Hence, cultures are not static but dynamic, and situational.

Social scientists have considered culture as hegemonic. Van Binsbergen protested against such a claim, and entitled his 1999 inaugural lecture 'Cultures do not exist' (*'Culturen bestaan niet'*, original Dutch title). With this lecture, van Binsbergen sought to go against the grain of taking culture for granted. He tried to deconstruct culture as a given, as a self-evident and hegemonic overall realm, in which individual identity can be defined. Individuals have multiple identities, or rather, multiple roles, and are not only defined by their ethnic group. People are not only Bemba, Lozi, or whatever ethnic identity, but are also urbanites, church-goers, teachers, fathers, sons, *etc.* A claim on culture or difference in cultures usually means that individuals are considered only as belonging to an ethnic group, which indicates that they are reduced to a particular ethnic group, which would be different from another ethnic group or culture. Hence, following this line, we do not see people as individual human beings, as urbanites, as mothers, daughters, students, *etc.* Moreover, cultures are mixed everywhere, especally in the current globalized world. For instance, in urban Africa, many marriages are mixed, Zambians go to a Western-based church, watch Mexican soap opera's, eat Western types of food, *etc.* In addition, particularly in urban areas', people take over certain aspects, ideas, customs from other ethnic groups, either entirely or partly. In this process, these ideas and customs are changed, while at the same time, the 'traditional' culture, as derived from the village, is only partially brought to town. Hence, cultures are fragmented, partial, and not hegemonic.[6] With these theoretical concepts of globalisation and ethnicity, we can now explore the concept of virtuality, which will be the topic of the next section.

The concept of virtuality

Van Binsbergen (1997, 1998a) claims that globalization is accompanied by virtuality. The terms virtual and virtuality have a well-defined history (van Binsbergen 1998a: 876).The term virtuality was already used in the Aristotelian philosophy, upon which Western Europe based its intellectual life. Around 1800, virtuality became an established concept in the field of physics. The theory of the 'virtual image' had already been formulated a century earlier by optics: the objects shown in a mirror image do not really exist, but are merely illusory representations, which we see at the end of the light beams in the surface of the mirror. In the twentieth century of information technology the term 'virtual' has gained a new meaning, derived from the meaning given to the term in optics (van Binsbergen 1997: 9 ; 1998a: 876).

Virtual reality has become a common placeof globalised postmodern daily life, including in contemporary Africa, where electronic media, such as television and video, are increasingly being used. Therefore, this form of virtuality can also be seen here (van Binsbergen 1998a: 876).[7]

---

[6] See van Binsbergen 1994a, 1999b, see also Landman 1999.

[7] In one of his most recent publications (van Binsbergen 2016: 85-168), van Binsbergen has published to substantially rewritten, greatly expanded version of his 1997 short book on *Virtuality* (Editor).

However, the applicability of the concept of `virtuality' extends further. Foucault (1966) applies the notion of `virtual discourse', while Jules-Rosette uses the notion of virtuality for a verbal discourse in a specific discursive situation which she calls the `symbolic vindications of modernity's broken promise' (Jules-Rosette 1996: 5), which plays a central role in the construction of post-colonial identity. Van Binsbergen (1998a) applies the concept of virtuality further concerning beliefs and images that are provoked and inspired by the use of electronic equipment, as well as in a more social and ritual context. It is in this context that he notes that virtuality refers to a specific relation of reference between elements of human culture.

> 'This relation may be defined as follows: 'once, in some original context C_1, A virtual referred to (that is, derived its meaning from) Areal. This relationship of reference is still implied to hold, but in actual fact Avirtual has come to function in a context C_2 which is so totally dissimilar to C_1, that Avirtual stands on itself; and although still detectable on formal grounds as deriving from Areal, has become effectively meaningless in the new context C_2, unless for some new meaning which Avirtual may acquire in C_2 in ways totally unrelated to C_1. Virtuality, then, is about disconnectivity, broken reference, de-contextualization, yet with formal continuity shimmering through' (van Binsbergen 1998a: 877).

In this definition, van Binsbergen considers Non-Locality as Given, Locality as an Actively Constructed Alternative, Virtuality as the Failure of Such Construction. He further states that:

> 'Applying this abstract definition, we may speak of virtuality when, in cases involving cultural material from a distant provenance in space or time or both, signification is not achieved through tautological, self-contained, reference to the local, so that such material is not incorporated and domesticated within a local cultural construct, and no meaningful contemporary symbolic connection can be established between these alien contents andother aspects of the local society and culture.' (van Binsbergen 1998a: 877)

So, important aspects related to virtuality are locality and meaning, or, more precisely, the individuals' production of meaning. The concept of virtuality is used to analyse the construction of meaning in today's urban Africa. Meaning can no longer be seen as encountered and manipulated in a context far way, in time and space, from the social context of production and reproduction; it is no longer local and systemic, but fragmented, virtual, odd, maybe even absent (van Binsbergen 1998a: 886). These concepts of locality and meaning will be elaborated below.

Virtuality, locality and meaning in globalised Africa

Before the development of communication technologies such as the telephone and means of transport, the similarity between social space and geographical space was obvious. In pre-industrial African societies, where such technologies did not exist, social space and geographical nearness continued to be similar, that is: the geographically near is the local, or even the local culture. Although elements of local culture are intertwined and depend on each other, there is actually a large number of separate cultural aspects, and material cultural objects exist among different groups of people involved, within a rather limited space and time. Hannerz (1987) uses the term creolization, which does not mean that the systemic nature of local culture has been abandoned or destroyed by outside influences, but that it accounts for appreciably less than the entire culture: a considerable part falls outside the system. Creolization can be seen as a specific form of virtuality, as a departure from the systemic nature of local culture. As Appadurai (1995) has described: If culture produces reality in the consciousness of the actors, then the reality produced under conditions of such departure is, to the extent to which it is virtual, only virtual reality. Virtual reality has become a common concept in post-modernism (Kapferer 1988).

In Africa, virtuality today manifests itself through the incomplete systemic incorporation of foreign cultural material which circulates primarily in urban areas. Examples of this form of virtuality are found all over Africa today, such as the Kazanga festival in West-Zambia (van Binsbergen, 1992a, 1994) and the Zambian kitchen parties, as described by Rasing.[8]

These rites and ceremonies go together with forms of social organization which create the socially local within the global perspective. The local is here created through the formation of new categories and groups, the boundaries around them, and the positioning of objects and symbols through which these boundaries are manifested. Such groups are no longer geographically local, as they are not a bounded geographical space, while the population consists of people belonging to different social groups or organizations. These groups or organizations are ethnic associations, churches, political parties, schools, *etc.* They ac-

---

[8] See Rasing 1999, 2000, 2001a, 2001b and 2009.

tually have a limited geographical area, but within that area, the majority of inhabitants are non-members, and therefore, they do not constitute contiguous social spaces.

Even though this is typical for the urban setting, social and ritual life is derived from village life. As van Binsbergen claims (1998a: 881): 'virtuality presents itself in the form of an emulation of the village as a virtual image.'

In his publications on virtuality, he used female initiation rites in urban Zambia, - the topic of my study -, as a case study to illustrate this theory. Before examining these rites in the face of virtuality (see other sections in this article), it is needed to investigate meaning in rural and urban settings. This will be done in the following section.

Social life and meaning in an urban – rural context

Even though there is much evidence that in the distant past African cultures have mixed with other parts of the world, as goods and ideas have been transported from other parts of the world, especially the Middle East and South East,[9] the previous anthropological image of `the' African culture and of the African village was wrongly presumed as holistic, self-contained, locally anchored, based on ethnic identity. As van Binsbergen (1998a: 882) claims, this image was deliberately constructed so as to constitute a local universe of meaning, which actually is the opposite of virtuality. Since such a culture was thought to form an integrated unity, all its parts were supposed to be part of that same coherence, which provided an illusion of localized meaningfulness.

The focus of globalization studies is the urban setting, in which one has easily access to international lifestyles mediated by electronic media. Today, however, African villages are affected by western ideas and goods, and therefore, are now also globalized. According to van Binsbergen, when focussing on meaning in terms of virtuality, we should examine globalization from the starting point of the African village and its internal processes of meaning. African towns have always been a context for cosmopolitan meaning which does not stem from the

---

[9] See van Binsbergen 1996a, 1996b, Casely-Hayford 2012.

villages in the rural regions surrounding the town, but reflects, and is reflected in, the world at large. While initially, the world view of urbanites strongly referred to that of the village, at the same time it was invoked with many changes. Therefore, also the meanings changed (van Binsbergen 1998a; 874).

In Zambia, towns have been established since the 1930s, when, after the development of the copper mines, towns in these areas were founded and grew rapidly. The area became known as 'the Copperbelt', and became the largest urbanised area in Zambia. This area attracted many people from all over Zambia, especially the Bemba and Bemba related people from Northern Zambia, who started living in the area, in which originally people from the ethnic group Lamba resided. In this process, the Bemba identity, undergoing considerable transformation and expansion, became dominant in these towns, and the 'town Bemba' became the *lingua franca*. Hence, towns comprises of a heterogeneous migrant population, and have social, political and economic structures that are different from the village context. Until about the 1970s, for the urban migrants on the Zambian Copperbelt, the village society was the context in which many-urbanites were born. Today, even though many inhabitants in urban settings have been born in town and never visited the village, there is still some reference to village life and culture. In the 1940-1980's studies, the dichotomy between town and village was a central issue. In current anthropological studies, however, this dichotomy has lost importance. Urbanites have increasingly less contact with the village, especially those who were born in towns and have never been to the village. At the same time, however, for villagers town becomes increasingly near, as travelling urban areas is rather easy today. Therefore, goods and ideas are transported from town to the village, rendering the difference between urban and rural only gradual.

However, van Binsbergen (1998a: 884) claims that the dichotomy between town and village remains relevant in so far as it influences individuals' conceptualizations of their life-world and social experience. The image of the village is an imaginary context, as it is either idealised but more often considered by urbanites as obsolete or backwards.[10]

It is worth mentioning that the construction of the twentieth century village

---

[10] See Powdermaker 1962, Turner 1968, van Velsen, 1971, van Binsbergen, 1981, 1992b, and 1998a.

should be seen in the context of huge social, political, economic and religious changes in the nineteenth and twentieth centuries, in which people were confronted with the erosion of the village model, as well as with other structures and political and religious ideas.[11] As a consequence, the Zambian population, as elsewhere in Africa, have struggled in combining various ideas and forms of organizational, ideological and productive innovation that were imposed from outside with local practices and ideas to reconstruct a new sense of community.

In addition, the old village order, and the ethnic cultures under which it was usually subsumed, may in themselves have been largely illusory, or constructed by the ideological claims of elders, chiefs, colonial administrators and missionaries, open to the cultural bricolage of invented tradition by these actors claims that the village still represents a model of viable community among African villagers and urbanites today, as this model refers to a collective idiom pervading all sections of contemporary society. This may have been true in the 1980s, when he wrote this, but I dare to contest this for today, the twenty-first century.

From migrants to urbanites

The issues van Binsbergen raises are inspired by the ideas of the researchers from the Manchester School / Rhodes-Livingstone tradition, who conducted research on the Zambian Copperbelt in the then recently constructed towns, and who considered the population of the urban Copperbelt as migrants, who, born and bread in the villages, were supposed to return to these villages after several years. They only had the village as reference point.[12]

However, the majority of today's urbanites have been born in urban areas, and even their parents and grandparents have. Many urbanites have never been to the village, and do not even know their relatives who might live in the village. Therefore, urbanites should predominantly be seen as urbanites. As Gluckman (1960: 57), already put it more than fifty years ago: `the African townsman is a towns-

---

[11] See Richards 1959, 1969 and 1971, van Binsbergen1981, Hinfelaar 1989, Oestigaard, 2014.

[12] See Gluckman 1945, 1960, Mitchell 1956, 1969, Epstein 1958, 1867, 1969, 1978, 1981, 1992, Powdermaker 1962, Mayer and Mayer 1974.

man'. In other words, the African townsman is not a displaced villager. Also Mitchell (1956) emphasised the influence of townlife on migrants, who tried to establish their life in town according to new structures, while Hannerz (1980), too, considers townsmen as townsmen, as many of them were born in town.

Currently, there is a new anthropological interest of studying urban life (*e.g.* Van Dijk, Spronk), but these studies mainly focus on the middle and upper class.

Even though there is much difference between social classes in towns, - the urban poor living in small house in compounds, with little education and or or poorly paid jobs whereas the rich occupy large houses, have well-paid jobs and have a high level of education - the majority of towndwellers consider themselves as urbanites, modern, educated, Christianised, as reference to the modern, Western world, and being part of that word. Although expectations of urban life did not come true, and being part of the global world may lead to alienation (Ferguson, 1999), urbanites consider themselves modern and globalised.

Despite claims on ethnicity, and some slight differences in the ethnic background of the urbanites, concerning matrilineal or patrilineal kinship system, emphasis on cattle, the marital system, cosmological ideas, the rural background and culture of urban migrants is not all that different. Through their rather similar cultural background, they can share meaning.

Moreover, urban migrants develop a new common idiom, both for communication as well as for the patterning of their daily relationships (van Binsbergen 1998a: 888). In addition, urbanites have to deal with issues related to the capitalist mode of production, which structures their economic participation and hence their experience of time, space, personhood and social relations; which may lead to alienation (Ferguson 1999). In addition, Christianity provides organizational forms and ideological orientations that appeal to the urban population. Also, cosmopolitan consumer culture, ranging from fast food shops to luxury furniture stores displaying a middle-class life-style, contributes to create identity of a modern, urban person.

The cosmopolitan sets of meaning differ considerably from the meaning embedded in the rural setting. They are derived from a western context, but are present in urban African as mutually competitive, fragmented, and optional.

Yet, together they constitute a realm of symbolic discourse that, even though internally contradictory, prevails over the local repertoires of meaning of urbanites. Urbanites have their specific ways to interact with and conceptualize, construct, keep apart, and merge cosmopolitan and rural idioms (van Binsbergen 2004). Many urbanites today, experience that the urban setting as a social space lacks the coherent integrated structure which could produce a system of meaning (van Binsbergen 1998a: 890).

As the idea of wholeness of a culture appeared questionable, the postmodern approach of a fragmented and incoherent multiplicity of repertoires of meaning became common. The postmodernist view may be combined with the notion of articulation of modes of production, - a notion on which van Binsbergen has extensively published (see van Binsbergen 1981, Geschiere and van Binsbergen 1985), -in order to better explain ideas of incoherent and segmented urban culture, fragmentation and contradiction of meaning, resulting in alienation.[13]

In addition, Hannerz (1980, 1986, 1989, 1990, 1992) has explained processes of cultural production, variation and control that appear in African towns from the perspective of the modern world as a unifying, globalizing whole. However, Hannerz does not problematize the concept of meaning, but takes meaning rather for granted and focuses on the social circulation of meaning (Hannerz 1992: 17, 273).

According to van Binsbergen (1998a: 891), urban immigration in Africa, with new urban migrants living together with their relatives from the village, suggests that this migration 'tends to be partial' as 'towns display apparently rural-derived elements, and tend towards high levels of virtuality / discontinuity / transformation'. Therefore, as he (van Binsbergen 1998a: 891) claims, 'it is important to look at meaning in African towns not only from a global perspective but also from the perspective of the home villages of many of the urbanites or their parents and grandparents.' I agree with this statement that urban areas have both elements derived from the village and from the foreign, global world, and subsequently we should look at meaning from both perspectives, but I think there is a difference between migrants and urbanites of whom many are the second or third generation in town, and between the lower class in the urban compounds, and the middle and upper class occupying

---

[13] See van Binsbergen 1981; van Binsbergen and Geschiere, 1985, van Binsbergen, 1993a, Ferguson 1999.

urban compounds, and the middle and upper class occupying low density areas, who, through high education, are much influenced by the globalisation process, as they have more access to Internet and other means of communication and travel abroad.

Female initiation rites in urban Zambia

I will now focus on present-day female initiation rites in urban Zambia, as has been a topic of my research since 1992. According to van Binsbergen (1997 and 1998a), these initiation rites illustrate particular forms of virtuality as part of the globalization process. I will examine whether his statement is true, arguing that this is too narrow an interpretation, and discuss what other interpretations can be applied to interpret these rites.

Female initiation rites are one of the dominant kinship rites (even more so than male initiation rites). These rites are similar all over South Central Africa.[14] In these rites girls are taught about adulthood and womanhood. Therefore, these rites are performed when a girl becomes matured, that is, when she starts menstruating.[15] In these rites girls are taught about their productive and reproductive tasks, gender relations, sexuality and other norms and values. Until the turn of the twenty-first century, this was done in detail, using clay figures and wall paintings (*mbusa*) and songs. Today, however, these rites have been severely curtailed; many aspects are omitted, and are often postponed to the wedding ceremony. This is due to various reasons, such as Christianity, western education, urbanisation and modernity, as will be elaborated on later in this article. Yet, despite the waning and curtailment of these rites, they are still performed, especially in rural areas, but also in urban settings.

Both the urban areas of the Copperbelt and Lusaka were the focus of my research on female initiation rites, where I have observed many of them, from the

---

[14] See van Gennep 1909; Richards 1939, 1945, 1956 / 1982, 1984; White 1953; Gluckman 1962; Turner 1964, 1967; Hoch 1968; Mayer 1971; Mayer & Mayer 1974; La Fontaine 1972, 1986; Jules-Rosette 1979; Corbeil 1982; Hinfelaar 1989; van Binsbergen 1987, 1992b, 1993b; De Boeck 1992; Rasing 1995, 2001a, 2001b, 2001a, 2003b, 2003c, 2004a, 2004b.

[15] See Richards 1956, Rasing 1995, 2001a, 2001b.

early 1990s up to today. In addition, I conducted research in rural areas, predominantly in the Northern (now Muchinga) and Luapula Provinces, and to a lesser extent in the Eastern, Western and North-Western Provinces. This provided me with a good picture of the performances and the changes in these rites, as well as with the differences between rural and urban initiation rites today. As these initiation rites are now often postponed to the wedding ceremony, and as such the initiation rite is curtailed and waning, there is still some sort of initiation rite performed when girls become mature. A major part of my research was to examine initiation rites as they are performed by Christian women, particularly in the context of the Roman Catholic Church.

The performance of female initiation rites in urban Christianised Zambia

Churches have played an important and interesting role both in the performance and the abolition of female initiation rites. The first missionaries who came to Zambia, in the late nineteen century and early twentieth century, did not know much about these rites. In the 1930s, certain Catholic priests studied these rites without judgement (see Labreque 1931 and Etienne 1948). In the 1940s-1960, attempts have been made to abolish these rites, both in Protestant and Roman Catholic circles in Zambia, with different types of sanctions(Verstraele-Gilhuis 1982, Rasing 2001a).Yet, at the same time women organised themselves to ask the clergy to stop these sanctions and to continue performing these rites (Verstraelen-Gilhuis 1982, Rasing 2001a). This has resulted in some attempts to construct Christian initiation rites (see Cox 1998). From the late 1960s onwards, after the Second Vatican Council, the Catholic Church, has changed its attitude towards these rites (Rasing 1995, 2001a, 2004a, 2010).This was in line with the policy of inculturation (Dondeyne 1967), in which priests studied cultural aspects, including initiation rites, in order to understand more about Zambian culture and rituals, and to format them into a more Christian rite. Attempts have been made to combine these rites with Christian rituals such as Confirmation, and to bless the novice in church after her initiation, but these attempts have mainly failed, because women preferred to keep these rites separate from the church.

Therefore, it is remarkable that today, these initiation rites have been approved and are encouraged by certain, although not all, Roman Catholic priests. The Roman Catholic Church, that can be considered a major agent of globalization, has sought to impose its particular ideas, religion, hierarchy, sanctity, and meaning, including that of human reproduction, and human life rituals, on the African population.

Imprinting Christian ideas was partly done by installing women's lay groups (and, to a lesser extent, male lay groups). These groups have played an important role in church, as well as in the social life of women, especially in towns. There are several lay groups, each with its own uniform and paraphernalia, formal structure within the overall church hierarchy, weekly meetings and prayers, and topics of attention such as caring for the sick, looking after the poor, the struggle against alcoholism, and cleaning the church. The organizational form and routine of these groups, and the social embededness this offers to its socially uprooted members, provides a structure of social locality and assistance among women, providing 'a place to feel at home'. This might be of greater relevance than the religious ideas and practices, while it is also a way to engage in formal organization. Within these lay groups, in which meaning is given to the members, at the same time the bedding and boundaries are created, within which the flow of goods, images and ideas as conveyed by globalization, provide an identity.

Some women's lay groups have been founded especially with the aim of 'teachings girls', that is, performing initiation rites in a rather traditional way, but somehow adopted to Christianity (Constitution Nazarethi, Rasing 2001a). The first group founded for this aim was the Nazarethi, founded in 1956 in Mufulira, a small Copperbelt town, by the then priest Mutale, who later became bishop. Like all other lay groups, this group has spread all over Zambia. Another group founded with this aim is the St. Ann that has spread to many parishes in Zambia, especially in Lusaka and Eastern Province. In addition, a men's lay group, St. Joachim, has been founded and spread in Lusaka and Eastern Province, aiming at teaching young men who are getting married. So, today, in the urban Copperbelt the Nazarethi, and in Lusaka, (particularly in Bauleni and St Francis parishes)both the Nazarethi and St. Ann, perform female initiation rites with the consent of the priests. In addition, in Lusaka, the Nazarethi and St. Ann use the church premises to have weekly rehearsals for these rites in the presence of the priest. This not confined to urban churches, but also in the rural district Chinsali, where I conducted part of my research, and in the Northern Province town Kasama, church lay groups now perform these rites in the context of and with open support from the Catholic Church.

Ii is even more remarkable, that initiation rites are performed for Catholic priests and sisters of certain congregations, when they are ordained or take their vows (Rasing 2004a).

The lay group's symbolic repertoire for initiation rites, as performed today in urban and rural church circles, are very similar to and have incorporated a lot of symbols of the 'traditional' initiation rite. This is also the case in the so called 'Christian' rite that is a rite performed by other Christian or Catholic women. The same 'traditional' clay models, wall paintings and songs are used, even though certain symbols are interpreted in a Christian way. Here, the issue of meaning comes in.

Elsewhere (Rasing 1995 and 2001a) I have shown that these rites remain very important for Christian women, as they teach about adulthood, femininity, agriculture, reproduction, sexuality, and other norms and values that a woman is supposed to know, thereby imprinting knowledge about adulthood and womanhood to the novice. In contrast with the previously rather hidden per-formance of these rites, the women lay groups have used the church and their authority as a context within which to perform initiation rites. This shows a remarkable change and gain on the side of the women.

Would the term 'folklore' be a proper term to interpret these initiation rites, as the priest involved in these rites told me? Indeed, I think the rites for priests and sisters, as well as the weekly rehearsals in church premises, might be seen as folklore. Con-cerning initiation rites for girls who are matured, would it be more appropriate to interpret these rites in terms of virtuality, as van Binsbergen does?

Theoretical concepts to interpret urban female initia-tion rites

While claiming that there has not been an adequate framework to interpret these rites, van Binsbergen illustrates several approaches that have been made to interpret them (van Binsbergen 1998a: 893). I will summarise them below, while adding my remarks.

One interpretation is that in terms of socialisation: people are socialized and grew up with the inertia of culture: even if urbanites adapt to new forms of social and economic life, in childhood they have been socialized into a particular (rural) culture which they seek to continue in their lives, especially concerning existential issues. In this way, performing a rural initiation rite in town would restore or perpetuate a cultural orientation which has its focus in the village, which actually means the rural residential group in towns. In terms of socialisation, it is interesting to note that Heather Monachonga (2009) in her comparative study among Zambian girls initiated in a 'traditional' way as compared to girls taught at school about 'modern' sexual reproductive matters found that the latter had little influence, since those girls were raised or socialised in their families or kin group, in which they had frequent 'lessons' and discussions about sexual issues, practices of attractive 'dancing' or wrinkling their waist, and so on. Even when girls were not initiated in the 'traditional' way, they knew a lot about traditional sexual and reproductive matters, as they were told informally by female relatives.

Another interpretation is that in terms of long-term cultural orientation, in the sense of Bourdieu's habitus: girl's initiation deals with the inscribing, into and through the body, of a socially constructed and mediated personal identity which implies, as an aspect of habitus, a total cosmology, a self-evident way of positioning one's self in the natural and social world; in a layered conception of the human life-world. Such habitus situates itself at a deep and implicit layer, despite individuals' and groups' adaptations in the conjuncture of social, political and economic conditions. Monachonga's argument (2009), as mentioned above, is also in line with this notion of habitus and positioning one's self in the social world. In addition, in terms of long term cultural orientation, reference should also be made to one's (ethnic) history. Particularly in the face of globalisation, in which 'cultures' tend to be mixed and less distinctive, while adapting western life-styles, there is a need for or claim of being distinctive, as a separate (ethnic) group, and the need to know one's roots and history, often referring to ancestral power and an (assumed) geographic area of origin. This is especially so in the contemporary Zambian context, in which several rituals, especially those dealing with sexual issues, are considered 'harmful cultural practices' by the UN and subsequently, other international NGOs, who therefore try to do away with them (Rasing 2014a, 2014b, 2015a and 2015b).

A third interpretation may be that due to being socially and economically insecure, urbanites seek to create, through rituals, a basis for solidarity so that they may appeal to each other in times of crises such as illness, funerals, and unemployment. In town, the traditional ritual may help to engender such solidarity. This would create a fictive kin group, and can be created by any ritual, not only that of initiation rites (van Binsbergen 1998a: 893).

These interpretations of rural initiation rites assume that the urbanites involved are recent urban migrants and still have much relationship with the village. This argument cannot be held in contemporary urban Zambia, as many urbanites have been born in town and have never visited a village. Therefore, it can be claimed that urban rituals seek to create an urban community, as a new form of social locality open to world-wide influences (van Binsbergen 1998a). This interpretation, however, emphasises the (construction of a) community, rather than emphasising social ideas of adulthood. Moreover, in my opinion, this interpretation focuses too much on urbanites selectively adhering to rural forms in an urban context. There is, however, a rather mutual interpenetration and blending of these rites; many ethnic groups have adopted some rituals from other ethnic groups, thereby constructing a mixture of the initiation rite. More remarkable, however, in line with the rather dominant ethnic group of the Bemba, many other ethnic groups try to adopt so called Bemba rituals into the female initiation rites. This is especially so if the novice is supposed to get married to a Bemba or Bemba-related ethnic group (Rasing 2001a).

In addition, many changes in these rites can be seen as compared to their performance in rural areas. My research has shown that there is much more difference between initiation rites in urban and rural settings, both concerning the duration and the songs and symbols used, than as compared to the elapse of time. For instance, the rite as observed by Richards in 1933 (Richards 1956), is very similar to the rite I observed in the same village (chief Nkula's village) in 1995, while urban rites are much more curtailed, many aspects are left out, and new aspects have been added as adopted to modern towns life. [16]

Interpreting urban female initiation rites as virtual

According to van Binsbergen, these rites should be interpreted in terms of virtuality, as urban women do not belong, nor aspire to belong, to the village world from which the rite is derived. He states:

> 'For while the kinship ritual emphasizes reproductive roles within marriage, agricultural and domestic productive roles for women, and their respect for authority within the kinship structure, these urban women are removed from the model of rural womanhood upheld in the initiation' (van Binsbergen 1989a: 897).

However, many of these women have their urban gardens, however small. Indeed, many urban women have little effective ties with a village, and their sexual and reproductive behaviour is different than what is officially taught in the rite, as many have multiple sexual relationships, before and during marriage, they are often without effective and enduring ties with a male partner, and are often female heads of households. Yet, officially the initiate is taught about kinship structures, taboos, reproductivity, sexuality, and authority of the husband, in-laws and her own kin, while at the same time women discuss their problems in marriage, their sexual escapades, rebuking their husbands and so on, in an unofficial way during the rite. In this way, the novice learns both about the construction of womanhood, social and kinship ties, and proper behaviour in an official way, as well as in an unofficial way about problems in social and daily life in today's society (Rasing 2001a).

This urban initiation rite is concerned with the construction of meaningful social locality out of the fragmentation of social life in the Copperbelt and Lusaka, both in the compounds and in low density areas.

Girls are initiated by the ritual leader (*nacimbusa*) that belongs to the same ethnic group as the girl. It frequently happens that these ritual leaders travel long distances, in order to initiate a girl from their ethnic group. This means that ethnic ties are still important. Also in the multi-ethnic church groups, the

---

[16] See Rasing 1995, 2001a 2001b, 2003a, 2003b, 2003c, 2004, 2014a, 2014b.

rites are performed according to the ethnic group the initiate belongs to. Yet, since the Bemba are dominant on the Copperbelt, and all attendants of the rite should assist in performing, it often happens that Bemba aspects are included in the rites for non-Bemba girls. In this way, both ethnic ties – whether constructed or not - are shown to be important, while at the same time a community of women is created.

The communal identity which is constructed through the initiation rite is that of a community of women, both as 'traditional' women, who live in town and therefore consider themselves as modern women, while at the same time they also consider themselves as Christians, and therefore leave out certain aspects of the initiation rite which they would not stick to in their daily life.

Even though only few urban women work in their fields, they do not stick to many of the sexual aspects taught in the rite, do not observe the rules of conduct and the taboos to which they were instructed at their initiation; and who in many cases will not have a formal marriage with their male sexual and reproductive partners, it is the ideal which they should know, the ideal that has been taught through many generations. Therefore, symbols, songs and clay models that have been used for many centuries in the initiation rites, continue to be used, as a proper way to install personhood and womanhood in the initiate. This might be explained as virtual. Yet, there is more to it.

As I have written elsewhere (Rasing 2001a) the social construction of womanhood, and personhood in general, is such a subtle and profound process that foreign symbols (as mediated through the Christian church) are in themselves insufficiently powerful to bring about the bodily inscription that produces identity. Even though girls are socialised in this process of adulthood in daily life, there is need for an official rite in order to establish and officially announce that the girl is now an adult, and subsequently knows the secret knowledge of adulthood. This is related to the ritually imprinted knowledge of adulthood on the physical human body, as mediated through the rites. Moore et al (1999) have shown that bodily aspects are important to establish personhood, gender and culture. Hence the bodily aspects as used and transformed in initiation rites, are needed to mediate female personhood. This comes close to Spronk's ideas about sexuality, intimate relationships and bodily sensations (2009 and 2014a). Spronk claims that body-sensorial knowledge can be used to understand the relation between the social significance of sexuality and erotic sensations, arguing that the sensual qualities of sexuality are mediators and shapers of social knowledge that help to understand how causal relations, such as the reconfiguration of culture, gender and sexuality in postcolonial African society, are registered in people's self-perceptions.

The fact that alien symbols are insufficiently powerful to establish adulthood can also be seen from the fact that many elderly (and also young) people complain about the claims by the UN and international NGOs, supported by teachers, that initiation rites are 'harmful cultural practises' and should therefore be abolished (Rasing 2014b).

So, we may say that initiation rites appear as virtual, as a lack of connectedness between the urban day-to-day practice of womanhood today and the ideological contents of the initiation. Yet, I do not think women hark back on a village model, but this model that happened to be performed previously in the common setting, which was the village, is now performed in what happens to be the common setting today, the town. In this way, we can speak of a long-term habitus.

The creation of gender and adulthood

In addition to the above mentioned analyses, these rites focus on the creation of personhood, meaning female adulthood, and the construction of kinship ties. Both construct the girl's gender, and social and reproductive relationships, which creates a sense of belonging.

These rites create adult women and install adult normative behaviour on the novice. Concerning this gender aspect, contrary to what is often believed, these rites do not install a merely submissive attitude on women towards their future husband or men in general. Traditionally, in these rites the girl was taught about her important role and tasks in the household, the extended family, relationships with the ancestors, and so on. Gender relations were based on mutual respect and complementarity (Richards 1956). Yet, the novice is told how to manipulate her future husband in such a way that it would seem as if he takes the decisions, while actually she will make sure that she gets what she wants. Moreover, female tasks such as household chores and cooking are considered important aspects of a 'proper woman',[7] as women are supposed to look after their domain of the interior house and the health of her family and relatives. (In this regard, the common idea that existed in the past that health is related to fire and sex,[8] has become obsolete and is therefore no longer taught in the rites). Yet, the idea that cooking and sex are related (Richards 1933, 1956, 1984, Rasing 1995, 1997, 2001a) still exist; a woman will only cook for her husband or relatives, and no man will eat together with a woman with whom he does not have sex with, unless it is a close relative. It is remarkable that among the middle class, a maid can do the cleaning, but is never allowed to do the cooking for her boss' husband. The relation between food and sex has also been described by Nootermans (1999).

While cooking is usually considered a typical woman's job, hence related to gender, I agree with Arnfred (2011: 254-255) who claims that in the African context, the concept of gender cannot strictly be applied, indicating that there should be a more flexible notion of the genders, i.e. of 'woman'. 'Certain social positions carry more power that the others, and, depending on the context, these positions may be occupied by biological men or biological women' (Arnfred 2011: 255). The same can be applied to the Zambian context. This can also be seen in the 'traditional' names that do not distinguish between male and female.

Sex is another important part of personhood, and was therefore an important aspect of the teaching in initiation rites. This is not only so in the context of procreation as the girl is supposed to produce offspring for the matriliny,[9] but also to maintain a proper and pleasurable relationship with her husband. As sex is supposed to be an important aspect of this relationship that should be done frequently and should be pleasurable for both the woman and man, the girl is taught how to deal with sex and how to make sure that both she and her future husband will enjoy sex, ranging from attracting a man, the foreplay, sexual intercourse and many variations of the sexual poses. It should be noted here, however, that today, due to pressure of world-wide organisations such as the UN, international NGOs and several churches such as Pentecostal and Protes-

---

[7] See Ogden 1996, Rasing 1995, 1997, 2001a, 2009.

[8] See Richards 1956 and 1984, Rasing 1995, 2001a.

[9] See Douglas 1971, Richards 1933, 1956, 1984, Rasing 1995, 2001a.

tant including the United Church of Zambia (UCZ), in many initiation rites these teachings about sex are minimized or omitted, with some exceptions of rural areas in North-Western and Eastern Zambia (Rasing 2014b).

However, currently in Zambia, girls learn about sex and sexuality in a more 'modern' or rather western way; at school, through television and Internet, and from their peers. Such ways to learn about sexuality are commonly used, and often illustrate rather woman-unfriendly ideas, subtly referring to a submissive position of women. This is contrary to the teachings in initiation rites.[20] More-over, the access to information about sex (as part of the globalisation process by means of communication technology), and the postponement and curtailment of female initiation rites, as a result of global influences such as churches and worldwide organisations (UN, NGOs) have resulted in a huge increase of teen-age pregnancies. This, together with a lack of respect for elderly people, has left the elderly and people with authority, in a rather awkward position. Against this background, there is a longing for the past, a cry to re-install initiation rites (both for girls and boys) in a rather traditional context (Rasing 2014b).

In addition, in contemporary initiation rites in urban settings, there is attention for 'romance' and 'love'. Even though I have heard these words in the initiation rites I observed in the 1990s, they are now more frequently used. As we do not know if these concepts were used in the past, and if so, what would be meant by them (see Stephens 2013), it can be assumed that these concepts have gained more importance (see Lindholm 2006, Spronk 2009 and 2014a, van Dijk 2013 and 2014). This might be so because they are (assumed to be) derived from western notions, and are applied mainly by middle class urbanites. Van Dijk (2014: 9) claims that 'romance has become an African reality, making the inven-tion of romance not exclusively a Western import product, although unmis-takably global images of how romance should or can look like are certainly a source of inspiration in Africa as well.' Whether 'romance' is a western product, is open to debate and could easily be contested. Nevertheless, the notions of romance and love have recently gained more attention by western scholars.[21] These studies focus on middle class urbanites that are obviously part of the

---

[20] See Rasing 1995, 2001a, 2014a, 2014b, 2015a, 2015b.

[21] See Lindholm 2006, Cole and Thomas 2009, Spronk 2009, 2014 and Van Dijk 2013 and 2014.

urban globalised world. This applies in particular for the studies conducted by Cole and Thomas (2009) who examine changing ideas of love relationships and how they are mediated by incorporating consumerist style as expressions of being modern, and by Spronk (2009, 2014a), who examines sexual and love relationships in the context of personhood and globalisation.

It can be observed that, influenced by western life styles, there is more attention to romance, especially among the youth, but also among adults. This is reflected in the initiation rites, and especially, in their more extensive form the wedding ceremonies that are a prolonged form of the currently curtailed initiation rite. Notwithstanding the discussions about shamefulness that were provoked when the two main Zambian newspapers, the 'Zambian Post' and the 'Zambian Times' (2007) showed a picture at their front page of then president Mwanasawa kissing his wife on the mouth, it shows that romance has become more common in Zambia. Yet, there are different ideas about what romance is (see van Dijk 2014). But surely, the significance of sexuality, intimacy and relationships in Africa has changed over the past decades, and have been impacted by the HIV / AIDS crisis (Rasing 2007a, 2007b, Spronk 2009, 2014a, Moyer et al. (2013).In addition, in Africa, love relationships are mediated by incorporating consumerist style as expressions of the modern (van Dijk 2014, Spronk 2009, 2014a).

Although I do not fully endorse van Dijk's (2014: 9) statement that 'ideas about romance, - particularly by the younger generations in Africa can be seen as a kind of culture critique vis-à-vis their elders and the way they want to maintain certain traditions in marital arrangements; opting for romance thus becomes an expression of protest',

I do agree that, in some ways, the romantic turns into a battleground of self-direction, self-styling and self-assertion (van Dijk 2014: 9). Although it differs what one considers romance, it mainly has to do with shaping relationships, their fascination for globally circulating models of love, and the manner in which they want to insert themselves into consumerist appetites, using global images of how romance should look (Spronk 2009, van Dijk 2014).

The idea of romance has also partly been imprinted by Western churches, such as during sermons, among the Catholic Marriage encounter, as well as in the courses for marriage preparation that young couples have to follow before they have their marriage blessed in church.[22]

Particularly middle class style of living is now being communicated from urban to rural settings via weddings (Van Dijk 2013: 142). Van Dijk shows how urbanites have their wedding ceremonies in the village, investing lots of money for cloths, drinks, food, and ritual activities, while in fact afterwards being left alone, meaning that now the couple is in a new state of adulthood and looking after themselves, but at the same time feeling they have been 'used' for the benefit of the relatives in the village. So, there is still a flux of goods and money from the town to the village. Van Dijk asserts the village benefits from the economy of the young married people in towns. He shows the connections that are established and pursued in wedding ceremonies: between the couple, their future children, their parents, both families, and between other relatives and people in the community. Although this is not new to anthropological analysis, Van Dijk (2013) claims that while a wedding ceremony is a ritual of connectivity, it is at the same time a ritual of disconnectivity; the couple are now considered adults, are independent from their parents household, have to look after themselves. The same applies partly to initiation rites, after which the girl is no longer considered a child, but is expected to behave as an adult. Also, by opening up a connection perspective, it can be seen that the ritual communicates a polyphone message of urban economy, global styles and personal experiences of adulthood and relational attachments (Van Dijk 2013: 156).

Constructing a female domain through initiation rites

Female initiation rites have been adapted to new, urban conditions in a changed social, economic and religious context, which suggests the versatility of these rites (Rasing 2004a). Obviously, they had to be adapted to urban situations and modern conditions in order to survive. However, recently, these urban initiation rites are disappearing, and are severely curtailed. This is due to various reasons, such as the Western school system in which girls should attend class, pressure of Pentecostal and Protestant Churches, pressure of global

---

[22] See Rasing 2001a; see also Van Dijk 2013 and 2014.

organisations such as the United Nations, - which in their 1995 convention declared these rites as 'harmful cultural practices'-, which' ideas were taken over by various international and national NGOs. In addition, a lot of money is needed to perform proper initiation rites, such as for the ritual leader (*nacimbusa*) and three drummers, for food and drinks for all attendants during the time of the rite, for transport for the most important attendants (the ritual leader, the girl's grandmother and aunt), to prepare the symbols (clay figures, wall paintings) used in these rites, to buy new cloths for the girl, and so on. In the current economy and huge unemployment, many people choose to leave out the initiation rite and reserve the money for the huge wedding ceremonies, either 'traditional' ones and / or the so called 'white weddings' – that is church blessings of the marriage, including the rather new kitchen parties.[23]

But, perhaps more importantly, these rites are disappearing because the cosmological ideas have changed towards a more Christian cosmology or religion, leaving the old cosmological or religious ideas obsolete.

Hinfelaar (1989) who put these rites in the context of a religious realm and cosmology has explicitly explained how these rites were related to and imbedded in a cultural and religious context. He showed that these rites were already changing in the eighteenth and nineteenth century, as was the whole context of kinship, social relationships, agriculture, economy, and political power, initially due to the construction and overruling of the Bemba chiefs, thereafter by western missionaries, soon followed by colonisers.

While in rural areas these rites still remain and are performed during several weeks (Rasing 2001a, 2014b), in urban areas they have severely been curtailed andare postponed to the wedding ceremony which is actually the same as the initiation rite (Rasing 2001a).

The continuation of initiation rites in an urban setting seems to be at odds with the life of modern Christian women but urban women are continuing to perform these rites in what they assume to be a traditional way. Women, at the interface of local cultural traditions and globalizing conditions inducing social and economic change, appear to be continuing to construct their own domain

---

[23] See Rasing 1999, 2000, 2001a, 2009; See also Tranberg Hansen 1996.

of power and meaning in a precarious urban environment (Rasing 200: 219).

Even though urban initiation rites are disappearing and curtailed, some parts of the rite are still performed. This means that these rites are still important for woman, also in urban areas (Rasing 1995, 2001a). These rites deal with the construction of female adulthood, as well as creating a community. Therefore, I agree with Arnfred (2011: 188) who states that 'After all, initiation rituals may not only be about the construction of women, but more in a wider sense about the confirmation and recreation of a particularly gendered sphere of powerful female sociality: women together maintaining the gendered cosmology of a particular ethnic group.' Hence, by being initiated in this universe, the initiate gains identity in a specific group. It is precisely to maintain this 'confirmation and recreation of a particularly gendered sphere of powerful female sociality' (Arnfred 2011: 188) that initiation rites continue to be performed.

Conclusion

Initiation rites must be seen in the context of the socio-economic organization of the society in which they are performed. This structure in the contemporary urban situation in Zambia has dramatically changed in socio-economic, political and religious terms as compared to the previous village life. Cosmological views are fragmented and therefore urban rites seem to be paradoxical. As Mitchell (1956) argued, they cannot be understood as an extension of rural rituals nor as a product of a purely urban social system but only by situating them within the political economic relations linking town and village. This suggests a more complex social and cultural process than simply a transition from rural to urban. Even when urban practices seem to resemble traditional ones, they have a different significance in urban surroundings (van Binsbergen 1981, 2000). Wilson (1972: 188) also saw a problem in ascribing relevance to rituals in fast changing societies.

Can we explain the existence of urban initiation ritesby using the concept of virtuality, as van Binsbergen (1997b, 1998) claimed? To recall his ideas about virtuality: virtuality is obtained when a certain socio-cultural item, for example a ritual, is transferred outside the context in which it originally functioned and from which it originally derived its meaning. The socio-cultural item has come to function in a new context so totally dissimilar to the original that it becomes effectively meaningless in its new context, except for some new meaning that it may acquire in this new context in ways totally unrelated to the original.

'Virtuality, then, is about disconnectivity, broken reference, de-contextualisation, through which yet formal continuity shimmers through' (van Binsbergen 1997b: 11, 1998: 878).

Although van Binsbergen's views on virtuality in urban modern Africa were explicitly illustrated by a discussion of girl's initiation in that context, I cannot completely endorse his analysis. Urban initiation may deal with disconnectivity as he suggests, but the old meaning still applies in the urban setting, even though it was not invented there in the first place. In the initiation rite a girl is transformed into a woman, affirming the novice's fertility, relations between fertility and agriculture, the ancient world view and the *mbusa* with their meanings. These are still relevant. Only a few of the images represented by the *mbusa* are no longer relevant. Even though the rites have changed, they are not detached from socio-economic conditions. Rites contribute to structure by (re-)installing the boundaries between young and old, initiated and uninitiated. The creation of boundaries assists in the acceptance of new situations. These boundaries are conceptual collective ways of seeing reality and of classifying, *e.g.* fashion, sexual and marital roles, ideas about law and order, and about cosmology and causality in terms of 'old fashioned' as opposed to 'modern' (Rasing 2001a: 220). Rites create order in exchange for obeisance to the ancestors. In addition, as a rite bridges the gap between experience and social change (Cohen 1985: 92; Jules-Rosette 1979: 198), it can decrease the divide between alienation from traditional life and a modern Westernized urban lifestyle. Due to their symbolic and communicative character, rites continue to exist.

Although rites may lose their symbolical meanings, the initiation rite still aims to transform girls into adult women. The novices are taught what every woman is supposed to know about behaviour in adult and married life, and in society generally. According to my informants, as well as my observations during my long-term study of these rites, the behaviour of novices changes markedly. In her book *Chisungu*, Richards (1956: 125) claimed that people called this the magic of the rite. The girls themselves claimed to feel grown up. Experiencing the rite, participating in ritual acts, and acquiring specific knowledge inscribe an identity in the girl. She gains self-confidence and strength and her identity is constructed. (Rasing 2001a: 220).

As I have shown, urban initiation rites remain important for women, as they imprint knowledge on the initiate about female adulthood and behaviour, thereby constructing personhood, female adulthood, kinship ties and ethnic ties. Also, it constructs a community of women, a female domain, in which women are powerful. This 'gendered sphere of powerful female sociality' (Arnfred 2011: 188) as constructed in initiation rites, can be created and exist everywhere, either in rural or urban areas, either in a local or global context. By creating this female domain, powerful symbols are used, both ancient symbols derived from village life, and modern symbols. With this, women show both their roots and their ideas of being modern, Christian, women in a globalised, urban environment.

References cited

Arnfred, S. 2011, Sexuality and Gender Politics in Mozambique: Rethinking gender in Africa. Suffolk: James Currey. Paperback 2014.

Appadurai, A. 1990, 'Disjuncture and Difference in the Global Cultural Economy', in M. Featherstone (ed.) Global Culture: Nationalism, Globalization and Modernity, pp. 295-310. London / Newbury Park: Sage.

Appadurai, A. 1995, 'The Production of Locality', in R. Fardon (ed.) Counterworks: Managing the Diversity of Knowledge, ASA decennial conference series 'The Uses of Knowledge: Global and Local Relations', pp. 204-25. London: Routledge.

De Boeck, F. 1992, 'Of Bushbucks without Horns: Male and Female Initiation among the Aluund of Southwest Zaire', Journal des Africanistes 61(1): 37-72.Casely-Hayford, Gus (2012) The Lost Kingdoms of Africa: Discovering Africa's hidden treasures. London: Bantam Press.

Cohen, 1985, The Symbolic Construction of Community. London: Tavistock Publications.

Cole, J. and L. Thomas, 2009, Love in Africa. Chicago: University of Chicago Press.

Constitution of the Nazareth, 1954.

Corbeil, J. J., 1982, Mbusa: Sacred Emblems of the Bemba. Mbala (Zambia): Moto-Moto Museum; London: Ethnographic Publishers.

Cox J.L. ed., 1998, Rites of passage in contemporary Africa. Cardiff Academic Press, Cardiff.

Dondeyne, D. 1967, De juiste bevordering van de culturele vooruitgang. In: Vaticanum 2: De kerk in de wereld van deze tijd, schema dertien, tekst en commentaar. Hilversum: Uitgeverij Paul Brand, pp. 148-199.

Douglas, M. 1971, Is matriliny doomed in Africa? In: Douglas, M. and Kaberry, P.M. (eds.) Man in Africa. New York: Anchor Books, Doubleday & Company, Inc., pp. 123-137.

Durkheim, E. 1912, The Elementary Forms of Religious Life. Paris: Presses Universites de France.

Epstein, A. L. 1958, *Politics in an Urban African Community*. Manchester: Manchester University Press.

Epstein, A. L. 1967, Urbanisation and Social Change in Africa, *Current Anthropology*8 (4): 275-95.

Epstein, A. L. 1969, The Network and Urban Social Organisation. In: J.C Mitchell (ed.) *Social Networks in Urban Situations. Analyses of Personal Relationships in Central African Towns*. Manchester: Manchester University Press, pp. 77-117.

Epstein, A. L. 1978, *Ethos and Identity*. London: Tavistock Publication.

Epstein, A. L.. 1981, Urbanisation and Kinship: The Domestic Domain on the Copperbelt of Zambia. 1950-1956. London: Academic Press.

Epstein, A. L. 1992, Scenes from African urban life: Collected Copperbelt Essays. Edinburgh: Edinburgh University Press.

Etienne. P. L. 1948, Notes sur les coutumes indigènes, unpublished book, (1948), Ilondola (Zambia).

Fardon, R. ed., 1995 Counterworks: Managing the Diversity of Knowledge, ASA decennial conference series 'The Uses of Knowledge: Global and Local Relations', London: Routledge.

Fardon, R., van Binsbergen, W.M.J., & van Dijk, R., eds., 1999, Modernity on a shoestring: Dimensions of globalization, consumption and development in Africa and beyond: Leiden / London: EIDOS.

Featherstone, M. ed., 1990, *Global Culture: Nationalism, Globalization and Modernity*. London / Newbury Park: Sage.

Ferguson 1999, Expectations of Modernity: Myths and meanings of urban life on the Zambian

Copperbelt, Berkeley / Los Angeles / London: University of California Press.

Foucault, M. 1966, Les mots et les choses: Une archeologie des sciences humaines. Paris: Gallimard.

Friedman, J. 1995, Cultural Identity and Global Process. London: Sage.

Giddens, A. 1990, *The Consequences of Modernity*. Cambridge: Polity Press.

Giddens, A. 1991, *Modernity and Self-Identity*. Cambridge: Polity Press.

Gluckman, M. 1945, 'Seven-year Research Plan of the Rhodes-Livingstone Institute of Social Studies in British Central Africa', *Rhodes-Livingstone Journal*4: 1-32.

Gluckman, M. 1949, 'The Role of the Sexes in Wiko Circumcision Ceremonies', in M. Fortes (ed.) *Social Structure*, pp. 145-67. London: Oxford University Press.

Gluckman, M. 1960, 'Tribalism in Modern British Central Africa', *Cahiers d'Etudes Africaines*1: 55-70.

Gluckman, M. 1962, Les Rites de Passage. In: Gluckman, M. (ed.) *Essays on the ritual of Social Relations*. Manchester: Manchester University Press.

Hannerz, U. 1980, Exploring the City: Inquiries towards an Urban Anthropology. New York: Columbia University Press.

Hannerz, U. 1986, 'Theory in Anthropology: Small is Beautiful? The Problem of Complex Cultures', Comparative Studies in Society and History 28(2): 362-7.

Hannerz, U. 1987, 'The World in Creolisation', *Africa* 57: 546±59. Hannerz, U. (1989) 'Culture between Center and Periphery: Toward a Macroanthropology', *Ethnos* 54: 200-16.

Hannerz, U. 1990, 'Cosmopolitans and Locals in World Cultures', in M. Featherstone (ed.) *Global Culture: Nationalism, Globalization and Modernity*, pp. 237-51. London / Newbury Park: Sage.

Hannerz, U. 1992, Cultural Complexity: Studies in the Social Organisation of Meaning. New York: Columbia University Press.

Hansen K. Tranberg, 1996, *Keeping house in Lusaka*. New York: Columbia University Press.

Hinfelaar, H.F. 1989, *Religious Change among Bemba-speaking Women of Zambia*. Ph.D. thesis, University of London, Unpublished Ph.D. thesis. London: SOAS.

Hobsbawm, E. and T. O. Ranger, eds., 1983, *The Invention of Tradition*. Cambridge: Cambridge University Press.

Hoch, E. 1968, Mbusa: A Contribution to the Study of Bemba Initiation Rites and those of Neighbouring Tribes. Chinsali: Ilondola Language Centre.

Jules-Rosette, B. ,1979, 'Changing Aspects of Women's Initiation in Southern Africa: An Explanatory Study', *Canadian Journal of African Studies* vol. 13 (3): 389-405.

Jules-Rosette, B. 1996, 'What Money Can't Buy: Zairian Popular Culture and Symbolic Ambivalence toward Modernity'. Paper presented at the international conference on 'L'Argent: feuille morte: L'Afrique Central avant et apres le desenchantement de la modernite', Louvain (21-2 June).

Kapferer, B. 1988, 'The Anthropologist as Hero: Three Exponents of Post-modernist Anthropology', *Critique of Anthropology* vol. 8(2): 77-104.

Kloos, P. 1991, *Culturele Antropologie*: een inleiding. Amsterdam: van Gorkum.

Labreque, E. 1931, *Beliefs and rituals of the Bemba and neighbouring tribes*. Ilondola / Chinsali (Zambia): Language Centre. (first printed in 1931-1934). (1982).

La Fontaine, S.J. 1972, Ritualization of women's life-crises in Bugisu, in: La Fontaine, J.S. (ed.) *The interpretation of Ritual: essays in honour of A.I. Richards*, London: Tavistock Publications Lit.

La Fontaine, S.J. 1986, *Initiation*. Manchester: Manchester University Press.

Landman, J. 1999, Interview met Wim van Binsbergen. Culturen bestaan niet. *Bijeen*.

Lindholm, C. 2006, Romantic Love and Anthropology. *Ethnofoor*, vol. 19, no. 1, pp. 5-22.

Maxwell K.B. 1983, Bemba myth and ritual: the impact of literacy on an oral culture. Frankfort on the Main / New York: Peter Lang Publishing Inc.

Mayer, P. 1971, ''Traditional" Manhood Initiation in an Industrial City: The African View', in E. J. de Jager (ed.) Man: *Anthropological Essays presented to O. F. Raum*, pp. 7-18. Cape Town: Struik.

Mayer, P. and I. Mayer 1974 / 1961, Townsmen or Tribesmen: Conservatism and the Process of Urbanization in a South African City. Cape Town: Oxford University Press

Mitchell, J. C. 1956, The Kalela Dance: Aspects of Social Relationships among Urban Africans in Northern Rhodesia. *Rhodes-Livingstone Paper* No. 27. Manchester: Manchester University Press.

Mitchell, J. C. ed., 1969, *Social Networks in Urban Situations*. Manchester: Manchester University Press.

Moore, H.L., Sanders, T., Kaare, B. eds., 1999, *Those who play with fire: Gender, fertility and transformation in East and Southern Africa,* London and New Brunswick, N. J.: The Athlone Press.

Monachonga, H. 2009, The School and Home Discourses on Sex and HIV / AIDS among Selected High School Girls in Zambia's Lusaka Urban. Oslo: Oslo University College.

Moyer, E., Burchardt, M. and R. van Dijk (eds., 2013) Sexuality, Intimacy and Counselling: Perspectives from Africa. *Culture, Health & Sexuality* 15 (Supplement 4), Special Issue. London: Routledge.

Nootermans, C. 1999, Verhalen in veelvoud: Vrouwen in Kameroen over polygynie en christendom, Nijmegen: Valkhof pers.

Ogden. J.A. 1996, 'Producing' respect: the 'proper woman' in postcolonial Kampala. In: Werber, R. and Ranger, T. (eds.) *Postcolonial identities in Africa.* London: Zed Books Ltd., pp. 165-192.

Oestigaard, T. 2014, Religion at Work in Globalised Traditions. Rainmaking, witchcraft and Christianity in Tanzania. Cambridge: Cambridge Scholars Publishing.

Powdermaker, H. 1962, *Copper Town: Changing Africa: The human situation of the Rhodesian Copperbelt,* New York / Evanston: Harper and Row Publishers.

Ranger, T. O. 1975, *Dance and Society in Eastern Africa, 1890-1970.* London: Heinemann.

Rasing, T. 1995, *Passing on the Rites of Passage: Girls' Initiation Rites in the Context of an urban Roman Catholic Community on the Zambian Copperbelt.* Leiden: African Studies Centre; Aldershot: Avebury.

Rasing, T. 1999, Globalization and the making of consumers: Zambian kitchen parties'. In: R. Fardon, W. van Binsbergen and R. Van Dijk (eds.) *Modernity on a Shoestring: dimensions of globalization, consumption and development in Africa and beyond.* pp. 247-269.

Rasing, T. 2000,Een ritueel feest: een Zambiaanse kitchen party. *LOVA,* vol. 21, nr. 2, pp. 22-27.

Rasing, T. 2001a, *The Bush Burnt, the Stones Remain: Female initiation rites in urban Zambia.* Leiden: African Studies Centre / Munster / Hamburg / London: Lit Verlag.

Rasing, T. 2001b, Initiatieriten en genderverhoudingen in Centraal Afrika in historisch perspectief. *Groniek,* nr.151, pp.107-120.

Rasing, T. 2003a, Initiatie onderzoeken en ondergaan. *Kwalon* 21, vol. 8, no.1, pp. 17-21.

Rasing, T. 2003b, Initiatie in het veld: onderzoeken en ondergaan. *Lova,* vol. 24, nr. 2, pp. 11-18.

Rasing, T. 2003c, Initiatieriten in stedelijk Zambia. *Lova,* vol. 24, nr. 2, pp. 72-89.

Rasing, T. 2004a, The persistence of female initiation rites: Reflexivity and resilience of women in Zambia'. In: W. van Binsbergen and R. van Dijk (eds.) *Situating Globality: African Agency in the Appropriation of Global Culture.* Leiden: Brill, pp. 277-310.

Rasing, T. 2004b, Rituals in Central Africa. In: Salamone, F. (ed.). *Encyclopedia of Religious Rites, Rituals, and Festivals.* New York: Berkshire / Routledge.

Rasing, T. 2004c, HIV / AIDS and sex education among the youth in Zambia: Towards behavioural change. www.hivaidsclearinghouse.unesco.org

Rasing, T. 2007a, 'Sex, a pleasure to fear: Ideas on sexuality and gender in times of HIV / AIDS in Zambia.' Paper presented at the AEGIS biannual Conference, panel on Sexuality in times of AIDS, Leiden, The Netherlands, 11-14 July 2007.

Rasing, T. 2007b, Gender and Death: Gender aspects of mourning in Zambia. *Lova,* vol. 28, nr. 2, pp. 27-39.

Rasing, T. 2009, Kitchen parties in Zambia: normbevestiging of macht van vrouwen? *Tijdschrift voor Gender Studies* (Journal for Gender Studies)vol. 12, no. 4, pp. 59-64.

Rasing, T. 2010, Traditional, modern and Christian teachings in marriage. www.fenza.org

Rasing, T. 2013, 'Harmful aspects of culture: A moral debate around women and sexuality in Zambia. *Quest: Journal for Philosophy*.

Rasing, T. 2014a, Gender and culture in Zambia: From pre- to post colonial gender and cultural norms. *Word and Context, Voices on Jubilee*, pp. 56-64.

Rasing, T. 2014b, Traditional Initiation and its Impact on Adolescent Sexual Reproductive Health in Western, North-Western and Luapula Provinces of Zambia. Research report for UNFPA.

Rasing, T. 2015a, 'We enjoy having sex': Sexual rights and pleasure among adolescent boys and girls in Zambia. Presentation at the EAGIS / ECAS conference, Paris, 8-10 July 2015.

Rasing, T. 2015b, Revitilising women's roots and knowledge for development. Paper presented at the UCZU conference on Gender and Development, Kitwe (Zambia), 14-15 August 2015.

Richards. A.I. 1939, Land, Labour and Diet in Northern Rhodesia: An economic study of the Bemba *tribe*. London: Oxford University Press.

Richards. A.I. 1945, Pottery images or mbusa used at the chisungu ceremony of the Bemba people of north eastern Rhodesia, *South African Journal of Science*, vol. XLI, pp. 444-458.

Richards. A.I. 1956 / 1982, Chisungu; A Girl's Initiation Ceremony Among the Bemba of Zambia. London: Tavistock Publications Ltd. Faber and Faber. 1982: London: Tavistock Publications Ltd.

Richards. A.I. 1959, The Bemba of North-Eastern Rhodesia. In: Colson, E. and Gluckman, M. (eds.). *The Seven tribes of Central Africa*, Oxford: Oxford University Press, pp. 164-193.

Richards. A.I. 1969 / 1940, The political system of the Bemba tribe - North Eastern Rhodesia. In: (eds.) Fortes, M. and Evans-Pritchard, E.E., *African Political Systems* pp. 83- 120.

Richards. A.I. 1971, The concilliar system of the Bemba. In: Richards A.I. (ed.) *Councils in action.*

Richards. A.I. 1984, *Bemba Marriage and Present Economic Conditions*. Manchester: Manchester University Press. Rhodes-Livingstone Papers, number four.

Spronk, R. 2009 / 2012, Ambiguous Pleasures: Sexuality and middle class self-perceptions in Nairobi. Amsterdam: Spinhuis / University of Amsterdam.

Spronk, R. 2014a, Sexuality and subjectivity: erotic practices and the question of bodily sensations. *Social Anthropology*, vol. 22, no. 1, pp. 3-21.

Spronk, R. 2014b, Exploring the middle classes in Nairobi: from modes of production to modes of sophistication. *African Studies Review*, Vol. 57, No. 1, p.93-114.

Stephens, R. 2013, *A History of African Motherhood: The case of Uganda 700-1900*. Cambridge: Cambridge University Press.

Turner, V. W. 1964, 'Symbols in Ndembu Ritual', in M. Gluckman (ed.) *Closed Systems and Open Minds*, pp. 20-51. Edinburgh: Oliver and Boyd.

Turner, V. W. 1967, 'Mukanda: The Rite of Circumcision', in V. W. Turner (ed) *The Forest of Symbols: Aspects of Ndembu Ritual*, pp. 151-279. Ithaca / London: Cornell University Press.

Turner, V. W. 1968, *Schism and Continuity in an African Society: A Study of Ndembu Village Life*. Manchester: Manchester University Press (reprint of 1957).

Vail, L. (ed.) 1989, *The Creation of Tribalism in Southern Africa*. London: James Currey; Berkeley / Los Angeles, CA: University of California Press.

van Velsen, J. 1971, The Politics of Kinship: A Study of Social Manipulation Amongst the Lakeside Tonga of Malawi. Manchester: Manchester University Press.

van Binsbergen, W.M.J. 1981, *Religious Change in Zambia: Exploratory Studies*. London / Boston: Kegan Paul International.

van Binsbergen, W.M.J. 1987, 'De Schaduw waar je niet overheen mag stappen: Een westersonderzoeker op het Nkoja meisjesfeest', in W.M.J. van Binsbergen and M. R. Doornbos (eds) *Afrika in Spiegelbeeld*, pp. 139-82. Haarlem: De Knipscheer.

van Binsbergen, W.M.J. 1992a, 'Kazanga: Etniciteit in Afrika tussen staat en traditie', inaugural lecture. Amsterdam: Free University. (Shortened French version (1993) 'Kazanga: Ethnicite en Afrique entre ' etat et tradition', in W. J. M. van Binsbergen and K. Schilder (eds) Perspectives on Ethnicity in Africa, special issue on 'Ethnicity', *Afrika Focus* (1): 9-40; English version with postscript (1994) 'The Kazanga Festival: Ethnicity as Cultural Mediation and Transformation in Central Western Zambia', *African Studies* 53 (2): 92-125).

van Binsbergen, W.M.J. 1992b, *Tears of Rain: Ethnicity and History in Central Western Zambia*. London / Boston: Kegan Paul International.

van Binsbergen, W.M.J. 1993a, 'Making Sense of Urban Space in Francistown, Botswana', in: P. J. M. Nas (ed.) *Urban Symbolism*, pp. 184-228. *Studies in Human Societies*, volume 8. Leiden: Brill.

van Binsbergen, W.M.J. 1993b, 'Mukanda: Towards a History of Circumcision Rites in Western Zambia, 18th-20th Century', in J.-P. Chretien et al. (eds) *L'invention religieuse en Afrique: Histoire et religion en Afrique noire*, pp. 49±103. Paris: Agence de Culture et de Cooperation-Technique / Karthala.

van Binsbergen, W.M.J. 1994, 'Dynamiek van cultuur: Enige dilemma's van hedendaags Afrika in een context van globalisering', *Antropologische Verkenningen* 13(2): 17-33. (English version (1995), 'Popular Culture in Africa: Dynamics of African Cultural and Ethnic Identity in a Context of Globalization', in J. D. M. van der Klei (ed.) *Popular Culture: Africa, Asia and Europe: Beyond Historical Legacy and Political Innocence* (Proceedings Summer School 1994), pp. 7-40. Utrecht: CERES.

van Binsbergen, W.M.J. 1996a, 'Black Athena and Africa's contribution to Global Cultural History', *Quest Philosophical Discussions: An International African Journal of Philosophy* 9 / 2 / 10(1): 100-37.

van Binsbergen, W.M.J. 1996b, 'Time, Space and History in African Divination and Boardgames', in D. Tiemersma and H. A. F. Oosterling (eds) *Time and Temporality in Intercultural Perspective: Studies presented to Heinz Kimmerle*, pp. 105-25. Amsterdam: Rodopi.

van Binsbergen, W.M.J. 1996c, 'Transregional and Historical Connections of Four-tablet Divination in Southern Africa', *Journal of Religion in Africa* 26(1): 2-29.

van Binsbergen, W.M.J. 1997, Virtuality as a key concept in the study of globalisation: Aspects of the symbolic transformation of contemporary Africa, The Hague: WOTRO; working paper.

van Binsbergen, W.M.J. 1998a, Globalization and Virtuality: Analytical problems posed by the contemporary transformation of African societies. *Development and Change*, pp 873-903.

van Binsbergen, W.M.J. 1998b, *Anna's room: A case study on becoming a female consumer in Francistown, Botswana*. Paper presented at the session on women's studies, CODESRIA General Assembly, Dakar 14-18 December 1998 Conference proceedings.

van Binsbergen, W.M.J. 1999a, Enige filosofische aspecten van culturele globalisering: onder bijzondere verwijzing naar Malls interculturele hermeneutiek. Paper read at the Dutch-Flemish day of philosophy (theme: globalisation) 30 October, 1999, Catholic University Brabant, Tilburg, Philosophical Faculty, published in the conference proceedings.

van Binsbergen, W.M.J. 1999b, 'Culturen bestaan niet': Het onderzoek van interculturaliteit als een openbreken van vanzelfsprekendheden, inaugural lecture, Erasmus University Rotterdam, Rotterdam: Rotterdamse Filosofische Studies.

van Binsbergen, Wim M.J., 2015, Vicarious reflections: African explorations in empirically-grounded intercultural philosophy, Haarlem: PIP-TraCS - Papers in Intercultural Philosophy and Transcontinental Comparative Studies - No. 17, also at: http://www.quest-journal.net/shikanda/topicalities/vicarious/vicariou.htm

van Binsbergen, W.M.J. and P. L. Geschiere eds., 1985, *Old Modes of Production and Capitalist Encroachment: Anthropological Explorations in Africa*. London / Boston: Kegan Paul International.

Van Dijk, R. 2013, A ritual connection: urban youth marrying in the village in Botswana. In: M. de Bruin and R. Van Dijk (eds.) *The Social Life of Connectivity in Africa*. New York: Palgrave MacMillan, pp. 141-159.

Van Dijk, R. 2014, Faith in Romance, inaugural lecture. University of Amsterdam.

Van Gennep, A. 1909 / 1960, *Les Rites de Passage*. Paris: Librairie Critique Emile Nourry.

Verstraelen-Gilhuis, G. 1982, From Dutch Mission Church to Reformed Church in Zambia: The scope for African Leadership and initiative in the History of a Zambian Mission Church. Franeker: T. Wever.

White, C. M. N. 1953, 'Conservatism and Modern Adaptation in Luvale Female Puberty Ritual', *Africa vol.* 23(1): 15-25.

Wilson 1972, The wedding cakes: a study of ritual change. In: La Fontaine, J.S. (ed.) *The interpretation of ritual: Essays in honour of A.I. Richards*, London: Tavistock publications, pp. 187- 201.

Recurrent Indo-European ethnonyms

by Fred C. Woudhuizen

Abstract: In the present contribution Indo-European ethnonyms are discussed which are of a recurrent nature. It so happens that basically the same Indo-European ethnonym can be attested in widely differing regions of the Eurasian continent and in widely differing times. It will be argued that awareness of this fact in a number of cases helps to improve our understanding of the linguistic nature of the ethnonym in question.

Key-words: Indo-European, recurrent ethnonyms, linguistic analysis.

Introduction

For Wim, who is fond of arguments involving
large distances in space and time

The process of Indo-Europeanization, which affected the Euro-Asiatic continent from the Tarim basin in the east to the British isles in the west presumably from the end of the Neolithic period onwards, was primarily the result of the spread of migrating Indo-European tribes originating from the North Pontic and North Caspian steppes moving into the regions in question. Now, the number of tribes or ethnoi involved is substantial, but not limitless. Therefore, it so happens that specific Indo-European ethnonyms pop up in more than one region affected by Indo-European settlers.

In the following, I will discuss a selection of these recurrent Indo-European ethnonyms with a bearing on Anatolia (A) and on western Iran and the eastern limits of the Indo-European family as formed by Bactria and the Tarim basin along its northern border (B).

Please note that the recurrence of an Indo-European ethnonym does not necessarily mean that the various branches speak one and the same Indo-European language, as there are numerous examples of the circumstance that a group of invaders, while planting their ethnonym in the region, nonetheless adopt the language of the indigenous population in that particular region. At least this is what happened, for example, in the case of the Akhaian (= Mycenaean Greek) take-over in the region of Adana (= Cilicia Campestris) during time of the upheavals of the Sea Peoples, henceforward called *Ḫiāwa* after them, and their foundation of the royal house of Muksas or Mopsos: the representatives of this royal house, like Awarkus / Urikki and the regent for the latter's descendant who was apparently not yet of age, Asitiwatas, during the late 8th and / or early 7th century BC conducted their monuments not in Greek but in the local script and vernacular, Luwian hieroglyphic (see Çineköy and Karatepe texts in Woudhuizen 2015a: 168-185). The same holds true for the group of *Peleset* or Philistines which settled in the same period in the Amuq valley to the south of Cilicia Campestris, who did not preserve their of origin Pelasgian (= Old Indo-European) language but, according to the monuments of their kings dating from the Early Iron Age, went over to the apparently dominant language in the region at the time, Luwian hieroglyphic (Woudhuizen 2015b: 295-296; cf. van Binsbergen & Woudhuizen 2011: 330).

We shall divide our discussion in two parts. First (subsection A) we shall discuss the ethnonyms with a bearing on Anatolia; then, in the shorter subsection B) we hall discuss those ethnonyms that have a bearing on Western Iran and the Tarim Basin.

A. With a bearing on Anatolia

(A1) Scoti - Skythians

Sometime late in the 7th century BC, Anatolia has been invaded from the North

Pontic and North Caspian steppes by war-like horsemen known as *Skythians* or *Scythes* (Σκύθαι). They are reported by the Greek historian Herodotos (*Histories* I, 106) to have controlled the Anatolian plateau for 28 years (from *c.* 610 to 582 BC) during the reign of the Median king Kyaxares, who ruled from 625 to 585 BC. In the Assyrian sources, the Scythes are addressed as *Išguza*, derived from Avestan *iazu-* "great" (Sergent 1995: 206; Sergent 2005: 216), and in the Biblical Table of Nations they are referred to as *Ashkenaz* (Genesis 10:3-4; 1 Chronicles 1:6; cf. van Binsbergen & Woudhuizen 2011: 177).

As emphasized in the early literature on the Celtic languages and peoples, like the works by Lorenz Diefenbach, it has been argued that the Celtic ethnonym *Scoti* as known as a form of address of certain inhabitants of Ireland and Scotland (Diefenbach 1840b: 260; 285) is to be identified with that of the Scythes. Notwithstanding the fact that *Scythae* = *Scoti* became a literary *topos* among Medieval writers like Beda and his successors (Diefenbach 1840b: 260), which appears to be based on nothing but "Klingklang" etymology, the identification may well be considered as having genuine historical roots in the light of the fact that the classical Greek author Plutarkhos has preserved the memory of the composite ethnonym Κελτοσκύθας (Diefenbach 1840a: 219, with reference to Plutarkhos, *Lives, Caius Marius* XI, 5; cf. Strabo, *Geography* 11, 6, 2). However, the balance is tipped in favor of a genuine historical connection of the Scoti with the Scythes on account of the fact that the former are alternatively addressed in Cymric as *Ysgotiaid, Ysgodaid*, etc. (Diefenbach 1840b: 425), which strikingly recalls the Assyrian form of address of the latter as *Išguza*!

(A2) Cimbrī - Kimmerians

Just anterior to the Scythian invasion of Anatolia, the political *status quo* in this region was shattered by the incursion of the *Kimmerians* (Κιμμέριοι). The latter are reported by Herodotos (*Histories* I, 15-16; 103; IV, 1) to have been driven from their homes in the North Pontic region by the Scythes (a Kimmerian funerary memorial is located by Herodotos *Histories* IV, 11 along the river Τύρης or present-day Dniestr), coming into this region from the east (cf. sub A (1) above). The Kimmerian invasion of Anatolia turned out to be catastrophic for the various military powers in the region, as the Assyrian king Sargon II died in his attempt to ward them off in 705 BC, the Phrygian king Midas is reported to have committed suicide at the eve of destruction of his capital Gordion by drinking bull's blood in 696 BC, and finally the Lydian king Gyges, the first ruler of the house of the Mermnads, also died in his war against them in 652 BC. Only one of Sargon II's successors, the Assyrian king Ashurbanipal turns out to be able to defeat the Kimmerians, headed by Lygdamis (Assyrian *Tugdammê*), in a battle during the last mentioned year (Kammenhuber 1976-1980: 594-596; cf. Wikipedia, *s.v. Cimmerians*).

Now, the Kimmerians, who are referred to as *Gimirrāja* by the Assyrians (Kammenhuber 1976-1980: 594), already feature in Homeros (*Odyssey* XI, 14: Κιμμερίων ἀνδρῶν δῆμος) and presumably turn up in the Biblical Table of Nations in form of *Gomer* (Genesis 10:2-3; *cf.* Ezekiel 38:6; *cf.* van Binsbergen & Woudhuizen 2011: 178). But what primarily concerns us here: they are positively identified with the *Cimbrī* in a classical Greek source. These latter are reported to inhabit the region of Jutland and the part of present-day Germany along its southern border in Antiquity. The source in question is Stephanos of Byzantium, who in his *Ethnica, s.v.* Ἄβροι, states the following:

Κίμβροι οὕς τινές φασί Κιμμερίους ('Kimbri whom some call Kimmerians')

As far as language is concerned, it deserves our attention that the Cimbri, notwithstanding their habitat in northern Germania, are suggested by Diefenbach (1839: 120) to be of "Keltische Abkunft". This latter suggestion would nicely tie in with what we know about the language of the Kimmerians, as the name of their leader, *Lugdamis*, is clearly based on the Celtic GN *Lug-*. More in general, the ethnonym *Kimmerians* confronts us with a reflex of the Proto-Indo-European (= PIE) root *$\hat{g}^him$-, from which it may safely be inferred that their language, just like Celtic, belongs to the *centum*-group among the Indo-European family. Of particular importance in this connection is that related forms based on this same PIE root turn up in Indo-European (= IE) Anatolian, viz. Hittite *gimra-* "open country", cuneiform Luwian *immara-*, Luwian hieroglyphic *apára-*, Lycian *Hprã-*, and Carian *Imbros*, of which the Luwian ones, apart from the regular loss of the voiced velar [$\hat{g}^h$], bear testimony of the phonetic development of the geminated nasal [mm] into [mb] (cf. Woudhuizen 2015a: 348) – a development which also typifies our identification of Κιμμέριοι with *Cimbrī* and in this manner emphasizes the validity of this identification.

Also of interest to us in this connection is the fact that, as noted in the above, the Kimmerians used the river-name Τύρης for the present-day Dniestr (Herodotos, *Histories* IV, 11; 47; 51), because this river-name is related to Lusitanian *Doúrios* or *Durius*, Ligurian *Durias*, etc., and as such rooted in the oldest Indo-European layer we come across in studying the process of Indo-Europeanization in Europe and Anatolia (Woudhuizen 2016a: 83-84 and cf. sub A (4) below).

(A3) Venedī - (H)enetoi

The Ἐνετοι are attested for the region of Paphlagonia in Anatolia in Homeros' enumeration of the Trojan allies in their war against the Akhaians or Mycenaean Greeks, known as the Trojan War, which event to all probability took place c. 1280 BC (Homeros, *Iliad* II, 851:

> Παφλαγόνων δ'ἡγεῖτο Πυλαιμένεος λάσιον κῆρ ἐξ Ἐνετῶν; for the given date, see van Binsbergen & Woudhuizen 2011: 249-250).

Subsequently, in the work of Herodotos, who wrote in the 5th century BC, the same tribal name is no longer found in the Anatolian region of Paphlagonia, but reported for southern Illyria (Herodotos, *Histories* I, 196: Ἰλλυριῶν Ἐνετοί; cf. Witczak 1993: 166; this particular group of *Enetoi* is probably also referred to by Appianos, *Mithridatic Wars* 55: Ἐνετοὺς καὶ Δαρδανέας καὶ Σιντούς, περίοικα Μακεδόνων) and presumably along the Po estuary at the head of the Adriatic (Herodotos, *Histories* V, 9: Ἐνετῶν τῶν ἐν τῷ Ἀδρίῃ).

Now, the given form of the ethnonym represents its developed form, in which the original digamma has been dropped and been replaced first by a spiritus asper and later by a spiritus lenis, in like manner as the Trojan royal name *Walmus* develops through the intermediary *Halmos* into *Almos*, both latter forms being attested in Greek literary tradition with a bearing on non-Greek population groups of central Greece (see Pauly-Wissowa Realencyclopädie, s.v.), or the name of the wife of the Akhaian or Mycenaean Greek king Menelaos, *Helena*, originates from *Welena* (*Deini[s] tad'anetheke khari[n] Welenai Menelawo*, see Jeffery 1990: 90 [= pl. 7, 1]). Therefore, its original form is **Wenetoi*, which in writing variant *Veneti* we encounter as the ethnic designation of an Italic population group inhabiting the region of Venice (<*Venetia*) along the estuary of the river Po at the head of the Adriatic in the Early Iron Age (as we have just noted in the developed form without the initial *wau* these are presumably referred to by Herodotos, *Histories* V, 9: Ἐνετῶν τῶν ἐν τῷ Ἀδρίῃ). As duly indicated by Diefenbach in his treatment of this ethnonym, the ancient sources are uncertain about the fact whether the Italic *Veneti* should be grouped with the Celts or not (see Diefenbach 1840a: 123-129). Benjamin Fortson, in his introduction to the Indo-European language and culture of 2004, considers the language of the Italic *Veneti* as attested for inscriptions dating from the latter half of the 6th century BC to *c.* 100 BC basically of Italic type, but with some from an Italic point of view queer features most closely paralleled for Germanic (Fortson 2004: 406-407).

Next to the Italic *Veneti*, this same ethnonym is, as duly observed by Witczak, also recorded for a Gaulish tribe in Brittany (Witczak 1993: 166; see Map in Sigmund Herzog's edition of Ceasar's *De Bello Gallico* of 1895).

In his linguistic analysis of the ethnonym *Veneti-(H)enetoi*, Witczak suggests a relationship with the indication of the Slavic autonym **Vęti* "Slavs", Latin *Venedi / Veneti*, and Germanic **Windôz*, which ultimately derive from PIE **Wenh₁tói* "people on friendly terms, relatives" based on the PIE root **wenh₁-* "to love each other; to be relative of" (Witczak 1993: 166).

(A4) Cauci - Kaukōnes

The *Kaukōnes* (Καύκωνες) are mentioned by Homeros, not in his catalogue of the forces in book II of his *Iliad* but only later in the same work (*Iliad* X, 429; XX, 329), as allies of the Trojans at the time of the Trojan War (*c.* 1280 BC; for this dating, see van Binsbergen & Woudhuizen 2011: 249-250). According to Strabo, *Geography* XII, 3, 5 they already inhabited the region of Paphlagonia, a province to the east of Troy near the course of the Parthenios river, at that time. From Herodotos, *Histories* IV, 148 (cf. also I, 147), we may infer that the Kaukones ultimately originated from Greece, to be more specific Triphylia in the Peloponnesos, in the region where the places Lepreon, Makistos, Phrixai, and Pyrgo were later founded. From here they were driven at some unspecified date by the Minyans, also a pre-Greek population group (cf. Woudhuizen 2016a: 66-67).

We owe it to the merit of Diefenbach (1840a: 320-321) and Bernard Sergent (1995: 84-85; 205; cf. Sergent 2005: 88; 214) that the Greek and Anatolian Kaukones are linked up with the *Cauci* (Καῦκαι [Diefenbach 1840b: 383]) as reported for Ireland. If we may proceed from the commentary of Carl Müller to Ptolemaios' *Geographia* II, 2, 8, these Cauci were neighbors of the *Menapii*

(Μανάπιοι [Diefenbach 1840b: 383; cf. Mallory 2013: 253, Fig. 9.3]), who in turn ultimately originated from the region to the south of the mouth of the Rhine in Belgium (see Map in Sigmund Herzog's edition of Ceasar's *De Bello Gallico* of 1895; cf. Woudhuizen 2010: 76, note 70). In line with this observation, it might reasonably be argued, following into the footsteps of Diefenbach 1840b: 414-415, that the Cauci arrived in Ireland together with the Menapii, and likewise originated from the lower Rhine region in Belgium. However this may be, Diefenbach (1840a: 320; cf. Diefenbach 1840b: 265) and Sergent (1995: 205; cf. Sergent 2005: 214) both suggest that the Germanic ethnonym *Chauci*, by means of lenition of the initial velar, may be a reflex of our present Cauci.

If the suggestion that the Cauci as attested for Ireland ultimately originated from the lower Rhine region in Belgium holds water, it seems likely that we are dealing here with a population group belonging to the so-called "Nordwestblock" (Meid 1986, presenting a critical survey of the work of Hans Kuhn; cf. Woudhuizen 2016a: 61-62), of which the river- and place-names belong to the oldest extant Indo-European layer. *Mutatis mutandis*, the Anatolian Kaukones may likewise be considered representatives of what is perhaps most adequately defined as the Old Indo-European layer observable as the earliest layer in the process of Indo-Europeanization in Anatolian as well (Woudhuizen 2016a: 61-81).

(A5) Brigantes - Brugoi - Phrygians

About the Phrygians (Φρύγες) it is generally assumed that they migrated from the borderland between Thessaly and Macedonia in Greece to the Anatolian plateau during the time of the upheavals of the Sea Peoples at the end of the Bronze Age. In Anatolia, then, they settled in regions formerly inhabited or controlled by the Hittites. There is undoubtedly a kernel of truth in this view, as Phrygian tribes under the name of Βρίγες or Βρύγοι (cf. Hesykhian gloss: Βρέκυν τὸν Βρέκυντα, τὸν Βρίγα. Βρίγες γὰρ οἱ Φρύγες) indeed migrated in substantial numbers to the Anatolian plateau in this period of political instability. But, if we take the evidence from Homeros' *Iliad* III, 184-187 at face value, the Phrygians, whose auto-ethnonym according to an Old Phrygian inscription from Midas town was *vrekun* (W-01, § 2, see Waanders & Woudhuizen 2008-9: 195-196), or more in specific Phrygian forces under the leadership of Otreus and Mygdon had already mustered along the banks of the Sangarios river (= Hittite

Saḫiriya) at the time that their ally Priamos, the king of Troy, was still able to fight himself, so say about a generation before the Trojan War of *c*. 1280 BC, *i.e.* in the 14th century BC. The suggestion that Phrygians were already present in the region of Anatolia east of Troy sometime during the Late Bronze Age is further underlined by the fact that they are mentioned by Homeros in his catalogue of the Trojan allies at the time of the Trojan War, *c*. 1280 BC, as auxiliaries coming "from far away Askania", *i.e.* the region of the Askanian lake (= present-day İznik gölü) to the east of Troy (*Iliad* II, 862-863: Φόρκυς αὖ Φρύγας ἦγε καὶ Ἀσκάνιος θεοειδὴς τῆλ' ἐξ Ἀσκανίης).

In my treatment of the Argonauts' saga of 2012 (= Woudhuizen 2012), I have argued that the Phrygian encroachment on the southern Pontic littoral already started in an even earlier period, from say *c*. 1500 BC onwards, and that it was initiated by maritime excursions of the Phrygians among the pre-Greek population groups who inhabited various regions of the Greek mainland, amongst which Iolkos in Thessaly during the period of its "Minyische Schicht", *c*. 1600-1400 BC. At any rate, the *Kaška*, who successfully withstood Hittite control over the northern coastal zone of Anatolia, are not yet present in this region during the time of the Old Hittite Kingdom (*c*. 1780-1500 BC), to which the coastal city of Zalpa formed an integral part, and among the Kaskan place- and personal names we come across onomastic elements of Thraco-Phrygian nature (cf. Woudhuizen 2016a: 44). That the maritime adventures by the Phrygians from Greece proper indeed did bring them all the way to Kolkhis as the myth maintains is indicated by the place-name *Phrixopolis*, which is the old name of Ideëssa to the north of the river Phasis (Strabo, *Geography* XI, 2, 8), and the ethnonym Μόσχοι, a variant spelling of the indication of the Phrygians in the Assyrian sources, *Muški*, attested for the region between Kolkhis and Armenia and after whom Moskhike, the Moskhian mountains, and *Moschorum tractus* are named (DNP, *s.v. Moschoi*). Note in this connection that the ethnonym *Muški*, which is separated by Anne-Maria Wittke (2004) from that of the Phrygians notwithstanding the fact that the Phrygian king Midas is referred to in Assyrian texts as *Mita of Muški*, is likely to be analyzed as an ethnic formation in -*k*-, corresponding to the "Nordwestblock" or Old Indo-European suffix for this function (Meid 1986; Woudhuizen 2016a: 62), of the basic root *Mus*- also traceable in variant form characterized by *a / u*-vowel shift for the Hittite country name *Maša*. It is true that the ethnonym *Mysians* based on the same root occurs separately from that of the Phrygians in Homeros' catalogue of the Trojan allies at the time the Trojan War (*Iliad* II, 859: Μυσῶν δὲ Χρόμις ἦργε καὶ Εὔνομος οἰωνιστής), but the two ethnic groups may have cooperated to the extent that the of origin separate ethnonyms became interchangeable as a reference to one and the same ethnic identity. At any rate, in Luwian hieroglyphic the Phrygians are also addressed by forms related to Assyrian *Muški*, namely: the ethnic adjective *Masàkana*- as recorded for one of the earliest inscriptions from the Early Iron Age, Kızıldağ 4, § 2 (Woudhuizen 2015a: 33), and the ethnonym *Musàka*- as attested (alongside, it must be admitted in all fairness, the related ethnonym *Muśa*- "Mysian") for a text from Karkamis from the reign of the regent Araras, Karkamis 6, § 6 (Woudhuizen 2015a: 104).

What primarily concerns us here, however, is the fact that the ethnonym of the Phrygians is based on the same PIE root *$b^hĝ^h(i)$- "high" as that of the Celtic Brigantes (Βριγάντες) recorded for central Britain as well as for southeast Ireland, and is therefore similarly to be explained as "dwellers of heights".[1] The relationship of the Phrygians with the Celtic Brigantes can be further underlined by the fact that some Phrygian place- and personal names happen to be based on Celtic roots like, alongside *brig*-, *adrast*-, *nem*-, *mid*-, and *teuta*- (cf. Woudhuizen 2016a: 61; 74).

(A6) Lougoi - Lugii- Lukka

The ethnonym and country name *Lukka* is first recorded in form of *Rwqq* as an ethnic for the seal bearer of the Byblian king Abishemu II in an Egyptian hieroglyphic inscription on a miniature obelisk from the temple of the obelisks at Byblos, datable to the reign of the Egyptian pharaoh Nehesy, *i.e.* 1710 BC (Woudhuizen 2014). Next, it turns up in the form of *[L]uqqa* in the annals of the Hittite great king Tudḫaliyas II (1425-1390 BC) as the first, and therefore most southerly, member of the so-called Assuwian league, which was headed by a king of the royal house of Arzawa named Piyamakuruntas and entailed the entire zone of western Anatolia, from, as said, Lukka in the south to Wilusiya "(W)ilion" and Tarwisa "Troy" in the north. In later Hittite and Luwian hiero-

---

[1] *Cf.* Sergent 2005: 217; Mallory 2013: 253, Fig. 9.3; 266 "hillfort-dwellers"; note the fact that we have here a *centum*-reflex of the given PIE root as [g] which originates from *[ĝʰ], and that Phrygian accordingly is likely to belong to the *centum*-group among the Indo-European family – an inference for which definitive proof has been provided in Waanders & Woudhuizen 2008-9: 211.

glyphic texts it occurs in form of *Lukka* and *Luka*, respectively, as well as in the plural, kur.kur^{mešuru}*Luqqa* and *Luka^{utnai}*"the Lukka lands" (the last mentioned form in Yalburt, §§ 4, 5 [Woudhuizen 2015a: 17]). Furthermore, it features in Akkadian cuneiform in the El Amarna letters from about the middle of the 14th century BC and in texts from Ugarit dating to the final years of the Bronze Age, and in Egyptian hieroglyphic once more as *Rwkw* in the texts by the Egyptian pharaoh Merneptah on the Sea Peoples (for an overview, see Gander 2010: 16-64 and cf. van Binsbergen & Woudhuizen 2011: 216; 237-238; 326).

From the Greek side, it deserves mention in this connection that in Mycenaean Greek we come across the ethnic *ru-ki-jo* "Lycian" in texts from Pylos (PY Gn 720, etc.), likewise dating to the end of the Bronze Age (Ventris & Chadwick 1973: glossary, *s.v.*), whereas in Homeros the Lycians are staged in the catalogue among the allies of the Trojans at the time of the Trojan War (Homeros, *Iliad* II, 876-877: Σαρπηδὼν δ'ἦρχεν Λυκίων καὶ Γλαῦκος ἀμύμων τηλόθεν ἐκ Λυκίης, Ξάνθου ἄπο δινήεντος).

In their own inscriptions, however, which date to the 5th and 4th centuries BC and are conducted in a local Anatolian variant of the Phoenician alphabet, the Lycians address themselves by the ethnonym *Trmile-*. Even so, there can be no doubt that with Lukka from the given Bronze Age sources reference is made to the country known by the Greeks as Λυκία and an inhabitant of this country from the same category of sources as Λύκιος.

Now, the ethnonym Lukka is strikingly paralleled by that of the Celtic Λοῦγοι as reported by Ptolemaios in his *Geographia* II, 3, 12 for Scotland and that of the Germanic Λοῦγοιοι *Lugii*, situated north of Bohemia, as referred to by Ptolemaios in his *Geographia* II, 11, 10. According to Sergent (1995: 211; 2005: 224) the latter were, notwithstanding their habitat in Germania, a "peuple sans doute celtique".

If the comparison of the Anatolian *Lukka* to the Celtic Λοῦγοι holds water, we receive invaluable extra information regarding the ongoing discussion in the matter whether the country name *Luwiya* with its adjectival derivate *luwili-* "in the Luwian language" is related or even directly derived from the country name or ethnonym Lukka – especially of interest against the backdrop of the fact that the Lycian language belongs to the Luwian group of languages, also including cuneiform Luwian and Luwian hieroglyphic. Arguments in favor of such a relationship have been provided by Onofrio Carruba (1964: 268; 1995: 68-69; 1996: 28; 37) and Emmanuel Laroche (1976: 18), and these have been taken seriously by John David Hawkins (1983: 141) and Donald F. Easton (1984: 27), whereas such a relationship has been categorically rejected by Frank Starke (1997: 475) and Harold Craig Melchert (2003: 14, note 8; for the most recent overview of the relevant literature, see Marino 2010: 490). As it comes to specifics, Carruba maintains that the unvoiced velar [k] is simply dropped: *Lukka-* > **Lukki-ya-* > **Luhiya-* > *Luwiya-*. This line of reasoning, which implies that Lukka originates from PIE **l(e)uk-* "to shine" in like manner as, for example, Greek *Levkas* (so explicitly Sergent 2005: 221), has been rejected by Starke on phonetic grounds, which are made specific by Melchert: "However, one must with Starke (1997a 475[97]) reject the repetition by Carruba (1996 28 & 37) of the derivation of the name *Luwiya* from *Lukka*. Only voiced **g*, not unvoiced **k*, is lost in Luwian. The two names, both undoubtedly non-Indo-European, have *nothing* to do with each other." The line of reasoning by Laroche, on the other hand, affects graphic [k], but in effect has a bearing on the voiced velars [ĝ] (Hittite *mekkis* = Luwian *mais*< PIE **méĝh₂-* "great, many") and [ĝʰ] (Hittite *kimmaras* = Luwian *immaras*< PIE **ĝʰim-* "winter, snow"; Hittite *kesseras* = Luwian *issaris*< PIE **ĝʰes-r-* "hand"; Hittite *pargus* = Luwian *parris*< PIE **bʰ ĝʰ(i)-* "high"), which, as rightly stressed by Melchert, are regularly dropped in Luwian – and so already from the time of one of the earliest records of the language onwards, as exemplified by the onomastic elements *nana-*< PIE **-ĝenh₁-* "brother, relative" and *wawa-*< PIE **gʷow-* "ox" in the Kültepe / Kanesh texts (Yakubovich 2010: 211-212). Now, the comparison of *Lukka* to Celtic Λοῦγοι suggests that the velar in question, notwithstanding its fortition in Greek Λυκία, *actually originates from a voiced velar not distinguished as such in the Hittite and Luwian systems of writing!* If so, the PIE root on which this country name and ethnonym is based is not **l(e)uk-* "to shine", as formerly assumed, but **l(e)ugʰ-* "to bind" (not to be mixed up with the same form for the meaning "to lie", see Pokorny 1959, *s.v.* nos. 1. and 2.), from which, on the analogy provided by the Thracian GN *Bendis*< PIE **bʰendʰ-* "to bind", also the Celtic name of the sun-god *Lug* is derived. All in all, then, it may safely be concluded that the country name *Luwiya* is related to the ethnonym and country name *Lukka*< PIE **leugʰ-* "to bind" by the for Luwian regular loss of the voiced velar **[gʰ]*, and that hence both names are of IE origin. Note, however, in this connection that, as Lycian is a Luwian dialect, the irregular velar reflex of PIE **[gʰ]* in the country name or ethnonym

Lukka must be attributed to the influence of a non-Luwian but nevertheless Indo-European substrate (Woudhuizen 2016a: 83; 86; 88), which can positively be identified as Pelasgian (Woudhuizen 2016a: 90-95). At any rate, one may legitimately wonder whether it is merely coincidental that the political organization of Lukka consists of a city-league, *i.e.* a number of cities bound together by mutual oaths of allegiance.

B. With a bearing on Western Iran and the Tarim Basin

(B1) Tungrī - Teukroi - Tukri - Tokharians

The earliest mention of an Indo-European tribe we owe to the so-called "Kuthaean Legend of Narâm-Sin", the ruler of Akkad whose reign is dated in the 23rd century BC (2291-2255 BC). The *Kuthi* or *Guti* of this myth are staged as northern mountain dwellers, who successfully raided on the Akkadian capital Babylon and controlled it for more than a century, until they were eventually expelled by one of Narâm-Sin's successors, Utu-hegal of Uruk, whose reign is dated 2120-2114 BC. What primarily concerns us here, however, is the fact that the homeland of the *Kuthi* or *Guti* in the mountainous lower Zāb region of western Iran in a text from the time of Hammurabi of Babylon (1792-1750 BC) was also addressed as *Tukri*. This latter name, then, has been identified by Walter B. Henning with that of the *Tokharians*, who inhabited the Tarim basin along the western border of China at the time from which the documents in their language stem, *i.e.* from the 5th and 6th century AD onwards. This identification is of particular interest in view of the fact that the *Tokharians* designated themselves as *Kuči* (*<Guti*), while their language is addressed as *Tocri* (*<Tukri*) by an Uyghur scholiast. What we have here is not an incidental correspondence between two names, *Tukri* and *Tocri* "Tokharian", which might be dismissed as "Klingklang" etymology, but an interrelated set of two correspondences, namely that of *Tukri* with *Tocri* and that of *Guti* with *Kuči* which both point into the same direction, namely a relationship with the *Tokharians*, which cannot be dismissed as merely coincidental (Henning 1978; cf. Witczak 1993: 164-165). Against the backdrop of this identification, the Tokharians may reasonably be suggested to have moved at some specific point in time from the lower Zāb region in western Iran to the Tarim basin along the western border of China, passing *Tokhristan* in Bactria along the route, where, considering this

name in later Greek sources, they appear to have left some traces (Woudhuizen 2016a: 49-50).

According to Antoine Meillet, the language of the Tokharians is most closely related within the Indo-European family with the western group, Italic, Celtic, and particularly Germanic (cf. Witczak 1993: 163). This view, however, is not generally accepted anymore and the most recently held opinion in this matter is that Tokharian, like IE Anatolian, split off from the rest of the family at an early date (Fortson 2004: 352; for an overview, see Blažek 2011: 137-139). However, it does not share with IE Anatolian the preservation of the PIE laryngeal *[h₂] in form [ḫ] and in this respect Tokharian sides with the other members of the Indo-European family.

Whatever one may be apt to think about the exact relationship of Tokharian within the Indo-European language group, the geographic name *Tukri* and the form of address of the Tokharian language *Tucri* bear a striking resemblance to the Germanic ethnonym Τοῦγροι as recorded by Ptolemaios (*Geographia* II, 9, 5: Τοῦγροι; commentary: also Τούγγροι, Τοῦγροι). This appears in Latin form as *Tungri*, who are reported by Tacitus in his *Germania* (§ 5) to originate from the eastern side of the Rhine or Germania proper and to have settled in the Belgic part of Gaul, of which fact the Belgian place-name *Tongren* still bears testimony. Note that the doubling of the *gamma* in the Greek variant form which in Latin results in [ng] of *Tungrīis* paralleled in Hittite for the river name *Saḫiriya* being identical with the *Sangarios* of the classical sources and in Luwian hieroglyphic for the related river name *Sakaàra-* being identical with the *Sangaras* of the Assyrian sources.

As Krzysztof Witczak kindly communicated to me by e-mail, the ethnonym under discussion may have sparked off in the Anatolian theatre in the form of Τευκροί, the name of the inhabitants of the Troad in northwest Anatolia according to Greek literary tradition representatives of whom, referred to in the Egyptian hieroglyphic texts as *Tjeker*, took part in the upheavals of the Sea Peoples at the end of the Bronze Age and subsequently settled at the Levantine town of Dor (van Binsbergen & Woudhuizen 2011: 285-288).

(B2) Gothī - Guti - Kuči - Kydōnes

In the discussion of the previous group of related ethnonyms we have already noted that the *Guti* are the earliest Indo-European tribe to be referred to in the Mesapotamian sources, namely in the "Kuthaean Legend of Narâm-Sin", the ruler of Akkad whose reign is dated to the 23rd century BC, and that this particular name corresponds to the Tokharian auto-ethnonym *Kuči*. In this connection, it deserves our attention that Victor Mair speaks of

"nomadic, fair-skinned, light-haired Guti" (Mair 2012: 275)

and that Mario Liverani informs us about a Gutian king *Tirigan* (Liverani 2014: 155), which name strikes one as being of composite nature and likely to be characterized by the first element *tri-*, corresponding to the PIE numeral **tri-* "3" (cf. the epithet of the Greek goddess Pallas Athene, Τριτογενής).

Now, Witczak argues at length that the ethnonym *Guti* of the Mesopotamian sources is identifiable with that of the Germanic tribe of the Γοῦται or *Gothī*"Goths"featuring in Greek and Latin texts from the centuries before and after the year of the birth of Christ (Witczak1993: 163-168). It is clear from these latter sources that the Goths inhabited the region along the eastern side of the middle Weichsel. Nevertheless, there is a persistent tradition recorded by Jordanes that the Goths originated from the island of *Skandza* or Scandinavia (cf. Ptolemaios, *Geographia* II, 11, 16: Σκανδία Γοῦται), in which more northerly region the Swedish city of Gothenburg and the Baltic island Gotland still bear the testimony of their former presence, and that they embarked their three ships on the southern coast of the Baltic sea which henceforward was called *Gothiscandza* (Hachmann 1970: 109-110). This literary tradition is seriously taken into consideration by Rolf Hachmann in his monograph *Die Goten und Skandinavien* of 1970, but plainly dismissed by Peter Heather in his book on the Goths of 1998. Hachmann tries to find evidence of the Scandinavian origin of the Goths in the archaeological record, but his quest ends in a "non liquet" for the period on which he focuses, the last two centuries BC. If, however, the *Guti* from the Mesopotamian sources are indeed to be identified with the Germanic tribe of the Γοῦται *Gothī*"Goths", their migration to more southerly regions like the Pontic area and, crossing the Derbend pass, into the Caucasus and beyond to the lower Zāb region, must have started already before the 23rd century BC! Ergo: Hachmann's archaeological investigations need to be reiterated in accordance with this adjusted timeframe.

Finally, as Krzysztof Witczak kindly communicated to me by e-mail, against the backdrop of its variant Γύθωνες the ethnonym under discussion may be attested as well for Crete in the form of Κύδωνες, a population group inhabiting the region of Khania already during the Late Bronze Age (*cf.* Linear B *ku-do-ni-ja* and Linear A *ku-zu-ni*, see Ventris & Chadwick 1973: glossary, *s.v.*; Woudhuizen 2016b: 218).

As far as the linguistic analysis of the ethnonym *Gothī-Guti* is concerned, Witczak points out that it is derived from the Germanic root **gut(t)-* "boy" with a collective **guti-* "(young) men". This root also occurs with a formans *-an-* as **gut-an-* for the expression of an individual meaning, on which formation Greek Γύθωνες, Latin *Gutones*, Old English *Gotan*, and Old Norse *gotnar* "men" are based (Witczak 1993: 163). The corresponding PIE roots are according to Eric Hamp (see Witczak 1993: 164; 168) **g^hud-i-* (collective) and **g^hud-on-* (individual).

References cited

Blažek, Václav, 2011, Tocharian Studies, Works 1., Brno: Masaryk University.

Carruba, Onofrio, 1964, Ahhiyawa e altri nomi di popoli e di paesi dell'Anatolia , occidente. *Athenaeum NS* 42. Pp. 269-298.

Carruba, Onofrio, 1995, Per una storià dei rapporti luvio-ittiti. In: Carruba, Onofrio, , Giorgieri, Mauro, & Mora, Clelia, (eds.), Atti del II , Congresso Internazionale di Hittitologia, Pavia, 28 giugno - , 2 luglio, 1993. Pp. 63-80., Pavia: Gianni Iuculano Editore.

Carruba, Onofrio, 1996, Neues zur Frühgeschichte Lykiens. In: Blakolmer, Fritz, e.a. , (Hrsg.), Fremde Zeiten, Festschrift für Jürgen Borchhardt , zum sechzigsten Geburtstag am 25. Februar 1996. Pp. 25-39., Wien: Phoibos Verlag.

Diefenbach, Lorenz, 1839, Celtica I, Sprachliche Documente zur Geschichte der Kelten; , zugleich als Beitrag zur Sprachforschung überhaupt., Stuttgart: Druck und Verlag von Imle & Liesching.

Diefenbach, Lorenz, 1840a, Celtica II, Versuch einer genealogischen Geschichte der , Kelten. Erste Abteilung., Stuttgart: Druck und Verlag von Imle & Liesching.

Diefenbach, Lorenz, 1840b, Celtica II, Versuch einer genealogischen Geschichte der , Kelten. Zweite Abteilung: Die Iberischen und Britischen , Kelten enthaltend., Stuttgart: Verlag von A. Liesching & Comp. (früher Imle &, Liesching).

Easton, Donald F., 1984, Hittite History and the Trojan War. In: Foxhall, Lin, &, Davies, John K., (eds.), The Trojan War, Its Historicity and , Context. Papers of the First Greenbank Colloquium, , Liverpool, 1981. Pp. 23-44., Bristol: Bristol Classical Press.

Fortson IV, Benjamin W., 2004, Indo-European Language and Culture, An Introduction. , Blackwell Textbooks in Linguistics 19., Oxford: Blackwell Publishing Ltd.

Gander, Max , 2010, Die geographischen Beziehungen der Lukka-Länder. Texte , der Hethiter

27. , Heidelberg: Universitätsverlag Winter GmbH.

Hachmann, Rolf, 1970, Die Goten und Skandinavien., Berlin: Walter de Gruyter & Co.

Hawkins, John David, 1983, Review of Suzanne Heinhold-Krahmer, Arzawa, , Untersuchungen zu seiner Geschichte und den hethitischen , Quellen, Heidelberg: Carl Winter Universitätsverlag, 1977. , Bulletin of the School of Oriental and African Studies 46., Pp. 139-141.

Heather, Peter, 1998, The Goths. The Peoples of Europe., Oxford: Blackwell Publishers Ltd (Reprinted, First published , 1996).

Jeffery, Lilian H., 1990, The Local Scripts of Archaic Greece: A Study of the Origin , of the Greek Alphabet and its Development from the Eighth , to the Fifth Centuries B.C., Oxford: Clarendon Press.

Kammenhuber, 1976-80, Kimmerier. In: Edzard, Dietz Otto, (Hrsg.), Reallexicon der , Assyriologie, Fünfter Band: *Ia... - Kizzuwatna*. Pp. 594-596., Berlin-New York: Walter de Gruyter.

Laroche, Emmanuel, 1976, Lyciens et Termiles. *Revue Archéologique* 1976, Fascicule , 1: Études sur les relations entre Grèce et Anatolie offertes à , Pierre Demargne, 1. Pp. 15-19.

Liverani, Mario, 2014, The Ancient Near East, History, Society and Economy., London-New York: Routledge, Taylor & Francis Group.

Mair, Victor H., 2012, The Earliest Identifiable Written Chinese Character. In: , Huld, Martin, Jones-Bley, Karlene, & Miller, Dean, (eds.), , Archaeology and Language: Indo-European Studies , Presented to James P. Mallory. Journal of Indo-European , Studies Monograph Series 60. Pp. 265-279., Washington, DC: Institute for the Study of Man.

Mallory, James P., 2013, The Origins of the Irish., London: Thames & Hudson Ltd.

Marino, Mauro G., 2010, From the Lukka Lands to Išuwa: A Comparative Analysis of , Analogous Problems. In: Hazırlayan, Yayına, & Süel, , Aygül, (eds.), Acts of the VIIth International Congress of , Hittitology, Çorum, August 25-31, 2008. Pp. 489-502., Ankara.

Meid, Wolfgang, 1986, Hans Kuhn's „Nordwestblock"-Hypothese, Zur Problem , der „Völker zwischen Germanen und Kelten". In: Beck, , Heinrich, (Hrsg.), Germanenprobleme in heutiger Sicht, , 183-212., Berlin-New York: Walter de Gruyter.

Melchert, Harold Craig, 2003, Chapter II: Prehistory. In: Melchert, Harold Craig, (ed.), , The Luwians. Handbook of Oriental Studies, Section One: , The Near and Middle East, Vol. 68. Pp. 8-26., Leiden-Boston: Brill.

Pokorny, Julius, 1959, Indogermanisches Etymologisches Wörterbuch. I. Band., Bern: A. Francke AG Verlag.

Pokorny, Julius, 1969, Indogermanisches Etymologisches Wörterbuch. II. Band., Bern: A. Francke AG Verlag.

Sergent, Bernard, 1995, Les Indo-Européens, Histoire, langues, mythes. , Paris: Éditions Payot & Rivages.

Sergent, Bernard, 2005, Les Indo-Européens, Histoire, langues, mythes., Paris: Éditions Payot & Rivages (Nouvelle édition revue et , augmentée).

Starke, Frank, 1997, Troja im Kontext des historisch-politischen und sprachlichen , Umfeldes Kleinasiens im 2. Jahrtausend. *Studia Troica* 7. , Pp. 447-487.

van Binsbergen, Wim M.J., & Woudhuizen, Fred C. , 2011, Ethnicity in Mediterranean Protohistory. BAR International , Series 2256., Oxford: Archaeopress.

Ventris, Michael, & Chadwick, John, 1973, *Documents in Mycenaean Greek.*, Cambridge: At the University Press (2nd ed.).

Witczak, Krzysztof Tomasz, 1993, Goths and Kucheans: An Indo-European Tribe? *Lingua , Posnaniensis* 35. Pp. 163-169.

Wittke, Anne-Maria , 2004, Mušker und Phryger, Ein Beitrag zur Geschichte Anatoliens , vom 12. bis zum 7. Jh. v. Chr. Beihefte zum Tübinger Atlas , des Vorderen Orients, Reihe B (Geisteswissenschaften) Nr. , 99., Wiesbaden: Dr. Ludwig Reichert Verlag.

Waanders, Frits, & Woudhuizen, Fred C., 2008-9, Phrygian & Greek. Talanta, Proceedings of the Dutch , Archaeological and Historical Society 40-41. Pp. 181-217.

Woudhuizen, Fred C., 2012, The Saga of the Argonauts: A Reflex of Thraco-Phrygian , Maritime Encroachment on the Southern Pontic Littoral. In: , Tsetskhladze, Gocha R., (ed.), The Black Sea, Paphlagonia, , Pontus and Phrygia in Antiquity: Aspects of Archaeology , and Ancient History. BAR International Series 2432. Pp., 263-271., Oxford: Archaeopress.

Woudhuizen, Fred C., 2014, The Earliest Recorded Lycian. *Res Antiquae* 11.

Woudhuizen, Fred C., 2015a, Luwian Hieroglyphic: Texts, Indices, Grammar., Heiloo (On line: academia.edu).

Woudhuizen, Fred C., 2015b, Some Suggestions as to the Improvement of our , Understanding of the Recently Discovered Luwian , Hieroglyphic Text Aleppo 6.*Ancient West & East* 14.Pp., 293-300.

Woudhuizen, Fred C., 2016a, Indo-Europeanization in the Mediterranean, With particular , attention to the fragmentary languages. PIP-TraCS – Papers , in Intercultural Philosophy and Transcontinental , Comparative Studies 16., Haarlem: Uitgeverij Shikanda Press.

Woudhuizen, Fred C., 2016b, Documents in Minoan Luwian, Semitic, and Pelasgian.

Yakubovich, Ilya S., 2010, Sociolinguistics of the Luvian Language. Brill's Studies in , Indo-European Languages & Linguistics 2., Leiden-Boston: Brill.

Our Lodge of Being: Language

by Bongasu Tanla Kishani

Our only lodge in Being-Language baffles how to say
Nothing else except God's award of signatures on the dot
for a gratuitous lifespan as mine as much as yours, speaks
best in chorus of endless volumes, many a voiceless ego casts
to bait a kindred half empty shoe of Being-Nothingness-Language.
Together ordinary-extraordinary tautologies ever walk arm in arm!
Time's triplicates breed flotsam, restless twin decadence-development!
The paralinguistic voids infix the unavoidable tongues in Being-Language!
Being-Nothingness-Language drives on as the plural kin and kith in rival rivers
along rainbow seasons with chained cribs in free creeks, the next of kin inherits.[1]

---

[1] This poem takes up the central theme of Kishani's philosophical manuscript on Being and Language, now being published in the series Papers in Intercultural Philosophy / Transconti-

The snail hints: Being sacredly houses its automatic Self,
replete in its entirety with Nothingelse, but its apostrophes!
Being ever fully begets typical bodily folds on its set self-calls.
It resets the woefully adrift at career crossroads carefully adroit.
Being drifts wholly downstream rounding off a risky risk upstream.
It wields to anchor, unearths to bridge hammocks in full spring tides.
Being ever creeps otherwise still with a fully creepy engendering agenda
of taboo voids, dreamers hush up in between the most upright cum ugliest.
Being as uniquely just glutton equally ratifies itself as uniquely intimate dieter.
To as from or up as down, Being spins self-looms in ways a no-one else best rivals
to x-ray in or out of litmus tests, ants or bees take turns to turn a Queen's evidence.
Being-Nothingness-Language daily synergizes its triple kin citizenshipon full time bases!²

Being-Language single-handedly lives up
to its true-life story, entirely all over kin ties
of the enigmatically self-making, selfsame Self.
Being-Language apostrophizes an egg or a seedling
like Nothingness, its wholly, seamless flesh and blood.
As sample wise palm wine, its tongue-in-cheek still tags
all at once, as ever intact, otherwise, sweet, sour and toxic;
its in vivo bin as its in vitro pin fully thrives in its self-portraits.
Being-Language exerts automatic voicesinto its whispering rainfall,
shedding tears, sobs of Christmas joy to poke its crackling sun shine,
minds harvest as the Nothingness, Being-Language Itself ever crops or tills.
Being-Language dips into top Nothingness to plait res nullius primi possedentis.
The Big Bang once reset the unique breach of integrity in declensions of Dasein.
Plural civil laws as nemo est judex in causa sua or lucem demonstrat umbra, lure all.³

---

nental Comparative Studies (PIP-TraCS) under the editorship of Wim van Binsbergen.
Res nullius primi possedentis: What belongs to nobody belongs to the first owner.

² *Nemo est judex in causa sua*: Nobody judges his or her own law-suit.

Today, time eclipses a basketful
of our witty findings into sales talk
for a resilient auction of self-reliance,
even for weevils to crop and sow grains.
Today, one stores its ruins on those ledges
where a fertile soil treasures to enrich a fruit
that one neither harvests nor sows in bits alone
preferring to treasure an ancestral bulk of dreams
parents freely hand over wholly as live future assets
in mutual bags or broken calabashes of cloven fortunes
such as a precious necklace of saàkinciy jewellery or pearls, *
yodelling or stammeringnames out ablaze in dazzling-flames. ⁴

Being-Language at once overly crawls
on all fours, walks on two but acrobatic
on three as stocks of riddles change hands.
We lock and unlock riddles, walking at dawn,
on all fours, at noon on two legs but at night fall
on three! To rename a ripe sequel as shadowy time
on a roll call silently knits and unknots grosser stories.
Our selfsame triplet: Being-Language-Nothingness at home
as abroad whistles as the referee re-enacts the match at play
inside the phenomenal rainbow where players still seek justice.
Otherwise, lifespan voicelessly clocks up with its blink of the eyes.
Being-Nothingness-Language avidly knits timescales as cyclonic yarns!

---

³ *Lucem demonstrat umbra*: There is no shadow without light.

⁴ *Saàkinciy:* A string necklace of blue and whitish decorative beads nobles of most ethnic groups in the Cameroon Grasslands wear as their classical tradition. Among the Nso', these beads symbolize wealth and nobility.

We credit a star of kindness and luck
in the warfare of its marketable kola nut
foreigners, daughters or sons, in the planets
swap for trails to open senses, mind and heart.
We oft set out to peck morsels of news in tongues,
one speaks in a poorer or richer land per unit of time.
Being at its pace quietly drums its synergies of Language,
whereby its in-vitro oils its in-vivo into ripe sibling rivalries.
Being evenlyradiates paradigms mirroring oily economic units!
Language namelessly auctions off as shares of paralinguistic names.
Yet, enigmatic Being-Language thrives on, the fittest of Godly Particles.
The Nothingness that is, intricately still shells its shares in Being-Language
within snail-like Shells together as its shadowy Selfsame Self in all its ways.
A twin decadence or development sets parity with Being-Nothingness-Language!

Protohistory, Presocratics, and philosophy

by Sanya Osha

Abstract: Wim M. J. van Binsbergen's work, *Before the Presocratics* (2012) presents a kaleido-scopic assessment of regional and global epistemic traditions and configurations before the advent of ancient Greek thought. He is concerned about interrogating worlds that relate to Afrocentricity employing an impressive assemblage of specialties namely, protohistory, archae-ology, comparative ethnography, comparative mythology, comparative linguistics and genetics. His central thesis is that rather than viewing different regional epistemic formations as singular and distinct, it is more appropriate to understand them as being part of a global and historical continuum of knowledge traditions that are perpetually subject to migration and transforma-tion – in short, all the elements of transplantation and dispersal. In this light, the strict separa-tion between regional and ethnic knowledge becomes misguided and often preposterous.

The reach and implications of van Binsbergen's work are too immense to attempt to arrive at a definitive conclusion quickly. It deserves to be read and analyzed diligently in order to do justice to its daunting scope, scholarship, and depth. At this juncture, what is of immediate concern is its discomfort with the general and specific aspects of the Afrocentric project. Van Binsbergen hopes his work would assuage Africa's doubts regarding its participation in trans-continental passages of global knowledge production. This article reflects on this specific angle as well as some of the central preoccupations of van Binsbergen's major work.

Introduction

Wim M. J. van Binsbergen's work, *Before the Presocratics* (2012) presents a ka-leidoscopic assessment of regional and global epistemic traditions and configu-rations before the advent of ancient Greek thought (see also 2011a–d; 2012b–f;

2013). He is concerned about interrogating worlds that relate to Afrocentricity employing: an impressive assemblage of specialties namely, protohistory, archaeology, comparative ethnography, comparative mythology, comparative linguistics and genetics. His central thesis is that rather than viewing different regional epistemic formations as singular and distinct, it is more appropriate to understand them as being part of a global and historical continuum of knowledge traditions that are perpetually subject to migration and transformation - in short, all the elements of transplantation and dispersal. In this light, the strict separation between regional and ethnic knowledge becomes misguided and often preposterous.

Convincing as van Binsbergen's arguments are, the messy phenomenon of race can undermine their appeal within the contexts and scripts of subalternity. Racial violence is not merely the abuse and denigration of subject peoples. It means more importantly; the total annihilation of consciousness, which of course touches on questions of the intellect. Racially abused peoples are never taken seriously intellectually. This is an angle completely absent from van Binsbergen's work as much as he attempts to advance a supposedly Afrocentric perspective.

Van Binsbergen calls into question the widespread perception held by many important philosophers - such as Heidegger and Gadamer – that the Presocratic thinkers started what is considered Western philosophy and that Empedocles initiated 'the system of four elements as immutable and irreducible parallel components of reality and in doing so, . . . laid the found for Modern science and technology, and the Modern World System at Large' (van Binsbergen 2012: 31). Afrocentrists' attempt to establish the primacy of the African continent and African cosmologies, often in direct opposition to outright racist objection. Van Binsbergen's project seeks to overcome this age-long 'paradigm of oppositionality' for a broader outlook of interconnectedness between human knowledge and epistemic traditions. Thus:

> as well as the rise of a vocal counter-hegemonic trend in scholarship all over the world, have ushered a new era, where the transcontinental continuities of the present invite us to investigate transcontinental continuities of the past, and to overcome such divisiveness as hegemonic interests of earlier decades and centuries have imposed on our image of the world and of the cultural history of humankind, and to help free Africa from the isolated and peripheral position that has been attributed to that continent in present-day World System (van Binsbergen 2012: 32).

Van Binsbergen also reminds us that he has conducted 'counter-hegemonic, transcontinental research for over twenty years now' (van Binsbergen 2012:32). This places his Afrocentric credentials to the fore even while interrogating the radicality of those same credentials, merely because he has taken up a project whose theoretical composition includes a far-reaching incorporation of genetic science, archaeology, linguistics, comparative mythology, comparative ethnography, and empiricism, in short, a range of radical methodologies that would end up signalling a whole new academic genre.

On the Pelasgian hypothesis

According to accepted paleoanthropology, archaic Homo sapiens evolved to anatomically modern human beings in sub-Saharan Africa as early as 200,000 years ago, and then dispersed to other continents. This view is termed the 'Out-of-Africa' (OOA) hypothesis (or 'recent single-origin hypothesis' (RSOH), 'replacement hypothesis,' or 'recent African origin model' (RAO) by experts in the field). There is also the 'Back-to-Africa' hypothesis, according to which human beings developed elsewhere, and then returned to Africa bearing new genes, religious and cultural practices, and new knowledge pertaining to science and technology. Van Binsbergen terms this migration back into Africa 'Pandora's Box.' He mentions some central hypotheses that he returns to frequently in his work, notably, the Borean hypothesis, as formulated by and Harold C. Fleming (1987; 1991) and Sergei Starostin (1989; 1991), which, as described by van Binsbergen, holds:

> all languages spoken today retain, in their constructed language forms, substantial traces of a hypothetical, reconstructed language arbitrarily termed 'Borean' and supposed to have been spoken in Central Asia, perhaps near Lake Baikal, in the Upper Paleolithic (c. 25 ka BP) (van Binsbergen 2012: 34)

On the other hand, says van Binsbergen, Stephen Oppenheimer (2001) argues, using the Sunda hypothesis, which postulates:

> considerable demic effusion of cultural traits took place from South East Asia to Western Eurasia (and by implication to Africa) as the South Asian subcontinent was flooded (resulting in its present-day, insular nature) with the melting of polar ice at the onset of the Holocene (10 ka BP) (van Binsbergen 2012: 34).

Van Binsbergen adds that to understand prehistorical and protohistorical philosophical thought, it is necessary to move beyond the philosophical enterprise as conceived as a narrow academic discipline and instead take in the study of the language, culture, and the social context in which Presocratic thought evolved. Accordingly, this methodological imperative necessitates a multiplicity of disciplinary competencies. In relation to philosophy itself, he states that he does not offer a clear-cut argument per se, but instead presents a 'historical and transcontinental-comparative *prolegomena* to an ontological philosophical argument on cosmology and the structure of reality'(van Binsbergen 2012: 41). Van Binsbergen labels his approach as 'counter-paradigmatic' inasmuch as it seeks to 'chart intellectual *terra incognita*' (van Binsbergen 2012: 43).

While conventional Global Studies deals with specific cultures, van Binsbergen's course is very much concerned with entire continents and the concept of globality itself. Thus, he begins from the Upper Palaelithic Age as a spatial construct while at the same time tracing 'a particular intellectual cultural complex characterized by such features as cyclicity, transformation and element cosmology' (van Binsbergen 2012: 43), thereby bypassing 'the highly presentist and localist perspectives prevailing in social anthropology ever since the *classic*, fieldwork-centered tradition in that field was established in the 1930s–1940s' (van Binsbergen 2012: 43). In addition, he learned that, within a given social context, cultural meaning is not only produced by social, political, and economic factors alone – he considers this a largely reductionist perspective – but also by symbols capable of retaining meaning and relevance across several cultural and geographical divides.

Karl Jaspers had propounded the notion of *Achsenzeit* (Axial Age: the period from 800 to 200 BCE, during which, according to Jaspers, similar new ways of thinking appeared in Persia, India, the Sinosphere and the Western world,[1] which, barring its overt Eurocentric connotations, as van Binsbergen reminds us, is central for an understanding of the concept of transcendence that became entrenched in human thought after the convergence of writing, the state, organized religion, and the monetary economy as key factors in the organization of society. Due to different waves of proto-globalization, these crucial features of organized society found their way into different regions of the globe such as the Aegean by way of Iran and China via Northern India. Those transformative bursts of proto-globalization were powered by chariot, horse-back, and water transport.

---

[1] See Jaspers 2010.

Van Binsbergen argues that certain cultural traits from the Upper Palaeolithic Age found their way into the African continent and he first became aware of this when conducting fieldwork in Francistown, Botswana, where a supposedly indigenous divination system, displayed strong similarities with

> 'an Islamic astrologically-based divination system that was established in Iraq around 1000 CE that in the meantime spread not only to Southern Africa but also to the entire Indian Ocean region, West Africa, and even Medieval and Renaissance Europe' (van Binsbergen 2012: 44).

This widespread family of geomantic divination systems, and other similar diagnostic and therapeutic traditions, all has a formal character that facilitates their transmission across several spacio-temporal contexts. Similarly, it is possible to study the correlations between cultural features-such as animal symbolism (such as the leopard and its spotted pelt), myths, and games belonging to the mancala (a board-game) variety – from a largely transcontinental perspective (see van Binsbergen 1995).

Transcontinental Studies, van Binsbergen points out, have led to significant shifts in anthropological research and the global politics of knowledge, fostering in the process the rise of disciplines such as postcolonial theory, Afrocentrism, Mediterranean Bronze Age Studies, and Egyptology. In this regard, the work of American sinologist, Martin Bernal is central – especially his thesis elucidated in *Black Athena I-III* (1987–2006).

Van Binsbergen then defines

> 'strong Afrocentrism as a theory that considers Africa the origin crucial phenomena of cultural history' (van Binsbergen 2012: 46).

This aspect immediately connects with Dani W. Nabudere's notion of Afrikology, which essentially regards Africa as 'the Cradle of Humankind,' and Afrocentric theorists such as Molefi Kete Asante's, whose notion of Afrocentrism possesses quite a number of arresting subtleties quite distinct from the usual ethnocentric affirmation of Africa's cultural primacy. Van Binsbergen is always anxious to affirm his Afrocentricity; one of the ways in which he accomplishes this is by attempting to debunk 'the Eurocentric and hegemonic myth that philosophy started in Europe in historical times' (van Binsbergen 2012: 47).

In advancing what he terms the Pelasgian hypothesis, van Binsbergen argues that as a result of the Out Of Africa exodus, Africans settled all over the world, bearing along with them specific sociocultural features such as marriage, kin-ship systems, and divination practices. In addition, during this global dispersal, myths and other products of the collective subconscious from Africa found their way into other regions of the world. Once out of Africa, these cultural manifestations became embedded in what he terms 'Contexts of Intensified Transformation and Innovation,' which led to 'new modes of production (both within and beyond hunting and gathering), and of new linguistic macrophyla' (van Binsbergen 2012: 49).

Contrary to the Out Of Africa hypothesis, the 'Back-to-Africa' hypothesis is claimed to have occurred 'in the last 15 ka' (van Binsbergen 2012: 51), during which Asian peoples migrated to Africa carrying cultural attributes with them. These attributes pertained to kingship, ecstatic cults, divination systems, and language, as, for example, van Binsbergen claims there are Austric similarities in Bantu. It is suggested that the return to Africa most likely happened through (1) North Africa and the Sahara and (2) along the Indian Ocean from the Ara-bian peninsula or a more southern point of departure through the Swahili coast, Madagascar, or via the Cape of Good Hope through the Atlantic West coast ending up in the Bight of Benin and West Africa. As a result of this migra-tion, an Indonesian / South East Asian influence (including East and South Asian) otherwise termed as the Sunda influence - can be discerned at a trans-continental level which includes Africa. Van Binsbergen argues that it is possi-ble to trace the emergence of mancala board games in Africa to an Asian origin, with world religions such as Buddhism and Islam serving as platforms for their dissemination. 'Sunda' traits such as agricultural crops, xylophones, ecstatic cults and kingship structures, it is mentioned, can also be observed in West Africa. He further suggests that

'Sunda-associated, Buddhist-orientated states were established in Southern and South Central Africa around the turn of the second millennium (Mapungubwe and Great Zimbabwe are cases in point)' (van Binsbergen 2012: 64).

It is also possible to trace the history and movement of geomancy at the trans-continental level. One of the oldest textual and iconographic attestations of geomantic representational apparatus is of Chinese origin. Another ancient geomantic attestation springs from the Arabian context. It is claimed that these two geomantic systems in fact share 'semantic, symbolic and representational correspondences'

and hence

'a common cultural environment' (van Binsbergen 2012: 68).

Apart from Sino-Tibetan and Arabian geomancy (divination by the earth) which bear remarkable similarities with each other, there is also the same family of systems to be found in ancient Greek and Latin, Hebrew, Indian and pre-modern African contexts. In Africa in particular, other systems of divination include the Malagasy *sikidy*, West African *Ifa*, and the Arabian *'ilm al-raml*. While many scholars have affirmed the influence Arabian geomantic practices across the coast of the Indian Ocean, many Afrocentric scholars have in turn rejected the Arabian origins of the West African geomantic system. Van Binsbergen recalls the derision and resistance which met his claim that similar geomantic systems exist outside West Africa at an Afrocentric discussion group. Van Binsbergen cites Robert Dick-Read, who asserts that there is evidence of Arab / Islamic influence in West African geomancy especially Ifa, which employs the names of Islamic prophets within its corpus. So it is not inconceivable that *Ifa*

'may have an Indian Ocean, circum-Cape background' (van Binsbergen 2012: 72).

Van Binsbergen concludes that West and South African practices of geomancy are directly indebted to Indian Ocean / Sunda influence coming through the Cape of Good Hope. Also noteworthy is, in parts of Africa, there exist simple configurations of geomancy which are likely to be derivations of more intricate forms that possess a non-African origin most probably Chinese. This view has not been welcomed by strong Afrocentrists. Van Binsbergen asserts that divination bowls from Venda and West Africa are likely to be variations of Chinese divination bowls or nautical instruments. The Sunda influence we are informed, can be discerned in the Persian Gulf, the Mozambican-Angola corridor, the Bight of Benin, the Austronesian population of Madagascar. On the other hand, when Africans surface in T'ang China, it is as slaves so much so that the figure of the black trickster became a familiar literary trope. All of this would obviously meet with the disapproval of Afrocentrists.

Martin Bernal, who has gained the attention of Afrocentrists for mixed reasons, is viewed by van Binsbergen to be 'wrong for the wrong reasons' (van Binsbergen 2012: 84). Bernal is also accused of imposing his subjective views as statement of fact and resorting to *ad hominem* tactics to assert his claims. In other words, van Binsbergen has much to fault about his work. Émile Durkheim is another Western intellectual that van Binsbergen exposes for shoddy work. Durkheim in *The Elementary Forms of Religious Life* (1912) makes propositions regarding Australian Aboriginals and totemism without so much as a visit to the site of study. As such, he had theorized and hypothesized about an entire group of people without any personally organized ethnographic evidence and without any acceptable implements of comparative analysis.

Van Binsbergen stresses that he is more concerned about establishing the linkages, continuities, and connections between different continents of the world and hence the timeliness and validity of the notion of transcontinentality. Movement, migration and exchange, he points out, have for millennia been part of the currency of human transactions. If such is the case, not only goods and people have been transported far and wide but also ideas. And so it is possible to trace the intellectual history of the world as sequences of interlinkages between diverse systems of knowledge of which mancala and geomancy are major examples. In addition, this absorbing history can tracked employing genetic, linguistic, archaeological, comparative-ethnographic and comparative-mythological modes of analysis.

Employing these given modes of analysis, it can be taken that the Presocratics were not really the inventors of element cosmology as granted credit by the official archives of history and philosophy but were merely clumsy and less inventive recipients of a handed down system, primarily, in van Binsbergen's view, from ancient Asia and Africa. His thesis therefore seeks to affirm 'the transcontinental complementarity of the intellectual achievements of Anatomically Modern Humans in the course of millennia' (van Binsbergen 2012: 86).

The Nkoya and cyclicity

Van Binsbergen commences the tracking of this – for the moment hypothetical – trajectory of the intellectual achievements of humanity by focusing on the Nkoya of Zambia who have been major subjects of his anthropological research. During the pre-colonial and pre-statal epochs, the Nkoya were categorized according to clans in a manner that had powerful political connotations. The clans played a key role in the use and management of natural resources, the economic life of communities, the deployment of rituals, and the performance of communal obligations. Within the context of precoloniality, leadership of the clans tended to be ritualized as opposed to being actually political and mostly consisted of women. However, with the advent of long distance trade and conflict among competing nationalities, the practice and visage of kingship were altered in which they became, for one, usually male-dominated and largely politicized as opposed to being merely ritual. Van Binsbergen hypothesizes that this development may also have been the mark of Sundan influence. Similarly, the existence within Nkoya culture of

> 'iron working, Conus shell ornaments, and the introduction of a new type of xylophone-centred royal music' (van Binsbergen 2012: 90)

is enough to suggest the influence some transoceanic involvement possibly stemming from South Asia. Van Binsbergen mentions that his research on Asian-African continuities gives him reason to believe that the transformation of Nkoya political institutions and aspects of their material culture most probably owes much to interactions with those from South Asia.

The Nkoya, being perceived by colonialist anthropological researchers as primitive, were susceptible to the biases of totemism studies, which tended to objectify the cultures of such so-called primitive tribes in terms of pairs, which are both oppositional and complementary. Accordingly, Nkoya clan names can be analysed according to binaries that are both complementary and are in opposition. Van Binsbergen goes on to posit that there is

> 'evidence of a more complex and more dynamic structure of *threesomes* that transcends recursive repetition' (van Binsbergen 2012: 94).

He mentions that he is not able to obtain the desired degree of certainty regarding many details about the clan names or their origins from native informants. Although he able to establish some continuities between the Nkoya and Taoist systems of binaries, he concludes that rather than maintain its structure and stability, the Nkoya system had degenerated 'into aberrant multiplicity' (van Binsbergen 2012: 106). In essence:

> 'in the Taoist system the same five elements always play, in turn, the role of the destructor, destroyed, catalyst, but in the Nkoya system those roles have become disconnected hence the number of elements, or clans, has multiplied from six to eighteen' (van Binsbergen 2012:106).

The Nkoya system is not unique. Instances of element cosmology can be found in Chinese Taoism, Egyptian cosmology, and in other systems found in Africa, North America, India, Japan, and Ancient Greece.

The enigma of Empedocles

In ancient Greece, the four-element cosmology or the quadripartite conception of the cosmos (*materiae primae*, prime matter) – earth, water, air, and fire – was first adopted by Empedocles, shaman and philosopher of Acragas, and historically credited with founding the scheme. Van Binsbergen debunks this view, claiming that the scheme was already in existence at least two millennia before the Empedoclean formulation. He argues that until the early twentieth century, Empedocles was considered responsible for the system in most of the literature when fresh readings of Homeric and Hesiodic texts affirmed the opposite.

Following Empedocles, Plato and Aristotle also adopted the scheme, which eventually became the foundation of

> 'Western natural science, astrology, medicine, psychology, literary and artistic symbolism and iconography including color symbolism until into Early Modern Times' (van Binsbergen 2012: 109).

The scheme can be found in esoteric practices of mystery cults, the Ancient Sabaceans of Yemen, the alchemy of the Qarmatians, Islamic occult sciences, and even digital media games within the context of postmodernity.

According to van Binsbergen, Empedocles 'merely codified and corrupted' a system that had been used all over Eurasia for several millennia. Thus, the influence of Asian shamanism on Greek rationalism, through its spread by Empedocles, has always been discernible.

Shamans are understood to be special individuals gifted with powers of healing and who are able to move along the celestial axis and within the Underground in search of medicine and information to heal needy and afflicted individuals and communities. Usually distinguished by a strange and bewildering disposition, it is believed that apart from Empedocles, other shamans (the healer-sage, *iatromantis*) within the classical Greek tradition include: Pythagoras, Parmenides, Abaris, Orpheus, Aristeas, Epimenides, and Hermotimus. The lives of these historical figures are usually marked by life-transforming journeys to retrieve lost souls, as in the case of Pythagoras, who ventured into the underworld to fetch the soul of Hermotimus, or Empedocles, who claimed to be able to control not only the weather but also had the power to rescue lost souls from Hades. Most of these figures were usually multi-talented, combining the roles of magician and naturalist, poet and philosopher, preacher, healer, and public counsel (Bruce J. MacLennan cited in van Binsbergen 2012: 112).

Within classical Greek rationalism, there is the traditional opposition between Apollo and Dionysus, where the former is supposed to act as a 'mitigating and balancing' influence on the latter's wild practices of ecstatic religion. Traces of this shamanic influence can be perceived in the ancient Near East and in Egypt a millennium before the emergence of the Presocratics. The process of the transmission of these shamanic influences into the Aegean was aided by the healers of ancient Iran and the Mesopotamian Magi in addition to the Scythians and Thracians.

Rather than the four-element model of Empedocles, the five-element system was more widely practiced in Greek and Roman antiquity, in which the fifth element is the quintessence. The five element model also exists in several parts of Eurasia, such as Japan and China, and in religions such as Hinduism (*panchamahabhuta*), Buddhism (*skandha*), and the Bon religion of Tibet. In addition, it can be found among the Daisanites and among the followers of Baroaisan of Edessa.

The transformation cycle of these elements and their correlative systems are important to understand:

> 'human-existential dimensions (the heavens, minerals, animal life, plant life, kinship, politics, colours, music, topography, etc.) so that the entire cosmos can be subsumed in a matrix whose columns define symbolic domains and whose rows define cosmological / existential dimensions' (van Binsbergen 2012: 118–119).

Correlative systems also provide invaluable information in the pursuit of divinatory arts and knowledge.

In the Chinese Taoist cycle of transformation, there is the belief that the difference between the elements is 'accidental and situational,' meaning that each may be transformed to the next within a couple of steps, thus making the distinction between them 'ephemeral and non-essential,' both of which establish the concept of immanentalism in Taoist thought.

Transcontinentality at large

The presence of four or more element systems can be perceived in sub-Saharan Africa, with the Nkoya, in van Binsbergen's view, being one of the most obvious examples. There is also an example to be found among the Yoruba of West Africa, whose model is complicated by the collective belief in an omnipresent sky god, Olokun, and the demiurge, Obatala. However, in spite of these two major attributes, van Binsbergen posits, 'the pantheon of Yoruba gods does have strong reminiscences of the Hesiodic and Egyptian cosmology' (van Binsbergen 2012: 123) and its attendant transformation cycle of elements.

The Yoruban pantheon of gods necessitates a few words. Olorun, who resembles the Nyan-kupon of the Tshis, the Mawu of the Ewes, and the Nyonmo of the Gas, is the embodiment of the sky. As the counterpart of the Egyptian god, Pet, he is viewed as being too removed from the affairs of ordinary human beings, and so he is not directly consulted by them. As such, he has no priests or shrines to his name, which is only invoked when lesser gods are unable to intercede on behalf of mortals.

Obatala, who was conceived by Olorun, stands in place for Olorun in overseeing the affairs of heaven and earth. He is also believed to be a sky god with human endowments and seen to be the counterpart of the Egyptian god, Ptah. Being a judge of human beings, he is also deemed to have qualities that match another Egyptian god, Osiris. Odudua, Obatala's spouse, bears similarities with Isis, the Egyptian goddess, but being a promoter of the ethos of love, she is akin to the Egyptian god Hathor. Odudua gave birth to male child, Aganju, and a daughter, Yemaja. Aganju represents land (earth) while Yemaja is a water goddess. The children of Obatala and Odudua married, bearing a son named Orungun, who proceeds to sire the following children with his mother:

> Dada, a vegetable god (WOOD); Shango, lightning god (FIRE?); Ogun god of iron (METAL) and war; Olokun sea-god; Olosa, lagoon god; Oya, Niger god (sic): Oshun, river-god; Oba river-god; OrishaOko, god of agriculture; Oshosi, god of hunters; Oke, god of mountains; AjeShaluga, god of wealth; Shankpanna, small-pox god; Orun, the Sun (FIRE?); and Oshu, the [M]oon. Oshumare, the rainbow, is a servant of Shango, and his messenger Ara is the thunderclap; his slave is Biri, the darkness [CHAOS].

> Shango hanged himself but did not die, for he went into the earth and there became a god (*orisha*) Ifa, god of divination, who causes pregnancy, and presides over births. Elegba, a phallic divinity; his symbol is a short knobbed club, which was originally intended to be a representation of the phallus. Circumcision and excision are connected to his worship. Ogun, the war-god. The priests of Ogun take out the hearts of human victims, dry and powder them, mix them with rum, and sell them to people who wish to acquire great courage (van Binsbergen 2012: 124).

In spite of this elaborate pantheon of gods, foursome features - bearing similarities to four-element cosmology or reflecting the accepted quadripartite separation of the cosmos - are noticeable in the Yoruba context in the form of the

> 'four estates, four winds, four days of the week, four walls of the Yoruba kingdom and a divine foursome Shango / Oya / Oba and Oshun'(van Binsbergen 2012: 125).

In other parts of Africa and its environs, traces of the four-element system can be found in Madagascar, which earlier scholars had attributed to an Indonesian influence, but which van Binsbergen ascribes to a broader Sunda incursion entering Africa through South East Asia.

Apart of the four-element system, other products and practices of Sunda transmission into Africa are said to include the xylophone, the gong, and breast harp, ancient Roman coins, cowry shells, geomancy, kingship rituals, and activities of ecstatic cults. These products and practices are most likely to have travelled via the Indian Ocean through the Cape of Good Hope and then entered into Africa. Heliopolitan cosmogony is also an attestation of the four-element system, in which instances of transformation comprising destruction and regeneration are frequent.

Van Binsbergen has been an energetic contributor to the *Black Athena* debate instigated by the work of Martin Bernal. Bernal argues that much of the ancient Greek civilization is indebted to ancient Egypt, a line of thinking promoted by Cheikh Anta Diop. Initially, van Binsbergen departs from the two opposing authors by arguing that the ancient Greek myth of Hephaestus and Athena owes much to Central Mediterranean and Anatolian influences rather than to Egyptian inspiration. He was later compelled to reconsider his position, agreeing with Bernal in the process, after a closing reading of Ovid and Virgil in relation to the contested myth. In other words, he reached an Egyptianizing conclusion. However, this conclusion is demonstrated to be only partial, as he goes on to argue, 'some rudimentary transformative cyclical element system could be a Pelasgian trait' (van Binsbergen 2012: 140), arriving from West to

Central Asia. This argument immediately debunks the conclusions of Bernal and long-established Afrocentrists such as Diop.

Van Binsbergen mentions several Pelasgian traces to be found in ancient Egypt, such as the composition of the pantheon, which appear to owe much to a West Asian derivation. Shamanic practices, leopard-skin symbolism, and the royal diadem, which became common in Egypt, are all, in van Binsbergen's view, attributable to a strong West Asian influence. But it is also not in doubt that the main cornerstones of Presocratic philosophy were received from the accomplishments of ancient Egypt, which was already in decline. It is not in doubt that Empedocles's vulgar appropriation of Egyptian cosmology did much to define the climate of Greek thought, especially with regard to the espousal of the four-element doctrine.

Presocratic Greek thought was preoccupied with arriving at the meaning of ultimate reality. Different thinkers came up with different explanations, such as Thales, who identified water as the primal matter, while Anaximenes posited it was air, and Heraclitus said it was fire. Xenophanes argued that it was the earth, and Empedocles mentioned four elements, (water, air, earth, and fire). The important demarcation in Greek thought between what came before Socrates and what came after was established by Hegel and Schleiermacher, while Aristotle is credited with asserting the significance of Empedocles in a manner that was to have a profound effect on

> 'Graeco-Roman, Arabic, Indian, and European natural science, astrology, other forms of divination, medicine, iconography'(van Binsbergen 2012: 150),

thereby creating the main epistemes of modern science and thought. A major shift occurred in the four-element system between the time it was conceived within Taoism and

> 'the ontologies of Graeco-Roman late antiquity, medieval Byzantine, Arabic, and Latin Science and their Early Modern Derivative' (van Binsbergen 2012: 150).

Under the later appropriations, the system became fixed, rigid, and standardized, through which the issue of transformation no longer mattered. As such, the elements no longer mutated from one substance to another, but remained unchanged, as established under the Empedoclean system (*rhizomata*), even when interacting with other elements, which is not the case within the context of the original Taoist system.

The transition from the fluidity of the element system in its Taoist phase to its standardized form in Empedoclean orientation coincided with the shift from an illiterate, oral mode of knowledge progressing to a written format, which, in turn, signalled the emergence of the state, science, and organized religion as major organizing factors within culture and society. It also implied the entrenchment of the notion of transcendence in place of the Taoist concept of immanentalism.

Van Binsbergen is also able to track the global distribution of flood myths from a protohistorical perspective (2008). In Judaism, the tears of Archangel Michael are believed to have created the Cherubim (a group of angels), just as in the Taino culture, Jamaica, rain is perceived to be divine tears. Both of these myths bear similarities to the Egyptian mytheme regarding 'human beings from divine tears.' The mytheme of divine tears producing human beings correlates to leopard symbolism, in which its spotted skin is akin to rain. Also of interest in this connection are the rites associated with the planting season, with its tropes of abundance and fertility, just as communal rites of sexual promiscuity and ancient observances in memory of the dead.

To return to the Empedoclean corpus: It was believed that the four elements that constituted the system completed a transformation cycle and this rendering was adopted by both Plato and Aristotle. Those who came after them, on the other hand, believed that the four elements consisted of four distinct, separate, immutable ontological substances, which, accordingly, privileged the concept of *rhizomata* over the contrary Empedoclean understanding and articulation of *effluvia* (the elments that enter through the sense organs).

Empedocles, to be sure, reflected upon the relations between mutability and immutability and sought a resolution between the two, thereby formulating a theory of element relations, as opposed to Parmenides, who adopted a stance of uncompromising immutability. Thus, in the Empedoclean and the Aristotelian tradition of the four-element system, fire, air, water, and earth are deemed to originate from each other, with each of them bearing a direct relation to each other. This much is made clear in Aristotle's *Meteorologica*.

Van Binsbergen reminds us that the Western intellectual tradition generally does not fully recognize the significance of the notions of transformation and cyclicity within Empedoclean cosmology, although this is gradually changing. A number of modern scholars interpret the transformative dimension of the cosmology as a perpetual transition between Love (*philia*) and Strife (*neikos*). In this instance, Love is symbolized as a spherical state embodying all the cosmic elements, which eventually disintegrates under the influence of Strife; then the cycle of spherical embodiment and tumultuous disintegration is resumed once again. This abbreviated modern interpretation of Empedoclean cosmology can, in turn, be compared with the simple movement of the pendulum, or, more appropriately, as moving in acyclical fashion, between production and destruction on one level, and fusion (love) and separation (strife), on another.

Van Binsbergen expends a great deal of effort in discovering the origins of the quadripartite division of the cosmos as couched in the Empedoclean doctrine. As mentioned earlier, he argues that Empedocles was not the originator or author of the scheme, even though Aristotle had claimed so in his *Metaphysics*. Van Binsbergen conducts his research beyond the scope of Graeco-Roman antiquityto focus on the Palaeolithic Age to find answers.

It is important to note that, apart from Empedocles, the Milesian school of philosophy upheld the Empedoclean doctrine, and to note that Heraclitus, the author of the doctrine, was aware of the significance of cyclicity within it. Nonetheless, the spread and influence of foursomes in the symbolism and iconography of different peoples has been tremendous, including, for example, the four trigons of the zodiac; the four humors of Galenus (blood, black bile, yellow bile, and phlegm); the four stages of man (gold, silver, bronze, and iron); and the four virtues, namely justice, fortitude, prudence and temperance (van Binsbergen 2012: 176). In addition, there are four cardinal truths in Buddhism, which are: the meaning of life is suffering, the source of suffering is attachment, there is a possible end to suffering, and there is a route available to the end of suffering. Other major foursome symbolisms and iconography include the four men of the Apocalypse; the four distinct suits of the deck of cards notably clubs, diamonds, hearts and spades; the four categories of Arabian music theory; the four main characters of Chinese opera; the four major castes of India; the books of the Veda with each consisting of four parts; the interlinked parts of the Kabbala; the four Archangels; and the four lights of Gnostic mysticism.

Van Binsbergen concludes that many of these foursomes are based on the four-element system, while some others including some not mentioned here, are not. He also mentions a couple of instances in the field of pure science devoid of human agency where foursomes play major roles, such as in physics, where there is a quest to arrive at a unifying theory through the combination of gravity, electromagnetic force, strong nuclear force, and weak nuclear force. There is also the four-color problem in mathematics, and the reality that each DNA and RNA protein comprises a blend of four different amino acids among other noted examples.

Van Binsbergen's major academic aim is to find evidence of the four-element system existing before Graeco-Roman antiquity. Therefore, he begins his search in the African Palaeolithic Age, which offers scant and perhaps unreliable information apart from the Blombos Cave red ochre block in South Africa and rock art in Zimbabwe. In carrying out this aspect of the work, he has only comparative linguistics, archaeology, comparative mythology, and comparative ethnography as disciplines upon which to rely. His quest leads to the recognition that the major Neolithic or Bronze Age Triadic Revolution, which culminated in writing, the state, organized religion, and (proto)science becoming the major factors determining the course of human thought, culture, and civilization was indeed a revolution in more than one sense. Accordingly, this triadic revolution established the Hegelian conception of dialectics as a major paradigm, which is regarded as an advancement on binary systems comprising for instance, in the cosmological sense, heaven and earth, or land and sea. A third agent is required to overcome the endless recursion inherent in the binary model that the triadic revolution transcended and which, in turn, transformed history, society, and culture.

However, van Binsbergen suggests that North American and African formal systems - with the notable exception of Egyptian Hermopolitan cosmology - continued to be characterized by twosomes and foursomes, and hence their delay in effecting the transition to the triadic paradigm and its attendant benefits. He adds that the Empedoclean four - element system was less advanced than the Taoist doctrine of transformation cycle, which incorporates a catalyst that endows it with triadic structure. It is implied that African divination systems, specifically Hakata, Ifa, and Sikidy belong to

'an Upper Palaeolithic Old World standard pattern' (van Binsbergen 2012: 209).

Van Binsbergen ultimately concludes that the element system could have emerged from anywhere.

Van Binsbergen then examines the work and impact of Albert Terrien de La-couperie, the French-British Sinologist who had posited a Western origin for part of the language and part of the population of China, thereby concluding that Europe and much of Asia shared a common ancestry in terms of civiliza-tion. Terrien produced an astonishing body of work that encompassed

> 'contributions to the history of Buddhism and of South Asian, Central Asian and East Asian writing systems and scriptures, the ethnography and linguistic description of Formosa, the archaeology of Korea, explorations in Assyriology, and the first recogni-tion of the striking similarities between the Indus valley and Easter Island (pseudo-)scripts' (van Binsbergen 2012: 218).

In addition, Terrien is regarded as being responsible for the emergence of pan-Babylonianism, the theory that all of civilization emanated from ancient Mesopo-tamia. The attention given to the Terrien's Sinology was accomplished by renewed focus on *Yi Jing* (*Book of Changes*), a Chinese classical text that propounds a cos-mology covering all facets of human society and the universe. Within the context of this divination system (cleromancy) that serves as an all - decisive oracle, a ran-dom generator – most notably a coin – is expected to accomplish a specific result drawn one of sixty-four possible combinations, each of which bears specific divina-tory significations concerning existence and beyond.

Terrien continues to elicit considerable interest in China and Japan long after in his untimely passing. His detractors are equally many, with a noted Afrocentri-cist, Runoko Rashidi claiming, in contrast to Terrien's conclusions, that black Akkado-Sumerians of Elam-Babylonia were responsible for the the*Yi Jing*. Cleromancy, the divinations system embodied in the *Yi Jing*, has been a source of considerable academic controversy. The text of the *Yi Jing* has been touted by local Chinese advocates to contain the kernel of all the major scientific break-throughs, but this has been contested by many Western scholars. Advocates of the text affirm that those who approach it with patience and diligence would be rewarded with knowledge hidden within its labyrinthine and obscurantist meanings. Meanwhile, others have concluded that it is a product of an obscure Central Asian dialect, the origins and status of which can no longer be traced.

Van Binsbergen's major aim is to attempt to trace continuities between differ-ent continents, regions, historical epochs with underlying, if not largely in-tended, implications for race. His findings lead him to conclude that Asia and sub-Saharan Africa were

> 'part of a *multi-centred* and *multi-directional* prehistoric and protohistoric system of
> exchanges in which an emerging global maritime network played an increasing role'
> (van Binsbergen 2012: 225),

thereby attesting to a process of proto-globalization in the Bronze Age.

Through modern archaeological and epigraphical developments, it is now possible to affirm the historical validity of Chinese ruling dynasties dating back to the period of early counterparts in ancient Mesopotamia and ancient Egypt, prompting Chinese scholars to ascribe an endogenous trajectory of development for their civilization rather than the formerly held thesis that they shared a common civilizational pool with Western Europe, which was located in ancient Mesopotamia. This development goes against Terrien's proposition, part of which ascribes a Mesopotamian origin to *Yi Jing*. Van Binsbergen points out that the old thesis espouses a non-ethnocentric and therefore anti-hegemonic slant, while the latest Chinese proclivity to privilege an endogenous path of development for its civilization betrays the same chauvinism that the old thesis had sought to avoid.

In a lengthy passage that reveals his long and often complicated relationship with Afrocentricity, van Binsbergen states:

> 'Since the 1990s I have repeatedly championed the cause of Afrocentricity. This was not
> in order to curry favour with my African friends and colleagues (although it did in fact
> endear me with them). Nor was it an attempt at Political Correctness, verbally com-
> pensating Africans as recognized and self-acclaimed victims of recent global history, by
> offering them the mere illusion of a glorious past. My defence of Afrocentricity also had
> to do with my awareness that once peripheral, subjugated or excluded groups – with
> whom I, admittedly, do identify, by birth, choice, and adoption – may have preserved,
> in their specific worldviews, knowledge of historical facts and relationships which oth-
> erwise have been expelled from collective consciousness by the hegemonic paradigms
> of dominant groups in the World System' (van Binsbergen 2012: 229).

Part of van Binsbergen's support for Afrocentricity has resulted in his attempt to discover certain 'dissimulated facts' that may have been preserved in certain group memories, but which have been ignored by the dominant global knowledge paradigms, of which his Pelasgian hypothesis is an example. The hypothesis advances the claim that during the Neolithic and Bronze Age, a markedly pigmented ethnic group possessing knowledge of proto-geomancy, early metallurgy, a fire cult, a solar cult, and formative element cosmology existed in

Western Asia (ibid.), which, in a way, supports a part of Rashidi's Afrocentric thesis that has been criticized for its scanty scholarship and lack of academic rigor, an accusation that has been made of much regarding Afrocentric discourse. Van Binsbergen suggests that during the OOA exodus, in which indigenes of the continent dispersed to other parts of the globe during the Middle Palaeolithic Age, highly pigmented people may have settled in Asia. This pigmented cluster is credited with knowledge of rudimentary metallurgy and proto-geomancy and the dispersal of these practices Westward. This heterogeneous mix of peoples is what gives rise to the Pelasgian hypothesis.

Van Binsbergen then advances another interesting proposition. The Mediterranean was populated with a broad genetic and linguistic assemblage of peoples during which the 'older layers of ethno-linguistic specificities' associated with highly pigmented people were thrust to the bottom of the social ladder while the more recent layers comprising of Indo-European and Afroasiatic speakers formed a dominant aristocratic stratum. Given this distinct social composition, the marginalized highly pigmented substratum were eventually shoved to the margins of the Old World constituted by sub-Saharan Africa, southernmost South Asia, and Australia / New Guinea, thereby giving rise to 'an inveterate, old and widespread racialism' that has subsequently denounced and denied transcontinental connections between Africa, Asia and Eurasia. This is quite an interesting hypothesis; one which van Binsbergen admits requires further exploration.

Adolf Leo Oppenheim (1966), an influential Assyriologist whom van Binsbergen obviously respects, makes an arresting point concerning the Mesopotamian origins of divination systems. Divination later became a prominent practice in Asia and other less significant contexts, notably Japan and Etruria. The technology of writing, in places such as Mesopotamia and China, facilitated the preservation of the methods of divination and its modes of interpretation. Oppenheim makes a claim that is bound to disconcert Afrocentrists, which is that ancient Egypt does not feature in the history of the arts of divination until its final dynasties. He then throws out a challenge to succeeding generations of scholars to attempt a reconstruction of Asian intellectual history, taking in the centrality of Mesopotamian accomplishments in science and astrology, the geomantic traditions of China, and the intricate horoscope of recent India among other major intellectual preoccupations. By extension, at the transcontinental level, it would be of considerable interest to interrogate the linkages of divinatory arts between ancient Greece, ancient Sumer, ancient China, and ancient Egypt, while also exploring the similarities in comparative mythological iconography, which constitute the central conceptual intent of van Binsbergen's project.

Van Binsbergen corroborates the widely held view that the Anatolian / Black Sea region is noted for numerous innovative developments in the history of civilization, namely the cultivation of food crops and the domestication of animals, the development of elaborate linguistic patterns, and the dispersal of the Flood myths. Accordingly, during the Neolithic period, the region was responsible for the emergence of a numerical, classificatory, and divination practice incorporating 'a protoform of the transformative element cycle,' which became distinctive in its structure, form, and properties. This particular crucial geomantic development was subsequently adopted in China, where it manifested itself as *Yi Jing* and the Taoist element system, and then later in Mesopotamia and Arabia, and subsequently in North Africa and sub-Saharan Africa. Afrocentrists have always claimed that geomantic divination is indigenous to Africa, but much of the evidence provided by scholarship points to the Arabian *'ilm al-raml* as the source of sub-Saharan African geomantic practices. Comparisons between sub-Saharan geomancies and other traditions, such as the Greek element system and Chinese divination practices, have been conducted and the Greek model was found to be very different from local African traditions, which were also discovered to exclude the transformative and cyclical features of the Chinese model.

Van Binsbergen draws attention to other controversial claims, such as one that holds that the *Yi Jing* most probably has a western Asian origin mediated by Hellenism and Hellenist Egypt with the Presocratics acting as agents of the transmission. On the basis of research into comparative linguistics, van Binsbergen suggests that the invention of the transformation cycle of elements occurred in the second millennium BCE and then spread to East Asia. He is aware of accusations of Eurocentrism arising from claims such as this, but is prepared to stand by his findings all the same. This paradigmatic invention, he claims, can be perceived in the Indo-Iranian fire cult and the Lycian cult of fire and metallurgy ascribed to the deity, Hephaestus, after which the Ionian philosophers and their acolytes formalized an element system that significantly reduced its transformative and cyclical features.

At this juncture, van Binsbergen makes a telling argument. It is possible to trace a genealogy of transcontinental continuities between different regions of the world spanning several thousands of kilometres and several thousands of years. As such, a major cultural invention with considerable paradigmatic implications at say, the extreme eastern hemisphere would eventually travel over time and space to the extreme western hemisphere, which provides a rough translation of what transcontinental continuities would mean. In relation to the global spread of element system cosmology, its prevalence seems to be attributable to its rather formalized – and therefore rigid – modes of transmission and retention. In virtually all the societies in which it was to be found, it was regarded as esoteric knowledge retained and transmitted through elaborate initiation rites organized by secret cults. This strict formalism guarded its essential character and facilitated its fluid transmission from one generation and millennium to the next and from one extreme region to the other.

The protohistorical legacy

The history by which the four-element system became established is quite fascinating. The Presocratics were primarily concerned with establishing what constituted primal matter, with different schools of thought selecting either earth, air, fire, or water. Within classical Greek literature, this contestation is granted dramatic form and effect, such as in the Homeric conflict between Achilles (earth) and Hephaestus (fire) against Scamander (water). Notably, Ovid's *Metamorphoses* is essentially concerned with the motif of transformation. The classical concept of transformation is, by extension, distinguished by the polarities of, on the one hand, killing or annihilating versus, on the other hand, creating or giving birth; or impeding versus to aiding, with both polarities underlining the importance of states of flux and transformation, in which the quest for definitive cosmic stability could only prove to be elusive. Empedocles, as we know, adopted the four-element system, retaining its essential transformative character but within the context of his millennial legacy, this crucial feature is missing, although, as van Binsbergen correctly suspects, the lingering Eurocentric presuppositions, which privilege West Asia remain.

Neolithic South West Asia, which is deemed to be a cradle of genetic, linguistic, and cultural diversity, is also regarded as the site of immense paradigmatic innovations. Here, in the technological field, innovations relating to metallurgy occurred, including the invention of the chariot. In the cultural sphere, the concept of transformation was established in addition to the triad and the division between Heaven and Earth. In the political realm, the state as a political entity was established. These major developments led to new levels of socio-political stratification, whereby racial (highly pigmented people) and linguistic (speakers of Khoisanoid and of Niger-Congo, especially proto-Bantu) underdogs were forced out of the dominant centres of global culture.

The concept of transformation enshrined in the element system doctrine is also associated with shamanism, which some scholars believe to have emerged as a practice in the Upper Palaeolithic Age of West and Central Asia (c. 20–15 ka BP). In order to understand shamanism, van Binsbergen enumerates two key questions; the first relates to its origins and the second concerns the possible connections between element cosmology, cyclicity, and transformation, on the one hand, and shamanism on the other. Research into the nature and possible history of leopard-skin symbolism from a global perspective provides a key starting point for this phase of the project. His view is that leopard-skin symbolism probably emerged from West to Central Asia (10–20 ka BP). The advent of shamanism is also connected with the emergence of naked-eye astronomy. Van Binsbergen lists the major symbols associated with shamanic practice notably: speckled nomenclature for leopard; speckled nomenclature for other species; ecstatic cult; therianthropy (human beings posing as animals); leopard therianthropy; leopard-skin symbolism; the notions of the Exalted Insider; the Sacred Outsider; and the Mother goddess (van Binsbergen 2012: 257).

Apart from the symbols listed above, the notion of transcendence is central to shamanism because a shamanic practitioner is believed to be able to enter ordinarily inaccessible spheres of reality most notably, Heaven and Earth, or the outer reaches of the galaxy and the Underworld, in search of esoteric knowledge and information usually pertaining to restoring good health in afflicted patients or at the communal level, order, and stability.

Van Binsbergen asserts that there are similarities between the Nkoya clan system and the Taoist transformation cycle – that normally includes a catalyst –

which can be attributed to three possible primary factors, namely

- the Upper Palaeolithic Age back-to-Africa migration;

- Pelasgian continuities during the Bronze Age; and finally,

- East Asian incursions into Africa in antiquity.

The first mentioned factor, which is the back-to-Africa migration, sounds convincing because the girls' puberty rites of the NaDene speaking peoples, whose languages fall under 'one linguistic macro - phylum,' notably Sino-Caucasian (as do the linguistic phyla of Sino-Tibetan, Caucasian, Barushaski, and Basque) are similar to those of Niger-Congo speaking Africans. It has also been established that there are similarities between the material cultures of Central (Mongolia), North America, and Bantu-speaking Africa, especially in basketry, fishing equipment, and basic house architecture (van Binsbergen 2012: 261). Experts working in the field of comparative mythology have been able to confirm that there are affinities in the mythological motifs of Bantu-speaking Africa and the Americas. Finally, in the field of linguistics, there are genetic connections between African macrophyla (Khoisan, Nilo-Saharan, and Niger-Congo), Eurasiatic (Indo-European), Austric, and Afroasiatic languages.

In continuing the project of tracing transcontinental associations in pre-historical modes of thought and esoteric practices, van Binsbergen isolates some of the distinguishing features and traits associated with shamanism, such as the elongated tooth shape and particularly circle and dot incisions, which are common in most regions of the world, most notably in parts of sub-Saharan Africa, Madagascar, ancient West Asia, and the Arctic and sub-Arctic regions of North America. Circle and dot incisions are associated with tremendous personal power and are prevalent usually where leopard-skin symbolism exists. Pardivested shamans (donned with leopard skin) existed in ancient Egypt and ancient Mesopotamia but van Binsbergen ascribes a West Asian origin to the practice of pardivestiture. This practice also became common throughout sub-Saharan Africa most notably, Southern Africa.

Similarly, van Binsbergen points out that the southern African divinatory four-somes bear close resemblance to Indigenous American games and divinatory models. He ascribes the global transmission of these features and practices to

Central Eurasia in the Upper Palaeolithic Age, a cultural transference that must probably occurred by virtue of the back- to- Africa-migration, during which there was a significant pattern of migration from West and East Asia into Africa via the Sahara and the Indian Ocean in the last 15 ka. Apart from bearing people, this substantial migratory trend entailed a diffusion within Africa of genetic markers, technological innovations, and cultural practices. Van Binsbergen concludes that native American and sub-Saharan African divinatory and game tablets can be traced back to a much earlier origin 'in Upper Palaeolithic Old World, more than 10,000 years before Empedocles' (van Binsbergen 2012: 274).

Van Binsbergen's Pelasgian hypothesis identifies Neolithic Bronze Age West Asia as a region replete with numerous technological developments and innovative cultural traits and practices, which were subsequently disseminated to the Mediterranean, East, South, and South East Asia, Oceania, ancient Egypt, and sub-Saharan Africa, mostly among the Nkoya. Accordingly, together with the Back-to-Africa hypothesis, the Pelasgian hypothesis provides van Binsbergen with the crucial conceptual opening to trace numerous pre-historical transcontinental continuities spanning millennia. In seeking to establish the intellectual validity of the Pelasgian hypothesis, van Binsbergen is able to list at least eighty Pelasgian features among peoples of different races and regions who fall within the Pelasgian realm. For instance, the mythical wagtail (*Motacilla*) attests to the importance of reed in different cultures across the globe from ancient Japan and ancient Egypt to sub-Saharan ethnicities and nationalities such as the Nkoya, the Zulu, the Yoruba, and even the natives of America. In these various cultures, the reed carries powerful cosmogonic resonances that can be traced and interpreted at an interlocking transcontinental level.

Van Binsbergen reveals that his initial fascination with Afrocentricity is what led him through a trajectory in which it is now possible to track transcontinental continuities that are the main subject of his work. This fascination, in some ways, must have led to disappointment about the claims Afrocentricity makes regarding its status and its notions regarding prehistoric Africa. However, the same disappointment is mediated by startling discoveries he makes while undertaking a most interesting intellectual journey. Here, he attempts to resolve the unexpected dilemmas he has with a problematic aspect of classical Afrocentricity:

'In recent centuries, Africa and Africans have been pushed to the periphery of the

World System and to the bottom of the global scale of prestige and power – resulting in their appearance as the outsiders par excellence. To counter this unfortunate and historically distortive situation, I have cherished, for decades now, the idea of Africa's continuity with other continents, even if this means that the intra-continental cultural initiatives and achievements to be attributed to Africa appear in a more relative light of transcontinental exchanges and common origins, thus blurring what Strong Afrocentrists have claimed to be Africa's inalienable contributions to global cultural history, *e.g.* geomancy. Now, although, I have often expressed my sympathy for the Afrocentric perspective, the painstaking analysis of empirical data as in the present argument yet brings me to admit that Africa has always been an integral part of global cultural history at large, but hardly, since the Upper Palaeolithic (30–12 ka BP), with decisive, pancontinental impact Afrocentrists have claimed for the African continent' (van Binsbergen 2012: 278).

Of theses and hypotheses

Van Binsbergen's conclusions deny the essentializations of African identities, which are usually discussed as instances of extraordinary exception when they are, in fact, part of a much broader transcontinental history linking different cultures, regions, and millennia with Africa, often receiving foreign innovations in relation to knowledge and technology rather than inventing them, but all the same, being able to adapt and transform them to meet local specificities and requirements.

Van Binsbergen had wanted to advance a strictly Afrocentric position until his findings unwittingly led him against the canons of Afrocentricity, which seek to address marginality, silence, denigration, and misrepresentation. Afrocentricity is also about the establishment of relations with the texts of W.E.B. Dubois, Chancellor Williams, Cheikh Anta Diop, Théophile Obenga, Joseph Ki-Zerbo, and Molefi Kete Asante. These authors embody a specific position in which the Eurocentric marginalization of the African subject is fervently contested and undermined. This has always been a significant characteristic of Afrocentric discourse. Furthermore, Afrocentricity, in its classical orientation, claims to be the Cradle of Humankind. No advocacy of Afrocentricity in its classical or radical orientation can be complete or credible without a consistent affirmation of this stance.

If, as van Binsbergen correctly suspects, strong Afrocentrists would have misgivings as to the Afrocentric potentials and intent of his project, most however, would applaud the courageous counter-paradigmatic turn of his approach in striking out for an area so vast and so intriguing in its possibilities as to seek to constitute an entire genre onto itself, if not a whole new discipline. This much must be admitted about his unique project.

Van Binsbergen's deflation of Afrocentricity's credibility as a discourse affirming the cultural and civilizational primacy of the black subject does not appear willful. In addition, he manages to marshal a staggering amount of evidence to corroborate most of his claims. It is now left to Afrocentrists to deploy an equally daunting academic arsenal to restore Afrocentricity's intellectual standing, thereby hoisting it up once again, as a discourse of radical critique at a safe distance from the shackles of marginality, on the one hand, and, in turn, providing a worthy discursive alternative to van Binsbergen's astonishing series of hypotheses, on the other. For Afrocentrists to accomplish this task, a mastery of several disciplines is necessary; comparative linguistics, comparative mythology, protohistory, and genetic science, among others. Indeed, much of Afrocentricity needs to rise above mere sloganeering and establish its much-needed foundations upon an array of discourses van Binsbergen has assembled in arriving at such unanticipated results and conclusions, which are contrary to his initial stance as an Afrocentric sympathizer and are, in fact, counter-argumentative.

This may not be exactly so, as the Afrocentric agenda is marked by different accents and aims. Afrocentricity seeks to establish the full subjectivity, creativity, and resilience of the black subject after the multiple traumas inflicted by slavery, colonization, and other forms of racial violence and subjugation, such as apartheid. It celebrates the freedom and agency of the black subject even in contexts of entrenched violence and negation. In critical terms, Afrocentricity operates beyond the simple proclamation of Africa being the Cradle of Humankind, as if this is all that is needed to soothe the injured psyche of the black subject.

Afrocentricity operates beyond the reclamation of ancient Egypt as the original site of black civilization, even though this is central to the Afrocentric agenda, as it seeks to wrest meaning, dignity, and redemption amid the fundamental violence of slavery, colonization, and racism. Afrocentricity, in the midst of these multiple forms of elemental violence, seeks to create an inimitable buttress of pathos to soothe broken communal psyches as well as embrace the future with renewed courage.

The reach and implications of van Binsbergen's work are too immense to attempt to arrive at a definitive conclusion quickly. It deserves to be read and analyzed diligently in order to do justice to its daunting scope, scholarship, and depth. But as mentioned earlier, what is of immediate concern is its discomfort with the general and specific aspects of the Afrocentric project. Van Binsbergen hopes his work would assuage Africa's doubts regarding its participation in transcontinental passages of global knowledge production. This hope may be cold comfort for ultra-Afrocentrists, who may choose to abide with their view of Africa as the Cradle of Civilization and then proceed to point out that Africa, once again, has been relegated to the peripheries of culture in a ruthless gesture of racialized and epistemic violence.

At a deeper level, the Afrocentric agenda seeks to come to terms with centuries of racial abuse, in which slavery is its culmination and most potent expression. The process of coming to terms with the horror of this enormous injustice and then discovering the resources by which to transcend it inflects Afrocentricity with a quite specific complexion as well as trajectory, which non-victims may never fully understand in spite innumerable well-intentioned attempts. There is a chasm of mourning that must be crossed; there is a necessity to acknowledge an immense sense of loss; there exists a physical as well as sense of collective psychic dispossession with which to contend. When Afrocentricity operates at these kinds of levels, these are the conundrums it grapples with and which shapes its aims and structures its relationship with its abiding burden of loss and finally directs its continual conversation with a past that inevitably lingers and which is impossible to forget.

If approached more critically, indeed the formidable protohistorical accomplishments of van Binsbergen's work pose serious questions to theories of blackness regarding the origins of humanity, especially if they choose to prioritize a reductionist agenda couched in a (pseudo)triumphalist proposition, in which Africa is cast as the Cradle of Civilization. This agenda would, in van Binsbergen's morally significant terms, be the replacement of one form of racial and cultural hegemony with another. But when Afrocentricity moves beyond such narrow conceptual objectives in order to grasp the haunting as well as transformative effects of the multiple horrors inflicted on the black race, that is, when it transcends its historic traumas while at the same time managing to enlarge its creative potentialities, then it succeeds in re-formulating the conceptual singularity of its mission and its moral validity.

Indeed, van Binsbergen intends (and largely succeeds) to establish a series of continuities across different continents, regions, races, and epochs. In other words, his project re-evaluates the conventional perceptions and assumptions regarding global history, in which unities rather than ruptures become significant. In Afrocentric terms, the project is likely to appear too general, ridding Afrocentricity of much-needed ammunition. Nonetheless, its overall academic deportment is admirable even when staunch Afrocentrists would tend to flinch from it.

The black subject in antiquity often constitutes an anomalous and marginal presence, be it in the form of the black Irish and similar instances in the Western extremity of Eurasia, or the Dallit, labeled 'Untouchables' in South Asia. So the black figure, contrary to Clyde Winters's (1980) assertion that the Xia and Shang Yin dynasties were established by blacks, has repeatedly appeared as an intruder, an unwelcome presence, according to van Binsbergen's findings and other similar archaeological and anthropological discoveries, that looms in opposition to dominant cultural, linguistic, and theoretical paradigms, thus making the 'outsider' designation fit a specific racialized pattern of reception and perception.

The characteristics that define the black presence in the Bronze Age East Mediterranean include proto-Bantu-speaking features, elongated labia, round house architecture, spiked wheel trap, mancala board games, and the worship of a single supreme deity, all of which represent a counter-paradigmatic cultural and linguistic presence.

In tracing transcontinental continuities encompassing board games, geomantic practices and traditions, shamanic manifestations, linguistic revolutions, global migratory patterns, technological innovations, leopard-skin symbolism, astronomical schemas, divinatory systems, clan structures, and toponymical systems across millennia, van Binsbergen has attempted to construct a global intellectual history of gargantuan proportions. Writing a global history of this nature cannot be a straightforward affair, especially if there are numerous earlier hypotheses to be either proved or debunked, theoretical models to be tested and cross-checked, paradigms to be re-evaluated in accordance with historical specificities, schools of thought to be re-assessed, various contestations with leading authorities in different academic fields and disciplines, attempts at resolving the intractable dilemmas of one's untested hypotheses, intellectual contradictions within one's own traditions, open anxieties about, and obvious gaps in, aspects of the project, and myriad other concerns of both personal and professional dimensions. All these problems and challenges are reflected in van Binsbergen's work. Nonetheless, he has made a noteworthy attempt to advance a series of theses and hypotheses that deserve painstaking attention for their sheer boldness, breadth, and versatility.

References cited

van Binsbergen, Wim M. J. 1995. 'Divination and Board-Games: Exploring the Links between Geomantic Divination and Mancala Board-Games in Africa and Asia.' Paper read at the 1995 International Colloquium on Board-Games in Academia, Leiden, April 9–13; published as 'Rethinking Africa's Contribution to Global Cultural History: Lessons from a Comparative Historical Analysis of Mancala Board-Games and Geomantic Divination,' special issue, Tatlana: *Proceedings of the Dutch Archaeological and Historical Society* 29: 221–54; revised version at http://www.quest-journal.net/shikanda/ancient_models/gen3/mankala.html.

van Binsbergen, Wim M. J. 2008. 'Transcontinental mythological patterns in prehistory: A multivariate contents analysis of flood myths worldwide challenges Oppenheimer's claim that the core mythologies of the Ancient Near East and the Bible originate from early Holocene South East Asia,'*Cosmos: The Journal of the Traditional Cosmology Society*, 23: 29–80.

van Binsbergen, Wim M. J. 2011a. 'Existential Dilemmas of a North Atlantic Anthropologist in the Production of Relevant Africanist Knowledge.'In *The Postcolonial Turn: Re-Imagining Anthropology and Africa*. Edited by René Devisch and Francis B. Nyamnjoh, 117–42. Bamenda, Cameroon: Langaa / Leiden: / African Studies Centre.

van Binsbergen, Wim M. J. 2011b. 'Is There a Future for Afrocentrism Despite Stephen Howe's Dismissive 1998 Study?' In van Binsbergen, ed. *Black Athena Comes of Age*, 253–82.

van Binsbergen, Wim M. J. 2011c. 'The Limits of the Black Athena Thesis and of Afrocentricity as Empirical Explanatory Models: The *Borean Hypothesis, the Back-into-Africa Hypothesis and the Pelasgian Hypothesis as Suggestive of a Common, West Asian Origin for the Continuities between Ancient Egypt and the Aegean, with aNew Identity for the Goddess Athena.' In van Binsbergen, ed. *Black Athena Comes of Age*, 297–338.

van Binsbergen, Wim M. J. 2011f. 'Matthew Schoffeleers on Malawian Suitor Stories: A Perspective from Comparative Mythology,' in 'A Tribute to the Life of Fr. Matthew Schoffeleers (1928–2011): Malawianist, Renaissance Man and Free-Thinker,' eds. Louis Nthenda and Lupeaga Mphande, special memorial edition, *The Society of Malawi Journal* 64, no. 3: 6–94.

van Binsbergen, Wim M. J. 2011d. 'Shimmerings of the Rainbow Serpent: Towards the Interpretation of Crosshatching Motifs in Palaeolithic Art: Comparative Mythological and Archaeoastronomical Explorations Inspired by the Incised Blombos Red Ochre Block, South Africa, 70 ka BP, and Nkoya Female Puberty Rites, 20th c. CE.' http://quest-journal.net/shikanda/ancient_models/crosshatching_FINAL.pdf

van Binsbergen, Wim M. J. 2012. Before the Presocratics. Cyclicity, Transformation, and Element *Cosmology: The Case of Transcontinental Pre- or Protohistoric Cosmological Substrates Linking Africa, Eurasia, and North America.* Leiden: African Studies Centre. *This volume has an excellent References Cited of other works by Wim van Binsbergen regarding this and related topics.

van Binsbergen, Wim M. J. 2012b. 'A Note on the Oppenheimer-Tauchmann Thesis on Extensive South and South East Asian Demographic and Cultural Impact on Sub-Saharan Africa in Pre- and Protohistory.' Paper read at the International Conference on Rethinking Africa's Transcontinental Continuities in Pre- and Protohistory,' African Studies Centre, Leiden, April 12–13. http://tinyurl.com/pszl8tt.

van Binsbergen, Wim M. J. 2012c. 'Production, Class Formation, and the Penetration of Capitalism in the Kaoma Rural District, Zambia, 1800–1978.' In *Lives in Motion, Indeed: Interdisciplinary Perspectives on Social Change in Honour of Danielle de Lame.* Edited by Cristiana Panella, 223–72.Studies in Social Sciences and Humanities 174. Tervuren: Royal Museum for Central Africa.

van Binsbergen, Wim M. J. 2012d. 'The Relevance of Buddhism and Hinduism for the Study of Asian-African Transcontinental Continuities.' Paper read at the International Conference on Rethinking Africa's Transcontinental Continuities in Pre- and Protohistory,' African Studies Centre, Leiden, April 12–13. http://www.quest-journal.net/shikanda/topicalities/Mwendanjangula_final.pdf.

van Binsbergen, Wim M. J. 2012e. 'Rethinking Africa's Transcontinental Continuities in Pre- and Protohistory.' Keynote paper read at the International Conference on Rethinking Africa's Transcontinental Continuities in Pre- and Protohistory.' African Studies Centre, Leiden, April 12–13. http://tinyurl.com/oyzb98n.

van Binsbergen, Wim M. J. 2012f. 'Towards a Pre- and Proto-Historic Transcontinental Maritime Network: Africa's Pre-Modern Chinese Connections in the Light of a Critical Assessment of Gavin Menzies' Work. http://tinyurl.com/oq2hpo8.

van Binsbergen, Wim M. J. 2013. 'African Divination Across Time and Space: Typology and Intercultural Epistemology.' In *Realities Revealed. Divination in sub-Saharan Africa.* Edited by Walter E. A. van Beek and Philip Peek, 339–75. Berlin: Munster/Boston: LIT.

Kobia's clash

Ubuntu and international management within the World Council of Churches

by Frans Dokman

Abstract: In this chapter, van Binsbergen's thinking about *Ubuntu* will be related with a domain which is usually less connected to him, namely management. The aim of this chapter is to explore the relation between African *Ubuntu* management and, Western dominated, international management. This permits us to argue that van Binsbergen's articulations of *Ubuntu* are both transcontinental and trans-disciplinary. The article presents a case study of the introduction of *Ubuntu* management by the Kenyan general secretary Dr Samuel Kobia at the World Council of Churches (WCC). Although sharing a Christian identity, there is a disharmony between Southern and Western colleagues of this international organization about religion and management in general and *Ubuntu* management in particular. Therefore the main question is: how is *Ubuntu* management dealt with in an international Christian organization with staff and management from the Southern and Western hemisphere?

Keywords: Ubuntu, international management, critical discourse analysis, World Council of Churches

Introduction

During his career Wim van Binsbergen has shattered many mindsets of African philosophers, intercultural philosophers, anthropologists, Africanists, religion scientists and cultural scientists. By doing so he offered valuable insights and contributed to academic steps forward in various disciplines. In this essay we will relate van Binsbergen's thinking about *Ubuntu* with a domain which is usually less connected to him, namely management. The aim of this essay is to explore the relation between African *Ubuntu* management and, Western dominated, international management. Consequently making it clear that van Binsbergen academic body of work is both transcontinental and trans - disciplinary.

Based on empirical research and personal involvement, Wim van Binsbergen discusses African religions both in a local and global context. He especially emphasizes the political dimensions of African religions. From these motivations the article will present a case study of the introduction of *Ubuntu* management, an African religious management concept, by the Kenyan general secretary Dr Samuel Kobia at the World Council of Churches (WCC). Although sharing a Christian identity, there is a disharmony between Southern and Western colleagues of this international organization about religion and management in general and *Ubuntu* management in particular.

Therefore the main question is: How is Ubuntu management dealt with in an international Christian organization with staff and management from the Southern and Western hemisphere?

To many Southerners organizations do have a religious dimension while for most of their Western colleagues religion is, at the most, relegated to the private domain. And so there are conflicting interpretations around Ubuntu- and international management, one being that religion plays a role the other that it does not at all.

In order to study the introduction and reception of Ubuntu management at the WCC, 'critical discourse analyses will be used as methodology. Critical discourse analysts assume that organizations are created, maintained or transformed through a political struggle between dominant and peripheral discourses that strive for hegemony (Fairclough: 1992:58).

International management and Ubuntu management

The dominant theories of international management science are rooted in the West. Despite their universal claim the Western management theories, however, do not connect to local practices of governing constructed in Eastern and Southern management theories, such as Dharma management (India), Guanxi management (China) and *Ubuntu* management (Africa) (Jackson, Amaeshi & Yavuz 2008; Xu & Yang 2009). The dominant Western management theories make a separation between work and religion, while for most Eastern and Southern approaches of management the connection with religion is evident.

The basis of Ubuntu is summarized in the principle,

> *'umuntu ngumuntu ngabantu'* (a person is a person through other persons).

Ubuntu is characterised by the individual's solidarity with the community. From an *Ubuntu* perspective culture is by definition religious. Within cultures there is no distinction between religion and non-religion. At most the degree varies in which religion is visible within the culture. From this perspective, it follows that organizations only vary in the extent to which religion is visible. The fact that they have a religious dimension is undeniable.

> Under the influence of modernisation the role of religion in international management is marginalized. The scant attention paid to religion in international management stems from a tendency to view the world and humankind from the angle of modernisation. Broadly the modern premise is that people worldwide are increasingly distancing themselves from religion as a result of growing insights. Thus being religious is regarded as not having gained proper insights (Kienhuis 2000). This contrasts with developments in the South where, despite a growing secularization (Shorter & Onyancha 1997), being religious is usually qualified as having the right insights (Bujo 1998; Nkafu Nkemnkia 1999). The South-West divergence corresponds with theories of religious studies showing a global resurgence of religion then again Europe being an exception (Berger 1999; Davie 2000).

Within the context of globalised, urban societies of South Africa, van Binsbergen warns for the dangers of *Ubuntu*. In a reaction to colonization, Apartheid and neo-liberalism *Ubuntu* has been constructed as a pan-African ideological identity. Van Binsbergen (2001:79) recognizes that *Ubuntu*:

> 'is in the first place born out of pain, exclusion, justified anger, and the struggle to regain dignity and identity in the face of Northern conquest and oppression'.

Some African scholars argue that colonisation has been exchanged for globalisation. Van Binsbergen states that the process of globalization is a cruel construction around the myth of universal limitless access. Van Binsbergen (2001:81) warns, however, for the power implications of launching *ubuntu* as a superior alternative for Western models of thinking because:

> 'Claiming an ethnographically underpinned superior insight simply means yet more Northern violence, inviting Southern counter-violence'.

In response to a neo-liberal global market economy *ubuntu* is a move to make use of Africa's community spirit in management. From the maxim that unity is strength important notions are shared property and collective decision making based on consensus. These notions of *ubuntu* are rooted in the sacred reality of life force: a universal, all-powerful energy around which all thought and action circles. There is an interaction of forces between a Supreme God and human beings, between human beings themselves, between humans and animals, and between human beings and inanimate matter. God is the creator and the source of all life force. After God come the spirits, the ancestors (the founders of the tribe, then the dead ranked according to age), then the living, and finally animals, plants and inanimate matter. One could call it a closed universe, in the sense that when one element gains more power, it is at the expense of some other element. This also poses a problem: that of witchcraft via 'bad spirits'. Be that as it may, all the participant forces – humans, spirits, animals and matter – refer to God. And because all elements share in God's life force, they are inter-related and form a religious community.

According to Lovemore Mbigi (2000) the spirits of the ancestors are to be used as social capital in the management of organizations and states. *Ubuntu* management positions organizations and colleagues within a pervasive religious life force. One of the consequences of globalisation is that a peripheral theory as *Ubuntu* management becomes more manifest. The result is tension within the international management between Western and Southern management theories.

In Southern and Eastern management theories religion is an important factor (Schiele 1990; Mbigi 1997). However Western researchers of Southern and Eastern management theories (Karsten & Illa 2001; Jackson 2004) hardly address the role of religion. This gap is remarkable in view of the global revival of religion (Berger 1999; Davie 2002) and the growing diversity in religious terms of international organizations. At the level of international organisations it is though recommended to be open to various management theories and the role of religion in it. Management of colleagues with a diversity of cultural and religious orientations is gaining urgency.

Ubuntu management

From an *ubuntu* perspective culture is by definition religious. Within cultures there is no distinction between religion and non-religion. At most the degree varies in which religion is visible within the culture. From this perspective follows that organizations only vary in the extent to which religion is visible. The fact that they have a religious dimension is undeniable. In the view of *ubuntu* management organisations are components of life force (Mbigi 1997; Mbigi & Maree 2005). If an organisation is successful, it is because human beings are in harmony with one another, with spirits, with ancestors and with God. If an organisation does not do well it is because there is no sense of community, no harmony with the spirits and with God. The journey of management and staff consists in restoring unity with one another, the spirits and God by means of ceremonies and rituals.

An essence of *ubuntu* management is that staff experience commitment and solidarity with the organisation as a result of the harnessing of religious resources. *Ubuntu* management makes staff members feel that within the organisation they are regarded primarily as religious beings rather than purely economic factors.

Spirits, associated with *ubuntu*, can help to transform an organisation: the supra-tribal hierarchy of spirits can be used to tap emotional and religious sources in an organisation. When that happens spirits are linked with various norms and values, as is evident in the following scheme, 'Spirits of management' (Mbigi 2000:38):

norms and values	spirits
Morality and dignity	Rainmaker spirit Gobwa
Performance and enterprise	Hunter spirit Shavi Rudzimba
Authority: know the truth	Divination spirit Sangoma
Power and conflict	War spirit Majukwa
Survival of oneself and one's Group	Clan / family spirit Mudzimu Wemhuri
Particular obsession, ability and creativity	Wandering spirit Shave
Bitterness, anger, revenge	Avenging spirit Ngozi
Cynism, negativity, destruction	Witch spirit Mutakati

Table 1.Norms and values, and attending spirits, in the cosmology of *ubuntu*.

Another aspect of *ubuntu* is advocating consensus. *Ubuntu* leadership is focused on the welfare of the whole organization, the collective, and not to the benefit of a small group of colleagues or an individual. However, the position of the majority is not pushed through, but one takes the view of the minority in decision-making. One takes into account the views of all participants. Leadership has of course the right to manage quick decisions, but on important themes and strategies is converging views.

The World Council of Churches and Dr Samuel Kobia, their first African general secretary

The main question of this article is: how is *ubuntu* management dealt with in an international Christian organization with staff and management from the Southern and Western hemisphere?

For an answer to that we look at the introduction and receipt of *ubuntu* management at the World Council of Churches.

The World Council of Churches (WCC) is an international faith-based organisation. It is a leading international Christian ecumenical body. Founded in 1948, its headquarters are in Geneva, Switzerland. Most Christian churches have joined the WCC. Currently 348 denominations from 120 countries are represented. The WCC represented some 80% of Christians, comprising about 380 million Protestants and 120 million Eastern Orthodox members. The Roman Catholic Church cooperates closely as a 'participant observer'.

The Assembly is the highest decision making meeting of the WCC, determines policies as well as appoints a Central Committee which serves as the chief governing body of the WCC until the next assembly. The Executive Committee (including the officers) is elected by the Central Committee and normally meets twice a year. The general secretary serves *ex officio* as secretary of the Central and Executive Committees.

The WCC is an ecumenical organisation. Ideally, Christians from many cultures and denominations do share one ecumenical religious identity. They feel called to the goal of visible unity in Christian faith. The WCC works for a better world.

Programmes are built around the themes of mission and evangelism, solidarity, international affairs, justice and peace. The staff comprises 200 workers reflecting the diversity of cultures and denominations.

At this moment the most important development in Christianity and also within the WCC is the southward shift of Christianity's centre of gravity (Jenkins, 2002). Western member churches who have mainly founded the WCC are now outnumbered by representatives of non-Western church organisations. The appointment in August 2003 of the African Dr Samuel Kobia could be seen as a symbol of this new constellation. The Kenyan Kobia was the first African general secretary.

At his first press conference Kobia announced his intention to run the WCC using the African management style of *ubuntu*. Both the appointment of Kobia as leader and the introduction of *ubuntu* management can be understood as symbols of a growing African influence within the WCC. This is most probably a sign that the position of Southern members has moved from the periphery to the centre.

Critical Discourse Analysis

In order to examine the introduction and receipt of *ubuntu* management at the WCC, 'critical discourse analysis' method has been used. Critical discourse analysts think that organizations are created, maintained or transformed through a power struggle between dominant and peripheral discourses that strive for hegemony (Fairclough 1992: 58). Discourses are all the expressions used by organization members to create a reality that frames their sense of identity (Mumby & Clair 1997: 181).

Grant and Hardy (2004:6) define an organization as an autonomous social community with its' own discourse. They enlarge the term 'discourse' to 'organization discourse' in which members, being subjects, produce and consume texts:

> "The term 'discourse' has been defined as sets of statements that bring social objects into being (Parker 1992). In using the term 'organizational discourse' we refer to the structured collections of texts embodied in the practices of talking and writing... that bring organizationally related objects into being as those texts are produced, dissemi-

nated, and consumed... Consequently, texts can be considered to be a manifestation of discourse and the distinctive unit... on which the researcher focuses."

Norman Fairclough (2005: 915) states that the study of discourse in an organisation is an important part of organisation studies, especially when the organisation is in a crisis situation. That is when individuals and groups devise plans to achieve their power goal. In that context data from critical discourse analysis reveal the manipulative strategies for creating and / or retaining power. Often this is accompanied by the exclusion of individuals and groups who do not see their views reflected in the new dominant discourse.

We draw on applications of critical discourse analysis in the management sciences. Critical discourse analysis can be used to determine the relationship between an organisation and the dimensions of power, culture and identity. Consequently critical discourse analysis is very useful to study what happens when an organization like the WCC after decades of Western dominance in leadership, finance and theology comes in contact with an African management style. When a Western, modern management discourse based on dualism (meaning a separation between management and religion) meets a Southern management discourse based on holism (meaning a union between management and religion), these management styles claim the dominant, central position in the WCC.

Below the discourse on *ubuntu* management will be studied, followed by Norman Fairclough's three dimensional model of discourse analysis. This model centres on language use and social change. Further theoretical and methodical considerations will be mentioned in the course of the analysis.

Linguistic practice

In harmony with the pragmatic turn, Fairclough (1992: 71) considers discourse to be a practice like any other practice. Discourse is a way of doing things. The only difference from other practices is its linguistic form. Thus the first dimension of analysis concerns linguistic features of the text such as vocabulary and metaphor (Fairclough 1992: 73-78).

The texts selected for detailed analysis focus on Kobia's discourse of *Ubuntu*- and religious management. We concentrate on Kobia's expressions because he embodies the discourse on *ubuntu* management in the WCC. The Central

Committee, other committees and staff (despite the fact that Kobia is a member of them all) are consumers of the *ubuntu* discourse and recipients of a religious management style. We look at the period of Samuel Kobia's leadership term as general secretary, from the time of his appointment in August 2003 up to December 2008 when he completed it. At the beginning of 2008 Dr Kobia announced that he would not stand as a candidate for a second term.

Public lecture

In his first speech after being elected as general secretary, on 28 August 2003, Kobia (2003b: 431-432) concentrates on African religion and the ideas of *Ubuntu*. According to him the general secretary's leadership, apart from the organizational program and finances, is mainly religious. Later Kobia (WCC, 2003) explains the meaning of *ubuntu*:

> 'Evoking the concept of *Ubuntu*, a Zulu word, Kobia explained that for Africans, it is 'that which makes human beings human'.

Kobia voices his intention to introduce *ubuntu* values into the WCC. Religion and humanity are at the centre of his view on management and based on such a relationship he hopes to achieve cooperation. Accomplishing the mission of the WCC is a journey you make together.

In an interview with Dutch newspaper *Trouw* (May 7, 2008) Kobia will emphasize that *Ubuntu* is of great importance for his leadership style: 'from the beginning I have been inspired by an African *ubuntu* proverb: 'If you want to walk fast, go alone. But if you want to go far, go together.'

Books

A dominant discourse in Kobia's book *The courage to hope* (2003a), concerns the conviction that African values are obstructed by colonialism and globalisation, based on Western hegemony in economic, cultural, political and spiritual power structures.

> 'For the African people, the relational dimension is accorded the highest value in measuring the quality of life in the community and the society at large. This relational dimension was greatly undermined and compromised by slavery and colonialism...' (Kobia 2003a: 83)

> 'Economic globalisation has gone even further and made finance and trade sacrosanct. Human beings and the value of life are subordinated to finance and not vice

versa.'(Kobia 2003a: 116)

In relation to the excesses of globalisation, including the neo-liberal market system and the spread of AIDS, *Ubuntu* is introduced as the traditional respect for the sanctity of life with an eye for humanity. Kobia (2003a: 125) calls on African theologians rediscover *Ubuntu*:

> 'Theology in Africa has to rediscover the positive attributes of our culture and the pride of *Ubuntu* for it to be able to fight globalisation.'

In his book *Called To The One Hope: A New Ecumenical Epoch*, Kobia (2006: 24-25) discusses the dominant Western influence on the organisational culture of the WCC. This reflects dominance at a theological, financial and bureaucratic level that is rooted in the past and does not correspond with the present multi-cultural and multi-denominational staff. Kobia sees Western church organisations as fearful of adapting to present-day circumstances in the ecumenical field, where many Christians are drawn to the vitality of the evangelical and Pentecostal movement and new charismatic churches, both in the North and in the South. Kobia's concern is that institutionalisation of the WCC is a development that not only harms its dynamics and spirituality, but also hampers agreement with a growing group of charismatic denominations (Kobia, 2006: xii). Kobia's (2006: 54) vision of the future is an ecumenical world in which dynamic churches, especially in the South, will predominate, having taken a distance from Western style of church:

> 'The shift of Christianity's centre from the North to the South is numerically, theologically and doctrinally and away from denominationalism, hierarchy and structure.'

WCC Meetings

In the perception of Samuel Kobia, business meetings are part of a religious process and not so much a worldly modus operandi. During his opening speech at the Assembly in Porto Alegre, Brasil 2006, Kobia (WCC Assembly Porto Alegre 2006) encourages participants to stop perceiving meetings as profane business-like:

> 'I am suggesting that we take a different approach to the 'business' of our meetings: our business is part of the process of spiritual discernment and is embedded in the festa da vida. Let us look at the assembly as a spiritual experience and not just as a business meeting that has to fulfil a constitutional mandate.'

In *The Courage to Hope* Kobia (2003: 161-169) makes a distinction between rationality and spirituality. He presents rationality as Western par excellence and spirituality as African. Kobia describes the European culture as aggressive, rational and non-spiritual. According to Kobia (2003: 163) the doctrines of Luther and Calvin support this type of European culture:

> 'Luther and Calvin succeeded in fashioning a new ethical statement for the West which was more in accord with the internal dynamics of Europeanculture. The doctrines that they developed supported the competitive, individualistic, aggressive, rationalistic, non-spiritual and detached behavior necessary for survival within the culture. There was no longer a question of emulating the New Testament portrait of Jesus.'

Discursive practice

Fairclough (1992: 71) states there's a dialectical relation between discourse as linguistic practice and discourse as social practice, and that discourse as discursive practice mediates between the two. By discourse as discursive practice he means the production, distribution and consumption of a text. Discursive practice focusing on the link between discourse and social context also looks at the intertextual aspect that is which texts inspire transformation of the discourse.

Production of the discourse

Kobia himself is a producer of the discourse on an African management style. In his book, *The courage to hope* (2003a), Kobia dwells on the ethical and religious aspects of African management, expressed in dealing with humanity. There is an intertextual relation between the ideas in the book and the discourse on *Ubuntu* management that Kobia conducts in the WCC. In *Called to the one hope* (2006) Kobia produces his own perception that institutionalisation of the WCC is to the loss of the organisation's spirituality.

But within the WCC Kobia is not the only producer of an *Ubuntu* discourse. Dr Hans Ucko, Dr Kasonga wa Kasonga and Dr Rogate Mshana appear to be also, next to Kobia, principal producers of an *Ubuntu* discourse in the WCC. In 2005 the WCC publishes *Worlds of memory and wisdom. Encounters of Jews and African Christians*. The book is dedicated to Dr Samuel Kobia. Hans Ucko, programme secretary of the WCC Office on Inter-Religious Relations and Dialogue, is one of the editors. Ucko (2005: 74) perceives *Ubuntu* essential to transform people to regain their humanity and according to a contribution by Kasonga wa Kasonga

(2005: 128), a theologian from Congo, *Ubuntu* has, influenced by modernity and Western values, been corrupted and declined meaning:

> 'In many ways, the practice of Ubuntu (bumuntu) is being corrupted nowadays when it is confronted with western values. In traditional African society, parents took seriously their responsibility to pass on 'bumuntu' during the socialisation of their children.'

The WCC has also a programme known as Agape (an acronym for Alternative Globalisation Addressing People and Earth) which is managed by Dr Rogate Mshana, WCC program executive for Economic Justice. *Ubuntu* is referred to in the sense of economic solidarity and cooperation (WCC PWE, 2007). The Agape Report *Consultation on linking poverty, wealth and ecology: African ecumenical perspectives* (2007) speak of *Ubuntu* in the sense of human fellowship and living in wholeness. *Ubuntu* is linked with abundant life (John 10: 10), with the accent on its affirmation of life, for instance through respect for nature.

Ubuntu also occurs prior to Kobia's appointment in 2004. In 2001 the WCC Justice, Peace and Creation (WCC Report WCAR, 2001: 20) section contributes to the UN World Conference against Racism in Durban. *Ubuntu* is offered as Africa's response to the global problem of racism:

> 'Africa must find courage and capacity to challenge the world to embrace as a basis for the war against racism, UBUNTU, the philosophy that I am because you are, and you are because I am. The drive for the renewal of this continent must not be driven merely by a desire to emulate the western world. It must be driven by a desire to put UBUNTU at the heart of Globalisation as an alternative philosophy for the world.'

These two aforementioned texts show that Kobia demonstrates the intertextuality of his texts with an existing discourse. This text contains no references to Dr Samuel Kobia.

Distribution of discourse

Samuel Kobia, Hans Ucko and Rogate Mshana are the principal producers of discourse on *Ubuntu* and *Ubuntu* as religious management. For maximum reception of the discourse time, place and target group are decisive factors. At what time and in what place do you, the 'producer', expect a particular group of 'consumers' to receive your discourse optimally?

Kobia decided to use his first press conference as general secretary elect in August 2003 to introduce and distribute his vision. Via the international press he brings his aspiration to operate according to an *Ubuntu* management style to the attention of both his organisation and the general public. Whilst reaching a larger public but directed mainly to the ecumenical world and the WCC itself, Dr Samuel Kobia wrote his books, *The courage to hope* (2003a) and *Called to the one hope* (2006). In *Courage to hope* he discusses *Ubuntu*. In *Called to one hope* Kobia advocates less institutionalised and more religious organisational structures and calls for transformation of the WCC. The book was published shortly after the WCC Porto Alegre Assembly.

Also addressed to an international readership, including colleagues at the WCC, Hans Ucko and Kasonga wa Kasonga distribute their views of *ubuntu* in *Worlds of memory and wisdom. Encounters of Jews and African Christians* (2005). Focusing more specifically on the WCC, Rogate Mshana brings the *ubuntu* discourse to our attention in the Agape programme (2007).

Consumption of the discourse

I gained further insight into the reception of the discourse on *Ubuntu* management from five key informants, both from the South and the West (Dokman, 2013). Three of them were members of the WCC management team and two worked for ecumenical organisations being often active in the WCC governing offices.

The first informant wondered whether it was possible to transplant *ubuntu* to a body like the WCC, being

> 'an organisation that is not a true community due to cultural differences and opposing interests. Besides, the WCC has no ancestors.'

According to this informant the question was not about a choice of management concepts but a choice of who is in charge. The Europeans who finance the WCC and want to make the decisions, but who in the Assembly meetings see themselves literally confronted with a majority of member churches from the South? The informant asked himself how serious Kobia was about *ubuntu*, since Kobia was familiar with the context of power games in the WCC. The informant raised the query: what chance does *ubuntu* stand in a context where it is ultimately a matter of money and power?

The second informant said that followers of *ubuntu* management and Western management are radically opposing each other. The informant perceived the conflict as between 'the political camp and economy camp'. The former makes more religious inspired decisions, the latter puts pressure on proceedings and financial (im) possibilities. The West, more particularly the German sponsoring member churches, are in the economy management camp. The informant finds it a pity that they do not rather draw their inspiration from theological princ- iples and referred to the (German) economy managers as the 'Western liberal Protestant culture of enterprise'. This leadership culture has predominated ever since the establishment of the WCC.

The third informant claimed that Kobia had no formal plan to introduce *ub- untu* in the WCC. What Kobia did as a leader was to form the 'Group of 14', comprising fourteen staff members who had to see to it that staff worked in a more integrated fashion and activities were more closely interrelated. Up to the time of Kobia's appointment the various programmes (Mission, Diakonia, Peace and Justice, *etc.*) worked in isolation and staff operated very much individually. Kobia wanted to change that and also created a Department of Planning and Integration at headquarters to change the movement to a more integrated ap- proach to programmes and interlink people and activities. According to this informant it is difficult to create a new organisational culture when the head- quarters of the WCC are situated in the West. A complete change in working methods, he said, would only be possible if headquarters were moved to the South, otherwise they will remain 'Northern shaped.'

The fourth informant speaks also about the dominance of the 'economy camp'and 'Western liberal Protestant culture of enterprise'. She experienced the Assembly as a religious process and appreciates Kobia's initiatives, contrary to the Western mindset, for a more religious management style. The decision making procedure at the Assembly according to consensus was received by this informant as religious community building. With decisions according to con- sensus, unlike the Western democratic decision model, the majority position does not win the day. The minority viewpoint is included in decision making. The views of all participants are taken into account.

The fifth informant compares the WCC with the United Nations as both or- ganisations with a great diversity of cultures, opinions *etc.* Within the WCC there are various perspectives on management styles. The reason that Kobia's plans received so much opposition, especially from Westerners, was because the African general secretary raised the power issue. It is well known that Christianity shifts from North to South but Kobia was the first one to connect this with, for instance, a more religious than rational, a more Southern then Western management style. This was difficult to accept for representatives of Western churches who see the WCC as a European institute. The informant emphasized that the appointment of Samuel Kobia as the first African general secretary filled the African church organisations with great pride. The informant labels the conflict around Kobia as a 'business disasters, causing a deep wound in the WCC.'

Kobia's clash

The last informant refers to the consumption, visible for the general public, of Kobia's policy namely the receipt of the general secretary's discourse by bishop Martin Hein from the German Evangelical Church and member of the WCC Central Committee. Bishop Hein voiced criticism in the media speaking disapprovingly of Kobia's way of taking decisions without consultation. This opposition was followed in February 2008 by news reporting on a non valid doctorate by Kobia.

As of February 2008 messages appear in the media that indicate instability within the World Council of Churches. Mid-February the German press agency EPD (Evangelical Press Service) brings the news that Kobia has received his doctorate of Fairfax University, not by the U.S. Government accredited educational institution. According to the news agency ENI is Fairfax University also nonexistent. Kobia indicates to be 'shocked and surprised' that Fairfax University is not certified. Since Kobia received twice an honorary doctorate, he may continue to pursue his doctorate.

These messages are broadcast just before the start of the meeting of the Central Committee of 13-20 February 2008. A meeting dedicated to the 60th anniversary of the World Council of Churches and the (re-)election of the Secretary-General.

Before the meeting the EPD publishes an interview with Bishop Martin Hein of the German Evangelical Church (EKD), in which this criticism has on the working method of Kobia. Hein considers that Kobia to travel a lot and criticises the Secretary-General because of taking decisions without consulting. Bishop Hein adds that the German Evangelical Church its financial contribution to the World Council of churches in the near future. That contribution accounts for a third of the income of the World Council of churches.

According to the fifth informant that title of the unaccredited Fairfax University was posted on Kobia's curriculum Vitae by the World Council of Churches and not by the Secretary-General himself. The ENI (Ecumenical News International) brought the message about the incorrect doctorate out which is highly peculiar as the ENI is a collaboration of church organisations, in which the World Council of churches has a majority. In fact, ENI placed her own boss Kobia in a bad light. The support for Kobia in the organization began to fall apart.

On 18 February the WCC pronounce that Dr Kobia has declared himself not available for a second term as general secretary for personal reasons. Dr Samuel Kobia has been the first leader to refuse a second term. A new general secretary was elected at the Central Committee meeting in September 2009. Two candidates had been nominated, Rev. Dr Park Seong-won of the Presbyterian Church of Korea and Rev. Dr Olav Fykse Tveit of the Lutheran Church of Norway. Dr Tveit won the election and he is general secretary of the WCC since then.

Social practice

The major question here is in what way discourses helped to stabilise or transform power relations and ideologies. Kobia's discourse is intended to transform the hegemony and the ideology in the WCC. What is the connection between his discourse and the organisation?

A principle of discourse analysis is that identity is a social construction with power as the main determinant. When connecting this principle to the WCC it was already noticed that the centre of gravity of Christianity has moved to the South. This changes the religion's dominant identity from a Western one to a Eastern and Southern (African, Asian and South American) one. The identity of Christian organisations like the WCC likewise follows this development. The appointment of the first African general secretary is a sign that in the WCC the identity of the South have moved from the periphery to the centre.

Kobia's first address as general secretary elect is focussed on changing the meaning of the organisation. After decades of Western-style management Kobia declares that he is inspired by *Ubuntu*, he links religion with management. In so doing the general secretary envisaged transformation of the WCC. His first address is not only informative but performative. It aims at affecting meaning and transforming the organisation. Kobia's speech to the Assembly, for a religious approach of the meeting, can be labelled similarly. The WCC should not be a business, rigid organisation but experience its management activities, according to u*buntu*, as part of a religious process.

The general secretary's discourse on *ubuntu* means to generate a new organizational identity (Kobia 2006: 106). In his opinion, the WCC needs to transform from an international Christian organization with a Western profile to one with a Southern profile. Following the numerical dominance by the South and East, Kobia considers it crucial to transform the WCC by creating a new organizational culture and identity. An *ubuntu* management style is a part and a expression of this transformation.

Ubuntu management is also a construction of African identity within the modern setting of the WCC, an identity formation by Christians from the South. Among African staff members there is a need for appreciation and a growing experience of their own African identity. Kobia's discourse praises African religion and management in terms of aspects like humanity and enhancement of community life. At the same time he polarizes by presenting Europe as an agent of colonialism and globalisation, an aggressive, rational and non-spiritual culture to which Luther and Calvin contributed with their doctrines.

Nonetheless, representatives of Western churches have always decided about the financial, economical policy and about the activities of the organisation. Bishop Hein represents their discourse of distrust which is publicly expressed against Kobia. This discourse is meant to stabilise power relations and maintain Western churches in the center of WCC. The clash of discourses reveals the conflict between the Western 'economy camp' which finances the WCC but experiences less influence and the Southern and Eastern 'political camp which wants to see their numerical dominance translated into greater say in policy making and a more religious inspired management.

Religion and management

Although the people of the World Council share an ecumenical religious identity there is a different view on the relation between religion and management. Two of the informants spoke about the so-called 'Western liberal Protestant culture of enterprise', also labelled as the economy camp. The other camp was called the political camp.

According to the followers of the 'political camp' the WCC is of God, a manifestation of religion without an opposition between business and religion. For his followers Samuel Kobia is literally leading God's business and *Ubuntu* management is the right reaction after years of Western management style. For Kobia (2006: xii) religion also manifests itself in organizational structures and he wonders if the WCC structures are inspired by God or the business world. Also in his view the organisation is an expression of God's presence. Within the WCC especially staff members from the South and East are supporting this management view. *Ubuntu* management is a demonstration of their identity and of *African Renaissance*.

This foregoing conception of a religious inspired management contrasts with the mainstream view in international management that religion may or may not play a role in the private sphere but it should be kept out of organisations. To the 'economy' managers the WCC is of the world with a separation between business and religion. The organisation gives expression to God's presence but has an autonomous, independent reality of a international enterprise. Business has its' own logic. On one hand there are ecumenical meetings and publications. On the other hand there is a practical type of administration.

The economy camp is more inclined to base itself on financial (im) possibilities and systematically puts pressure on proceedings. The West, more especially the German sponsors, are in the economy camp. Their style accords with the pragmatic and technocratic trends that can be seen in Western society, even in organisations with more idealistic goals. For them, the African management style of *Ubuntu* connected to spirits is controversial and as one informant states: 'The WCC has no ancestors.' According to mainly Southern informants the 'economists' do not rather draw their inspiration from religious principles.

These differences between the political and economy camp, or between orientations from the South and East, versus West on international management, features worldwide in many organisations (Dokman 2005). According to staff members with a more Western style, only secular principles and methods apply within management. While for staff members inspired by a Southern and Eastern perspective religion and management are fully integrated.

Conclusions

This essay examines, via Critical Discourse Analysis, how *ubuntu* management is dealt with in an international Christian organization with staff and management from the Southern and Western hemisphere. In this case study relations between colleagues are determined by the power factor and so conflicts do arise. Within the organization people recognize even a political and economy camp.

The first with Southern notions like *ubuntu* state that organizations participate in a religious context. The WCC is seen as an institute that shares in God's life force. In terms of this vision the success of the WCC depends on the degree of harmony between staff, management, partners, spirits and God.

The economy camp accords with the pragmatic and technocratic trends that Norman Fairclough (1992: 200-224) discerns in Western society, even in organisations with more idealistic goals.

Even so, in the end the WCC is caught up in a profound identity crisis.

Kobia's introduction of *ubuntu* management meant to generate a new religious and organizational identity. The WCC needed to change from an international Christian organization with a Western profile to one with a Southern profile, following the numerical dominance by the South. Kobia's vision and approach however did not correspond with what Western financers wanted, and so it was a matter of time to make clear that he no longer was considered as general secretary. Also seen from this experience one might question if it is possible for members of the WCC, though sharing an ecumenical identity, to rise above issues of sharing religion, power and money?

Kobia operated a polemical strategy by praising African religion and criticizing Europe as agent of globalization. In many of his speeches and publications

Kobia concerned the conviction that African values are obstructed by colonialism and globalisation, based on Western hegemony in economic, cultural, political and religious power structures. Kobia observed the same Western control in the institutionalisation of the WCC. Given the fact that the WCC's administration, finance and theology have been dominated for decades by the West, Kobia's discourse was very understandable. Van Binsbergen (2001: 81) writes:

> 'The point is that any social situation in which one truly, existentially takes part, breeds through the experience of such participation a subjective reality from which one cannot and will not distance oneself.'

From an *ubuntu* perspective one could have expected Kobia to seek for consensus. As leader being responsible for an international organization Kobia could have less emphasized his African point of view but presented a more inclusive perspective as consensus has connotations for instance in the Asian Sangseang. Furthermore, *ubuntu* management is focused on the orientations of the whole organization and not on a group of followers. However, Kobia's discourse expressed, and one can also say honoured, the experience of being a member of an excluded vision.

This study also shows the conflict between a management discourse based on secularism and a management discourse based on religion. Time will tell if there will be a consensus. For now, with the Norwegian Dr Tveit as general secretary of the WCC, the 'economy camp' appears to be more visible in the centre. This conflict between Western and African conceptions of management leaves the organisation deeply wounded. The case study shows that religion on the work floor is a sensitive issue with potential for a clash.

The case study proves as well van Binsbergen right. He welcomes non-dominating notions like solidarity and humaneness. But van Binsbergen warns for the power struggles between discourses that strive for hegemony. The launching of *ubuntu* as a superior alternative for Western models of thinking bears the risk of a continuous clash between critics and then again proponents. A way out is to critically discuss the violent dimensions of dominations, not only between people from the African and European continents but between people of all continents. We, including the WCC, can only transcend the patterns of center and periphery, dominance and margins, exclusion and inclusion if we confront ourselves with the outcomes of these misconceptions. According to van Binsbergen (2001: 82) *ubuntu* is above all an invitation to start this process.

References cited

Berger, P.L., ed., 1999, The desecularization of the World: Resurgent Religion and World Politics, Washington: EPPC.

Bujo, B., 1998, The Ethical Dimension of Community. The African model and the dialogue between North and South, Nairobi: Paulines Publications Africa

Davie, G., 2002, Europe: the exceptional case: Parameters of faith in the modern world. London: Darton, Longman & Todd

Dokman, F., 2005, The West and the rest of the World in theology, mission and co-funding. Nijmegen: NIM.

Dokman, F., 2013, De zevende dimensie: De rol van religie binnen internationaal management. Leiden: Quist Publishers.

Fairclough, N., 1992, Discourse and social change in society, Cambridge: Polity Press.

Fairclough N., 2005, Peripheral vision: discourse analysis in organization studies. The case for critical realism, Organization Studies 26 (6). London: Sage.

Grant, D. & Hardy, C., 2004, Introduction: struggles with organizational discourse, Organization Studies, 25 (1). London: Sage.

Jackson, T., 2004, Management and change in Africa: A cross-cultural perspective, New York: Routledge

Jackson, T., Amaeshi, K. & Yavuz, S., 2008, Untangling African Indigenous Management: Multiple Influences of the Success of SMEs in Kenya, in: Journal of World Business, 43 (4), 400-416.

Jenkins, P., 2002, The next Christendom: The coming of global Christianity, New York: Oxford University Press.

Karsten, L. & Illa, H., 2001. Ubuntu as a management concept, in: Quest, Vol. XV, No. 1-2, 91-112.

Kienhuis, N., 2000, Out of the boxes. A critical study into Hofstede's 4D model of culture, Nijmegen: CIDIN.

Kobia, S., 2003a, The Courage to Hope: A challenge for churches in Africa. Nairobi: Acton.

Kobia, S., 2003b, Remarks following the election of the WCC general secretary, in: The Ecumenical Review 55, 2003, 431-432.

Kobia, S., 2006, Called to the one hope: A new ecumenical epoch, Geneva: WCC.

Mbigi, L., 1997, Ubuntu: The African dream in management, Randburg: Knowres.

Mbigi, L., 2000, Managing social capital, in: Training & Development. Alexandria (USA): American Society for Training and Development, January, 36-39.

Mbigi, L. & J. Maree 2005, Ubuntu: The spirit of African transformation management, Randburg: Knowres.

Mumby D. & R. Clair., 1997, Organizational discourse, in: T.A. van Dijk (ed.), Discourse as structure and process: discourse studies, vol. 2. London: Sage.

Nkafu Nkemnkia, M.,1999, African vitalogy. A step forward in African thinking, Nairobi: Paulines Publications Africa.

Schiele, J., 1990, Organizational Theory from an Afrocentric perspective, in: Journal of Black Studies, Vol. 21, No. 2, 145-161.

Shorter, A. & Onyancha, E., 1997, Secularism in Africa: A case study: Nairobi city, Nairobi: Paulines Publications Africa.

Ucko, H., 2005, The story of Jews and Christians meeting in Africa, in Worlds of memory and wisdom. Encounters of Jews and African Christians, J. Halperin & H .Ucko (eds), Geneva: WCC.

van Binsbergen, W., 1999, 'Culturen bestaan niet'. Het onderzoek van interculturaliteit als een openbreken van vanzelfsprekendheden, Leiden: African Studies Centre.

van Binsbergen, W.,2001, Ubuntu and the Globalisation of Southern African Thought and Society. In: African Renaissance and Ubuntu Philosophy. Special Issue. Quest, an African Journal of Philosophy, Vol. XV No. 1-2, 2001.

van Binsbergen, W., 2003, Intercultural encounters. African and anthropological lessons towards a philosophy of interculturality, Münster: Lit Verlag.

WCC Report on the WCAR 2001, Making a fresh start: the urgency of combating racism, Geneva: WCC

WCC Report 2001, The Island of Hope: An Alternative to Economic Globalisation, Geneva: WCC

WCC Central Committee News Release, 'General Secretary-elect brings African touch to WCC', Geneva, 30 August 2003.

WCC Central Committee 2006, Gen.05, Geneva: WCC.

WCC PWE 2007, Reference Group on Poverty, Wealth and Ecology, Geneva: WCC.

WCC Agape 2007. 'Alternative Globalisation Addressing People and Earth (Agape): Consultation on linking poverty, wealth and ecology: African ecumenical perspectives.' Geneva: WCC

Wijsen, F., 2015, Christianity and Other Cultures: Introduction to Mission Studies. Wien: Lit Verlag.

Xu, S. & Yang, R., 2009, Indigenous Characteristics of Chinese Corporate Social Responsibility Conceptual Paradigm, in: Journal of Business Ethics, 93 (2). 321-333.

The Graeco-Egyptian origins of Western myths and philosophy

and a note on the magnificence of the creative mind

by Louise Muller

Introduction

He not busy being born is busy dying (Bob Dylan)

I have known emeritus professor Wim van Binsbergen since my early twenties as a passionate intercultural philosopher; he was the supervisor my MA thesis in Intercultural Philosophy. Somewhat later, as his student-assistant at the African Studies Centre in Leiden, I made sure that this professor did not run out of library books, which he needed to consult in order to write his ever-growing pile of self-written books. I always wondered what happened with all the books that I ordered for him and so did the librarian of the African Studies Centre. On a weekly basis, he was in need of more books, but he was reluctant to return any of them. I used to joke with the librarian by saying that he had eaten all of the books I had ordered for him. Clearly, Professor van Binsbergen needed his food for thought.

One of the books that I ordered for him and of which I, later on, received a copy as a token of his appreciation for my hard work was *Life against Death: The psychoanalytical meaning of history* by the American classicist Norman O. Brown. From the ceremonious way in which van Binsbergen handed me this book I understood it must have been very important to him. I tried to read it, but at the time I found the book quite impenetrable, incomprehensible, and generally not my piece of cake.

Later in life, I decided to read Brown's *Life against Death* again, and this time I read it in its entirety. I now understand that, according to Brown, human beings have not one, but two souls, which are the repelling magnets of nature.

Apollo and Dionysus

Every person is equipped with both the Dionysian or life force soul (in Greek *Eros*), and the Apollonian or death force soul (in Greek*Thanatos*). Dionysus was a Greek fertility god from *c.* 580 BCE associated with wine, music, and choral dance (Csapso 2016). In Attic art, Dionysus was often depicted as a slumping god on a ship, which had a vineover laden with grapes as a mast, surrounded by a sea with a pod of dolphins; the dolphins being the rescuers of sailors (life force) (Carpenter 1990). Dionysus, who was resurrected from death, represented hedonism, happiness, and the good life that he celebrated with a glass of wine. His half-brother Apollo was in many respects his polar opposite. Apollo was a cerebral god associated with the sun, light, and intellectual pursuit. Dionysus symbolized the ability of (wo) man to submerge him- or herself in a greater whole, the ecstatic, and the chaotic emotions. Apollo, on the contrary, symbolised his or her formally rational and reasoning mind.

The Dionysian and Apollonian natural forces are complementary and one needs both to be a balanced person; one needed to be capable of creating form and structure as well as being passionate and vital. Brown believed that a utopian society would primarily consist of such balanced persons who are at home in both the world of rationality and logic and of symbols and emotions(Brown 1959).

Dionysus and Apollo were two Greek deities, which suggest that the concepts of death and life force have a Greek origin and that the Greeks were the first to focus on the necessity of balancing natural forces within the human mind. When studied in renowned Egyptian myths, however, it becomes clear that as early as the historical period of the Old Kingdom (c. 2686 BC - c. 2181 BC), the Egyptians have been familiar with these naturalforces. Greek writers, such as Herodotus and Plutarch[1], borrowed these concepts from the Egyptians by studying their myths, such as the myth of Osiris; an ancient Nubian god that entered Egyptian mythology and the pyramid texts in 2350 BCE. Osiris, which was the first Egyptian god with a human appearance, is a pharaoh-god connected with life-giving power and the preserver of the ideal natural order (Maät). Amongst the Greeks in Alexandria Dionysus (death force) and Osiris (life force) were venerated simultaneously, as was observed by the historian Herodotus (Herodotus 2015 [first edition 440 BCE]: book II, 42).

Osiris' younger brother Seth, who did not inherit the throne, was associated with envy, death force, and the disorder and chaos of the universe. Seth was so jealous of his elder brother that he chopped Osiris' body into thousand pieces and dispersed all body parts over the earth. Since Osiris' body was no longer whole it appeared that Seth successfully prevented the resurrection of his brother. However, Isis, Osiris' sister and wife, found all pieces of her husband's body. She then transformed herself into a bird and used her magical skills to resurrect her brother. During the act, Isis and Osiris conceived their son Horus; the falcon-headed god. After Osiris passed away, Horus (life force) and Seth (death force) were involved in various struggles for political power. Their fight for the thrones of Upper and Lower Egypt ended with Horus winning both thrones. This fight symbolised the natural forces of order and disorder in the universe(Meeks and Favard-Meeks 1993).

It is the historian Herodotus, again, who comments that the Egyptian god Horus and the Greek god Apollo were one and the same. He remarks: 'In Egyptian, Apollo is Horus, Demeter is Isis, Artemis is Bubastis'(Meeks and Favard-Meeks 1993)(Herodotus 2015 [first edition 440 BCE]: book II, 156).The Greek mythology was thus strongly influenced by the ancient Egyptian myths.

---

[1] Until in the course of the 19th c. CE Western scholarship gained access to contemporary Egyptian sources for the myth of Osiris (from various Egyptian text fragments since the Old Kingdom, such as the pyramid and sarcophagus texts, hymns, the book of the dead, and descriptions of the Osiris festivals that were celebrated all over Egypt) this myth was only known in the version which the 2nd-c. CE Greek high priest at Delphi, Plutarch, presented in his *De Iside et Osiride*. *Cf.* Van den Berk 2015.

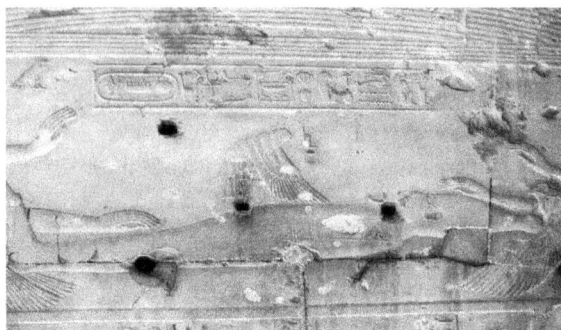

Figure 1: The procreation of Horus, son of Isis, Abydos Temple relief Sethos, Egypt.

Egyptian morality and a recent theory of higher consciousness

The ancient Egyptians believed that the natural forces of the universe (the macrocosm), *i.e.* the life and death force, were represented by deities (the microcosm). Ideally, the deities maintained the right balance between these forces to preserve the order of the universe. The Egyptian pharaohs were responsible for the preservation of the right balance between the aforementioned natural forces. They upheld the moral ideals of ancient Egypt and the laws of nature, represented by laws laid down by the goddess of Justice; Ma'at. The concept of Ma'at is used throughout Egyptian history and its meaning as an interrelated order of rightness, including the divine, natural and social, is repeatedly affirmed. Ma'atis that which confirms the order of the universe, the good, that which is opposite to disorder, injustice and evil (*isfet*). Ma'at stands for the totality of ordered existence in the universe and justice, right relations and duty in the socio-political communal domain and the following of rules and principles in the personal domain. It extends from the elements of nature into the moral and social behaviour of mankind. Whether a man has behaved righteous depends on a person's reputation with his pharaoh and fellow citizens. This does not mean, however, that a person has no personal conscience, but that conscience itself is a relational concept. What one thinks of oneself is based largely on the evaluation by significant others (Karenga 2004).

The existence of Ma'at in Egyptian society and its myths in the meaning of both the pharaonic and individual adherence to rules and principles to keep on the right path reveals that most Egyptians did have a good understanding of just and unjust social behavior. In terms of consciousness, this implies that Egyptians were self-reflexive; they were moral human beings capable of reflecting upon their own behavior over a period of time. This assertion is supported by the Italian neuroscientologist Antonio Damasio's theory of consciousness. In 'The feeling of what happens'(2000), Damasio makes a distinction between three cumulative forms of human consciousness:

1. *The protoself*: a person's bodily state, which is the most basic representation of self.

2. *The core self*: the awareness of the biological bodily state and emotions in the here and now, which is a more evolved form of consciousness.

3. *The autobiographical self*: a person's reflection on the awareness of emotions over a longer period of time. The autobiographical self is the third layer and most evolved form of consciousness. It draws on memory and past experiences which involve the use of higher thought processes. It requires a person to have a language, an autobiographical memory capacity, and reasoning ability.

Damasio believes that the autobiographical self is a necessary condition for both rational and mythological thinking. Therefore, to his mind, mythological thinking does *not* belong to a lower form of consciousness. Damasio stresses that myths are not the product of the core self but, similar to rational thinking, are the result of self-reflexive thoughts of the autobiographical self, which is both an individual and a group member. An adult constructs this self with its experiences, ideas, images, evaluations, likes, dislikes, achievements and failures. Although the autobiographical self is unique to a person, he or she shares narratives with members of the same peer group, community, or culture. This means that besides using our own experiences, we include the experiences, ideologies and beliefs we inherit from (deceased) members of our cultures, which makes us part of the larger narratives of mankind. The autobiographical or self-reflexive self is thus the result of mythological and logical individual thoughts of a person, whose consciousness is at the same time constructed by and part of the collective consciousness of humanity as a whole (Damasio 2000).

The Egyptian moral principle of Ma'at departs from the notion that autobiographical selves are constructed in a social context, which largely determines the social position of the individual self and the righteousness of a person's behaviour. The existence of Ma'at as a significant concept among the ancient Egyptians implies that these ancient Africans did not have a primitive mind. It demonstrates that, on the contrary, the Egyptians had a self-reflexive or higher order consciousness, which enabled them to create a sense of continuity of the self over space and time, a stream of contemplations and evaluations about themselves over longer periods of times and to develop ideas about the future. The Egyptians were moral human beings and the fact that they were capable of creating myths implies that they were also capable of thinking logically since both mythology (*mythos*) and rational thinking (*logos*) are products of higher order consciousness created by the self-reflexive autobiographical self.

A short history of theories of consciousness

My insight that the Egyptians must have had a high consciousness, as result of an autobiographical self, runs counter to the still mainstream idea among Western philosophers that the ability of human beings to think rationally and philosophise, *i.e.* to develop *logos*, did not exist prior to the emergence of the first philosophers, called the pre-Socratics, in ancient Greece (c. 500 BCE). Before that time, they argue, people were only able to come up with mythological explanations for social or natural phenomena, which were the creations of a primitive mind. The Egyptian myth of the boat journeys of the Sun god Ra in the sky, for instance, provided a primal explanation for sunrise and sunset. The pre-Socratics were allegedly the first who attempted to explain these movements of the celestial bodies by making empirical observations and thinking rationally.

The eighteenth-century German philosopher Immanuel Kant believed that once the Greeks had developed their ability to think rationally and to come up with proto-scientific explanations for the existence of the universe and social order, mythological thinking disappeared from Greek society. In his opinion, myths belong to societies that are deprived of the ability to think rationally, because they are less developed than societies that celebrate the use of the *logos* over *mythos* and value philosophers and other rational thinkers more than narrators and minstrels.

It is striking how obstinately the stage theory, which relegates myths to the cultural products of human beings with a lower level of mental development, is rehashed in today's academic circles. Especially, when one reads the oeuvre of 19th – century-born German philosopher Ernst Cassirer, who convincingly argued, following Aristotle some 2,300 years ago, that man is *animal symbolicum* and *animal rationale*. In his three-volume work, *The Philosophy of Symbolic Forms*, Cassirer explains that what distinguishes a man from an animal is not only his exact comprehension of the world (*Weltbegreifen*) but his ability to create and use symbols and to represent the world (*Weltverstehen*). In Cassirer's words

> 'the 'expressive' phenomena, like myths and religion, and the conceptual systems, like mathematics, are all the result of a *Weltverstehen*, a free activity of the mind'(Cassirer 1957 [first published in German 1929]).

To Cassirer's mind 'what distinguishes man from nature is the system of human activities of which language, myth, religion, art, science and history all are the constituents'(Cassirer 1944). In his opinion, scientific discoveries did not make the function of myths in societies defunct as they continued to offer an explanation for phenomena in society in a metaphorical language. Cassirer stressed that *mythos* and *logos* should be understood as complementary concepts. In his view, the creation of myths and the use of ratio are alternative coexisting ways to explain the existence of phenomena and situations in the social word in both (ancient) history and contemporary times.

However, for various present-day scholars, such as Watterson (2013) and Massey (2008), the ancient Egyptians and the bearers of other ancient cultures with a significant focus on mythology have remained to be primitive people with an underdeveloped mind. These scholars believe that the Egyptians lacked self-reflexivity and were not capable of rational thought. They found evidence for their point of view in the Egyptian relationship between body and mind, more precisely, the relationship of one body with a multiplicity of minds or souls.

In the next two sections, I aim to enhance understanding of the Egyptian concept of body and mind and to debunk this stage theory derived proposition.

The Egyptian concept of body and soul

For the Egyptians, the human body (*khat*) 🖼 was a model of the universe that resembled the macrocosm. The energies in the universe were considered to be similar to those inside the *khat*. The two sides of the body were considered to have different qualities due to the connection of these body parts with the entrance and departure of the personality soul (the *ba*). The *ba* was thought to enter the body into the right ear at the moment of a baby's birth and to leave the body via the left ear after a person had passed away, allowing it to start its journey through the underworld (Merkel and Joyce 2003). The right part of the human body was, therefore, associated with life force and light, and the left part with death force and darkness. Similar to humans, the gods were believed to have a light side and a dark side. The eyes of the pharaoh-god Horus, for instance, were associated with the sun (right eye) and the moon (left eye) and with the life and death force in nature (Littleton and Fleming 2004).

Figure 2: The weighing of the heart ceremony on 'The papyrus of Ani'. The British Museum (Rossiter 1974).

Besides a human body, the Egyptians believed that an individual was born with multiple souls that were loosely connected to a person's *khat*. In fact, the body was perceived as the temporary seat of five soul elements that formed the individual: the *ka* or life force 🖼, the *ba* or personality / bird soul 🖼, the *akh* or immortal combined *ka-ba* soul (the ibis or phoenix) 🖼, the *ren* or name soul, and the *sjoet* or shadow soul. These souls could permanently or temporarily dwell inside the *khat* until death brought a person into a liminal stage of existence. This stage would end after the *ba*-spirit had made its journey to the gate of Osiris, the Lord of the Netherworld (the *duat*) (Lamy 1989). The *ba*-soul was only allowed to enter this gate to be eternally united with Osiris in the case of a positive outcome of the verdict over the heart (*ib*), which was still inside the mummified body, during the weighing of the heart ceremony. During this ritual, the heart of a person was weighed against the feather of Ma'at to determine whether or not (s) he had lived righteously. Only when the heart of the mummified person was equal to or lighter than Ma'at's feather, the *ba* of the deceased gained access to the *duat* (Rossiter 1974).

During the opening of the mouth ceremony, the *ba* - soul would return to the mummified body of the deceased to enable him or her to eat and drink again in the afterlife. The Egyptians believed that a deceased person would need his or her body again during the afterlife. For this reason, depending on the social status of the deceased and his or her gender, all kind of attributes that the deceased could use in the Afterlife were added to the grave, such as combs, jewellery, or perfumes; the alleged sweat of the Sun-god Ra. Once the deceased person's senses had returned to the mummy and his or her *ba* had passed Osiris' gate and thus completed its dangerous journey through the underworld, the *ba* became immortal. This happened through the unification of the *ba* with the *ka*–soul and their transformation into the *akh*; the deceased's magical powered and enlightened immortal soul. The *akh* soul only emerged when the mummification process had been successful(Geru 2013).

Figure 3: Vignette of Spell 23 - the opening of the mouth ceremony from the Papyrus of Hunefer (New Kingdom - 19th Dynasty).

The Egyptians considered the relationship between body and soul to be non-dualistic. Although the *akh* was immortal, its existence depended on the wholeness of the body of the deceased. The body's completeness was meant to be guaranteed by its mummification. Neither in the social world nor in the *duat* were body and soul considered to be separated entities (a dualism). Instead, the Egyptian person was made up of a loose connection between a material body and five (quasi)-material soul entities that together made up the ancient Egyptian as a person (Merkel and Joyce 2003). The enlightenment of one of the *akh* souls was the positive outcome of the dangerous journey of the Egyptian souls through the *duat*, which symbolised the struggle of the soul(s) on its way to higher consciousness (the eternal light) and reincarnation. The ancient Egyptians are often misunderstood as people obsessed with death and death manuscripts and rituals, such as the book of the dead and the process of mummification. In reality, however, they were the venerators of life force and spiritual enlightenment. The ancient Egyptians considered the daily reincarnated sun to be the source of all life, including that of the reincarnated and enlightened individual soul (the *akh*). The Egyptian book of the dead was not a veneration of death force but a hymn to life force and the human ability to reach higher consciousness. For the ancient Egyptians, enlightenment and soul harmony were the positive outcomes of the moral struggle of a person's multiple souls (Geru 2013).To my mind, the ancient Egyptian's experience of a bright light and a higher consciousness prior to one's (near) death resembles those of contemporary patients with a near-death experience. In 'endless consciousness', the cardiologist Pim van Lommel (2011) describes the experiences of patients that he interviewed, who survived a cardiac arrest. He mentions their strong awareness of the aforementioned bright light and how; consequently, their personalities underwent a permanent change. After the arrest, his patients had become aware of the fact that love is most important in life and that the heart was the seat of their affection and higher consciousness. The patients informed van Lommel that they had reached a higher stage consciousness and had become more emphatic and sensitive towards the feelings of others. On the basis of his twenty years of research, Van Lommel concludes that the current mainstream materialistic philosophy of consciousness, which states that all consciousness is the result of brain activity, is too narrow to properly understand the aforementioned near-death experiences. These experiences demonstrate that the human consciousness does not always coincide with brain functions and can be experienced separately from the body. Did van Lommel empirically research the existence of a phenomenon that was part of the ancient Egyptian's intuitive knowledge?

Of the multiple souls, the ancient Egyptians considered the heart to be the most central of all quasi-material entities. They believed that the heart was the seat of both reason and emotional knowledge. The heart was the most important thinking and feeling human soul. For this reason, the heart was the only organ that was not removed from the body prior to its mummification. This organ was the seat of a person's morale and only those Egyptians who had lived righteous were allowed to undertake the dangerous journey from the land of the living to the Netherworld. The wrongdoers ended up as a delicious meal for the Nile crocodile Ahmit after which the body of the deceased would undergo a second death and disappear forever into oblivion. The living would, then, cease to remember the person, which heralded the end of his or her existence (Rossiter 1974).

Despite the ancient Egyptian's relatively advanced knowledge of surgery obtained by the mummification of bodies, the nervous system and of (brain-related) diseases they believed that the heart and not the brain was the most important organ of the body. The Egyptians did not have much empirical knowledge of this organ's blood circulating function but they often, symbolically, referred to the working of the heart as the flooding of the Nile. Similar to this long river, which was the source of Egyptian life, the heart consisted of channels that connected all organs to one another and delivered them the necessary fluids to function (e.g. blood, air, saliva, sperm, and nutriment). The brain was regarded as insignificant and was, therefore, discarded during mummification. In ancient Egypt, diseases were often explained by a human state of blocked organs, like irrigation canals through which the water could no longer flow (Finger 2005). Similar to Indian early civilisation, connected to the ancient Egyptians through various trade networks (van Binsbergen 2012), yoga-like exercises were prescribed to those patients who needed their heart channels to be opened up for the mentioned fluids to flow healthy through their bodies. Too much openness of the body was, however, regarded as dangerous, because malevolent spirits could more easily dwell inside an open body and unpermitted use it as their new home (Finger 2005).

It is very likely that the ancient Egyptian open or blocked organ theory is the origin of Hippocrates' system of medicine known as humoralism. This system detailed the makeup and working of the human body positing that an excess or deficiency of any of four distinct bodily fluids (humors) directly influenced a person's temperament and health. These fluids were: blood, phlegm, yellow bile and black bile. In the treatise *On the Nature of Man*, Hippocrates describes that health is primarily the state in which the fluids are in the correct proportion to each other, both in strength and quantity, and are well mixed. Pain occurs when one of the substances presents either a deficiency or an excess, or is separated in the body and not mixed with others (Mann and Lloyd 1983). The body fluids also represented the Four Temperaments or Humors (of personality) as they later became known. The medical theory of humoralism shows parallels with the central theoretical ideas of Ayurveda, which developed in the mid-first millennium BCE (Basham 1976, Comba 2001). As a medical theory, humoralism influenced the ancient Greeks and subsequent Western medical philosophy and Muslim scholars. Humoralism remained an influential theory until the nineteenth century (Wittendorff 1994). The ancient Greeks thus borrowed from ancient Egyptians and Indian medical wisdoms to gain an understanding of the human body and its connection to the soul.

The Greek concept of soul

Before the fifth century, the Greeks believed that only human beings had a soul (*psuchê*) and that the Netherworld was the sole domain of deceased persons (and thus not of e.g. the Egyptian Nile crocodile Ahmit). In the Homeric poems, the word *psuchê* is most often used in those cases where people's life was at risk and the spirits of death were close by, such as in the case of the life of Achilles in the Iliad (Fagles] 1998 [original c. 800 BCE]). In the sixth and fifth centuries BCE, the PreSocratics also started to use the word 'ensouled' (*empsuchos*) to refer to the condition of being alive, which was applied to human beings and other living things. The human soul was associated with courage and morality, and an entity engaged in thinking and planning. In ordinary Greek life, the soul was treated as the bearer of moral qualities, and as responsible for practical thought and cognition. In pre-Socratic philosophical thinking in this period, Empedocles, Anaxagoras, Democritus, and Pythagoras, also believed that plants and animals had souls. Pythagoras even believed that human souls could animate plants(Lorenz 2003). As mentioned before, the Egyptians already had a much longer tradition of attributing soul life to animals and plants. Since 3200 BCE they, therefore, buried various small and large animals, including the African wild bull, the leopard, the elephant and a hippopotamus. From 700 BCE, which is two centuries before the pre-Socratics, the Egyptians also embalmed many animals, such as cats and birds (Quirke 2014). If we add this to the fact that Pythagoras and many of the other pre-Socratics had studied and lived in Egypt, it is likely that the idea of *panpsychism* or animism was not of Greek but of African origin.

In both ancient Egypt and Greece, one believed in a divine life principle to which each and everything was connected. Body and soul were not radically different and considered to be part of the matter of the universe. The pre-Socratic Heraclitus, for instance, believed that the soul was bodily (material) but consisted of fine matter, such as air or fire. The Greeks made a distinction between a life soul, three body-souls (*thymos, nous* and *menos*) and a free soul (psyche, in Greek *psuchê*), which is comparable to the Egyptian ka-soul. *Thymos* was the active soul comparable to the Egyptian *ba*-soul, the source of intense emotion, passion and associated with breath, blood, heart, liver, and the human desire for recognition. One believed that *Thymos* was material - dependent on the body - mortal and only active when the body was awake. This soul was located between the heart and the sternum. *Nous* was the intellect or the faculty of the human mind necessary for understanding what is true or real and was located in the chest, and *menos*, meaning strength, was located in the head. The ancient Greeks believed that of all mentioned souls only the free soul or (*psuchê*) was immortal. When someone passed away, the (*psuchê*) would leave the body through a body opening, mostly the mouth or the person's nose. Unlike the Egyptian, the Greeks did not believe that the free soul (*ka* or *psuchê*) would return to the body during a later stage in the life of the deceased to unite with another soul (The Egyptians believed that the *ka* and the *ba*-soul became *akh*). The Greeks thought that the *psuchê* would lead its own life, away from human bodies. Therefore, the Greek ancestors could allegedly neither speak nor smile or walk, which is why they moved differently from the living. The Greeks did, however, not leave any exact descriptions about the wanderings and behaviour of the immoral souls of their ancestors and beloved ones (Bremmer 1987). The pre-Socratics thus only had some vague beliefs in the immortality of the *psuchê* separate from the mortal body.

To sum up, since the sixth and fifth centuries BCE, the ancient Egyptians and Greeks shared a belief in the human body as a model of the macrocosm and multiple souls of whom at least one was considered to be immortal.

Present-day views on the ancient Greek and Egyptian concepts of soul

Bremmer (1987), and other philosophers of consciousness have categorised the belief in multiple souls as 'primitive'. Brenner wrote,

> 'although the concept of *psyche*developed into the modern unitary soul, its 'primitive' character can be discerned in the so-called shamanistic traditions and the early descriptions of dreams'(Bremmer 1987:11).

Soul development goes together with the notion that in every 'body' houses just one rational and immortal soul (*psyche-logos*), which brings life to the body. This soul idea was developed by the Greek philosopher Plato, the intellectual successor of the PreSocratics, who introduced an anthropological radical form of dualism in the belief between body and soul. Although Plato's ideas about the relationship between body and soul are unclear, in his view, intelligence is a characteristic of the soul, but not of the body. The latter is dependent upon the soul for its movements. Although the body is thus conveyed by the soul, the soul is contained within the body, like a prisoner in jail (Olshewsky 1976). In *The Republic*, Plato offers a theory of soul (comparable to the Egyptian *ka*), which allows attribution of (in principle) all mental or psychological functions to a single subject, the soul as separated entity from the body (Lorenz 2003). And whereas the body, the sense organs, enables us to gain an understanding of the physical world, the world of matter, it is the soul that connects us to the world of the perfect forms and ideas. According to Plato, body and soul link us to two different and radically separated worlds, of which the metaphysical one is superior to the physical. To know the higher truth, one should not rely on one's senses but upon one's soul and its meditative connection to the perfect forms. It is the ideal world that is the source of the immortal soul, which pre-existed before a person's birth and which, after one passes away, will always return to the ideal world (D.J. Vaughan] 1997). Several scholars, such as George G.M. James (1954), believe that Plato derived his ideas of the immortality of the soul from Egyptian priests, who believed in the eternal *ka-soul*, *ba-soul* and the *akh*-soul. Plato received part of his philosophical education from these priests, with whom he stayed for a period of over thirteen years. Nevertheless, Plato is often considered to be the father of Western philosophy, because of his singular concept of the soul as the seat of rationality as one can find in Plato's *Phaedo*.

The idea of one soul as the seat of *logos*, rather than a multiplicity of soul enti-ties, is generally considered to be a step forward in the history of the philoso-phy of consciousness. The main idea is that only unitary souls are considered to be self-reflexive, which is a characteristic of higher consciousness. The reason-ing is that soul reflexivity can only exist if there is a continuation of a narrative within the soul over a longer period of time. Such a perpetuation of streams of conscious ideas about the self cannot occur in bodies with multiple souls flying in and out of a body because there would not be one single self to develop a narrative about the self. This train of thought implies that the consciousness of ancient Egyptians and pre-Socratics, who adhered to a belief in multiple souls, was less developed than that of the Greeks since Plato and successive genera-tions of (Greek) philosophers. It would also imply that rational higher order thinking would not occur before Plato, who is, therefore, the legitimate father of Western philosophy.

If this is the truth, the question remains, where this leaves the ancient Egypt-ians, the Presocratic Greeks and also contemporary Muslims and (Pentecostal) Christians in *e.g.* Africa. In her award-winning book *Dreams That Matter*, the anthropologist Mittermaier examines the encounter and engagement of pres-ent-day Muslims in Egypt with the Divine. She comes to the conclusions that many contemporary Muslims in this North African country believe that (reli-gious) knowledge is channelled through the gods and that various spirits enter the body to enlighten Muslim believers (Mittermaier 2011). Similarly, many Pentecostal Christians in various countries in Africa, including Nigeria, believe in malevolent spirits, which can dwell inside a person and can only be removed by church rituals of exorcism(Adogame 2012). Would all these ancient adher-ents of a concept of multiple minds and the abovementioned contemporary believers really be incapable of thinking rationally or is there something else at stake? By all means, what these ancients and contemporary Muslims and Chris-tians have in common is the belief in a central core soul among all souls (the heart or psyche), which remains in place regardless of the alien spirits' tempo-rary occupation of a person's body. To my mind, as long as this central core soul develops a narrative of the self and is thus self-reflexive, the multiplicity of soul theory is merely the result of a cultural difference than that it, in any way, affects the thought process of its adherents. In my view, the human capability to think rational (*logos*) is thus, for the most part, unrelated to people's philosophy of consciousness in past or present times.

By all means, historical knowledge and the remains of the pyramids themselves make it easy to refute the presumption that neither the Egyptians nor the pre-Socratics were capable of philosophising. Between 3300 BCE until 396 BCE, the ancient Egyptians wrote in hieroglyphs, which were native to Egypt. Concerning the hieroglyphs, the famous Jean-François Champollion (1790-1832) who deciphered the Rosetta stone, commented:

> 'It is a complex system, writing figurative, symbolic, and phonetic all at once, in the same text, the same phrase, I would almost say in the same word'.

Champollion, who held a chair in Egyptian history, did not consider the ancient Egyptians to be primitive people, whereas he did not live long enough to discover that many of the hieroglyphic writings contain literary works, poems, medical texts and mathematical treatises, that predated the Greeks by thousands of years (Parkinson 2002). The *Mathematical Papyrus*, for instance, which deals with arithmetical problems and geometry, was written during the reign of the Egyptian King A-User-Re (approx. 1650 BCE). This significant document in the history of maths thus far predates Plato's geometrically based theory of the ideal forms. The Egyptians were the first to invent the use of decimal points and fractions as well as methods of solving problems pertaining to volumes of pyramids and areas of cylinders, which have been adopted by modern mathematicians (Olela 1997). Egyptian sage philosophers were so well known for their mathematical knowledge that the Presocratic Pythagoras left his birth town Samos in Iona to study in Egypt, which was a more stimulating place to work (Brock 2004). Besides, the Egyptians also had a relatively advanced knowledge of surgery, which they used to cure patients with psychological or physical illnesses and to remove those organs that they wished to preserve after a person had passed away (the lungs, the liver, intestines, and the stomach). The heart was left in the body so that the righteousness of the in the heart located soul could be judged against the feather of Ma'at ⸾ (Finger 2005).

These characteristics of ancient Egyptian and pre-Socratic culture demonstrate that the assumption that rational thinking (*logos*) did not develop until the philosophy of Plato does not hold. It is, therefore, more useful to concentrate on how narratives explain the world (*mythos*) and descriptions of natural phenomena based on the sense experiences (*logos*) were used as alternative complementary explanatory models in the Graeco-Egyptian ancient world. The pre-Socratic philosopher Thales, for instance, believed that water was the underlying substance of all materialism in the universe and that all worldly material would reincarnate and return to an invisible form of water. Thales, who was educated in Egypt by sage priests, borrowed his ideas from the Egyptian wisdom that, together with the sun, the Nile was the source of all life. However, he also studied the Egyptian myth that the god of the ocean water (Nun) was the primordial chaos and darkness of which the Sun-god Atum emerged (Hart 1990). Thales took his ideas from the Egyptian belief in a divine form of consciousness as the source of all matter (in Greek *nous*). Both the ancient Greeks and Egyptians thus not only defined their gods as natural forces but also observed the natural phenomena as gods.

The Egyptian gods and their energies were considered to be part of nature. The Egyptians were very good at observing nature and they used symbols to decode the natural universe. They, for instance, empirically observed frogs and came to the conclusion that frogs were amphibians, who lived between water and land. The goddess frog Heket symbolised the transforming energy in these amphibians, which enables them to live in between water (the primordial chaos of which all life came into being) and land (the dry space that babies enter after they have left their mother's watery womb). Heket also symbolised a midwife. She accompanied Chnoem, the god of creation, in the conception of the virgin birth of the children of Ahmes; an Egyptian queen in the old Egyptian kingdom. The goddess Heket thus symbolised transformation. This does not mean, however, that Egyptians were not capable of understanding rationally how a frog could live under water and on land. Nevertheless, by visualising the energy of genesis in nature the ancient Egyptians enhanced a deeper phenomenological understanding of the natural world. Heket symbolised the transition from the stage of non-being to being, from chaos to order, from the existence of land and (human) life out of water due to the flooding of the Nile(Van den Berk 2015). As the pre-Socratic historian, Hecataeus of Miletus and the ancient Greek historian Herodotus put it: 'the birth of Egypt is a gift of the Nile' (Gwyn 1966, Herodotus 2015 [first edition 440 BCE]: Book II, chapter 5).

Another animal, which among the Egyptians also functioned as the head of a god, was the scarab (*scarabaeus sacer*). Because these beetles use the dung ball that they are rolling for themselves as a brood chamber for their eggs, it symbolised the reincarnation of the dead and the rising of the sun. Equal to the sun the scarab beetles rolled their ball from East to West and equal to Ra the young scarab beetles appeared to be capable of spontaneous creation as they emerged from their burrow(Van den Berk 2015). Twinge scarabs also characterised the human brain, which was similar in shape. The Egyptians associated the scarab and the brain with creative (sexual) force and spontaneous resurrection. Since the Middle Kingdom, the scarabs were placed on the heart of mummies to protect this alleged central organ during the body's entry in the afterlife. The bottom of the scarab contained an inscription from 'the Book of the Dead'. The inscription was meant to protect the deceased against confessing to any of the wrongs that (s)he might have committed during his or her lifetime once (s)he was brought before Osiris and the tribunal of the gods (the weighing of the heart ceremony)(Wilkinson 2008).

Figure 4: The god Khnum creates the child of queen Ahmes (Ihy) with the assistance of the frog goddess Heket. A relief from the mammisi (birth temple) at Dendera Temple complex, Egypt (south of Abydos).

The scarab-god Khepri was closely associated with the creator god Atum and the sun-god Ra. This demonstrates that the Egyptian belief was syncretistic, which means that the natural forces were often represented by a combination of gods, such as Atum-Ra or Khepri-Ra. The fact that the frog goddess Heket and Khepri symbolised transformation demonstrates that the Egyptians made use of a multiplicity of approaches to symbolically explain the working of the universe. They believed that specific energies, such as that of transformation, inhabited several beings. This enabled them to choose from a variety of symbols to represent these energies and to mythically explain the origins of the universe. For them, the analysis of empirical observations of nature was the foundation of the mythical explanations that were believed to sprout from a higher order of consciousness (Hart 1990). The Egyptians believed that the symbolic understanding of the natural energies was more important than the understanding of the empirically observable world. This does not imply, however, that the Egyptians were not capable of thinking logically nor that this preference for a mythological explanation for the existence of the universe was related to the Egyptian concept of mind. Instead, it was the result of a cultural orientation that placed mind over matter and believed that the heart was the centre of the soul.

Figure 5: A picture of Khepri, the scarab-headed god, the tomb of Nefertari (Dynasty XIX) in the Valley of the Queens.

The effect of this cultural orientation was that the Egyptians, who were well aware of the working of the human nervous system, used rational methods to cure their patients (*e.g.* by treating a nervous condition with herbs), but also invoked the appropriate gods, such as Chonsoe - the god of medicine. The combination of *logos* and *mythos* in the treatment of illnesses was, moreover, also common in the Greek society of the fifth century. After the aforementioned Hippocrates, the father of medicine, discovered that not the heart but the brain was the central organ of the human body, many neurological diseases - such as epilepsy- were no longer believed to be a punishment inflicted by the gods. Yet, Hippocratic physicians were not offended if their patients wanted to pray and were sometimes encouraged to invoke their gods. In the Hippocratic treatise 'On Regimen,' one can read that prayer is good, but while calling on their gods for help and comfort, humans should also treat each other (Hippocrates [460 BCE-380 BCE]). Because the Hippocratic physicians were influenced by the pre-Socratic vegetarian Pythagoras, these doctors were not allowed to dissect any human or an animal (Finger 2005).As a result, they knew very little about anatomy and physiology and focussed their treatment on patient care and prognosis rather than diagnosis. A French doctor, therefore, called the Hippocratic treatment 'a meditation upon death' (Jones 1868). In the third century BCE, for the dissection of bodies for medical purposes, Greek doctors went on an internship to Alexandria in Greek-ruled Ptolemaic Egypt. In Alexandria, physicians conducted ground-breaking investigations into internal human anatomy by dissecting human corpses. Under the reign of Ptolemy (323-30 BCE), the medical school at Alexandria grew quickly to become the medical centre of the Hellenic Age. The medical research that took place in Alexandria in this century, which built upon the knowledge of the ancient Egyptians of mummification and that of the Hippocratic physicians, was a unique event in the history of medicine and the Ancient World. In the West, dissection remained prohibited until the Renaissance and, consequently, anatomists were not allowed to practice on human corpses. This prohibition obstructed anatomic research until humans became the central focus of attention in the universe (Faulkner 2015).

Mythos and Logos

In the previous paragraphs, we have seen that many of the Egyptian myths focused on the maintenance of the balance between the death and life force in nature (the macrocosm) and in the individual (the microcosm). Several of the struggles between the gods, such as those between Seth and Osiris, and Seth and Osiris' son Horus were not only fights for the throne of Upper and Nether Egypt but also symbolical contests between light and darkness, morally better and worse behaviour. The boat journey of the Sun-god Ra through the Sky during the day and the Netherworld at night not only symbolised the daily reincarnation of the sun but the reincarnation of all life on earth. The individual journey of the deceased person's *ba* through the Netherworld clarified the connection between microcosm and macrocosm. The Sun-god Ra and the *ba* of a righteous deceased person made the same journey through the Netherworld as part of the eternal cycle of life. Both the sun and an individual person would be reborn and be part of the energy of the Sun-god Ra, the life force and light of the universe. The self-reflexivity of a person in this universe was his or her conscious knowledge of being part of the eternal circle of life, which included the ancestral spirits (the living-dead) in the Netherworld located in the reed beds of Egypt (Allen 2003).

The historical consciousness of that person was a consciousness of the eternal cycle of historical events of the rise and fall of the Egyptian civilisation and the birth and re-birth of the pharaohs and their subjects (the Egyptians created lists of pharaohs and annals of individual reign. All pharaohs were considered to be the reincarnation of the first pharaoh-god, the Jesus-like Horus, whose birthday was celebrated each year) (Quirke 2014). The Egyptians were raised with a cyclical time consciousness (*neheh*) and the concomitant circular perception of individual and social history understood as a series of events. In this consciousness, self-reflexivity is not the result of a linear process but of the awareness of the circulation of natural energies including those of the individual. Next to *neheh*, the Egyptians were also aware of linear time (*djet*) that they associated with the time of the land (Traunecker 2001). Cyclical thinking, which was the dominant way of thinking among the Egyptians, is not illogical but derives from multiple starting points. It allows thoughts to flow, to brainstorm about a topic in attempts to arrive at a solution in the process. Non-linear or right hemisphere thinking by the brain increases possible outcomes and enables the mind to zoom in and out to see details or the bigger picture there were necessary. It is as useful as linear thinking in thought processes and does not imply a lack of development of the *logos* but of an alternative way of dealing with challenges and dilemmas (Joseph 2012). As demonstrated before in this article, by no means did the right hemisphere dominance prevent the ancient Egyptians from being inventive in the fields of *e.g.* written language, mathematics, astronomy, geometry and surgery and above all morality.

Morality and the search for balance between better and worse forces in nature and mankind were at the heart of ancient Egyptian society. The ancient Egyptiansunderstood that good and evil are not one another's opposites but energies on a spectrum and that eventually no harmony can be reached in (human) nature without an underlying strife between the feminine and the masculine, the light and dark natural forces (Karenga 2004). The myth of the strife be-tween Seth and Osiris and Seth and Horus were necessary for the Egyptian society to remain healthy and harmonious. Despite all of Seth's dark energies, he was also venerated and beloved by the Sun-god Ra, who often supported him in his fights against Osiris and Horus (Littleton and Fleming 2004).

Morality, which focuses on the question of what is good, is also an important field in Western philosophy. Ideas similar to those of the ancient Egyptians about harmony after inner (moral) strife can be found in the works of the Presocratic philosopher Heraclitus (540 BCE) and the German philosopher Friedrich Nietzsche (1844-1900) (Nietzsche 1995 [first published 1871]). Moral philosophical thoughts cannot be realised without rational thoughts. And yet, in most twenty-first-century textbooks, it is not the Egyptians but the Greek Presocratics, who are considered to be the first proto-rational or philosophical 'Western' thinkers. However, most of the first Presocratics (Thales, Anaxi-mander and Anaximenes from Miletus; part of a Greek colony in Asia Minor) were not concerned with morality and reflecting on what is right on a social and individual level. Instead, they asked themselves what elements the world is made of and what the origins of the universe are. They were, therefore, recog-nised as (proto) natural scientists (the *physikoi*). What these Presocratics had in common was the notion that they themselves were capable of gaining knowledge of (human) nature and that knowledge was not channelled to them by the gods. This point of departure distinguishes the presocratics from earlier renowned Greeks, who contributed to literary studies, such as Hesiod. This poet felt that his main poetic work, the *Theogony*, was a gift of the gods. The ancient Greeks were, however, not unique, in their notion that not all knowl-edge had a divine origin. In fact, the source of the aphorism 'know thyself', which could be found in the temple of the Greek oracle of Delphi was the In-ternal Temple of Luxor. One this temple wall, one could find the aphorism:

'Man, know thyself, and you are going to know the gods' (Schwaller de Lubicz and Lamy 1978).

This aphorism reveals that some Egyptian priests believed that the source of human knowledge was mankind itself. Another similarity between the ancient Greeks and the Egyptians is that one of the creation myths shows the connec-tion between the gods and the elements of nature. According to the Egyptian Theogony of Heliopolis, the ancestor-gods of Isis, Seth and Osiris are natural elements. The god Nu, for instance, represents the element water, Shu stands for air, Tefnut for moisture, Geb for the earth, and Nut for the sky (Hart 1990). This Theogony reveals that equal to contemporary physicians, the Egyptians were searching for a so-called 'theory of everything'; a hypothetical theory that fully explains and links together all physical aspects of the universe.

Whatever van Binsbergen's attempts (2012b) to revisit the Presocratic Greek philosophers and reinterpret their thought as secondary and peripheral aberrations of a cyclical element transformational cosmology that had near-global distribution in the Neolithic and Bronze Age, the Presocratics are considered to be the first philosophers, the earliest proto-logical thinkers because they disbelieved or were critical about the belief in gods. Xenophanes of Colophon (c.570-475 BCE), for instance, was convinced that men anthropomorphized the gods, which is why

> 'Ethiopians say their gods are snub-nosed and black, Thracians that theirs are blue-eyed and red-haired' (Watterson 2013).

The classicist Whitmarsh is convinced, though, that atheism is a characteristic of every (ancient) culture. Atheism is thus not a distinctive feature of ancient Greek culture, but Greek society was, in most cases, exceptionally tolerant towards atheists (Whitmarsh 2015). Greek society brought forth great philosophers and proto-scientists, such as the Presocratics, Plato, Aristotle, the astronomer Ptolemaeus and the physician Hippocrates. Nevertheless, most Greeks were convinced that the immortal gods had supernatural powers and could interfere with the lives of the mortals. Apollo was, for instance, considered to be empowered with the gift of predicting the future. The Greeks believed that the wisdom of this god was channelled to the priestess (Pythia) of the Oracle of Delphi. Similarly, in the New Kingdom, the Egyptians attributed soothsaying powers to Seth, Isis and other deities and the ancient Egyptians could visit their oracles to seek advice (Shafer, Baines et al. 1991). Apollo's image was mixed with that of the Sun-god Helios and equal to the Egyptian Sun-god Ra, he travelled through the air in a vehicle (either a boat or a solar car) to enable the sun to shine during the day. Both Greek and Egyptian gods could turn themselves into an animal and use its energies to their benefit. According to the Book of the Dead (chapter 27), the local goddess Henen-Su, for instance, turned herself into a cat. She did so in order to please the Sun-god Ra and to kill the malevolent serpent spirit Apophis (or Apep) that fought with Ra and aimed to destroy his solar boat each night thereby threatening the rebirth of the sun the following day. Henen-Su's heroic deed was also carried out by the ancient Greek god Apollo, who slew the great dragon Python and by the Medieval dragon slayer Saint George (Remler 2010, McCullough 2013).

In summary, the ancient Egyptians and the ancient Greeks were combined convergent with divergent thinking, which resulted in the creation of myths and rational thoughts. Despite the proto-scientific discoveries of these ancient people, they were predominantly right brain hemisphere oriented. As discussed in this article, the myths, philosophies and the discoveries of the ancient Greeks about the origin and working of the natural world were in line with those of the sage priests of ancient Egypt of whom they borrowed ideas and on whose shoulder they stood to make their own contributions to both the humanities and the natural sciences.

Figure 6:The local goddess Henen-Su turned herself into a cat to kill the malevolent serpent spirit Apophis (or Apep) (Remler 2010).

Figure 7: An oil painting of the French Rococo Painter Ernemond-Alexandre Petitot (1727-1801), at: https://www.1st-art-gallery.com/Ennemond-Alexandre-Petitot/Ennemond-Alexandre-Petitot-oil-paintings.html.

Figure 8: Saint George and the Dragon Greek, Ionian Islands School Mid-17th century Tempera and gold on gesso and wood. The Temple Gallery, London.

Conclusion

This chapter aims to contribute to the understanding of the birth of Western philosophy (the *logos*). The author argues that rational thinking and philosophy are not a Greek invention. The ancient Egyptians, for instance, were perfectly able to think logically, which resulted in great discoveries in the fields of written language, mathematics, astronomy, surgery and medicine. The fact that the ancient Egyptians left us hieroglyphs with astonishing mythological figures and parts of stories, such as the myth of Osiris, does not reduce their capacity for rational thinking. Although the ancient Egyptians were predominantly right brain hemisphere oriented, they combined convergent and divergent thinking and valued both myths and logic. The article stresses that the ancient Greeks, including the Presocratics and Plato, were not so different from their Egyptian predecessors, whose rich and advanced culture they admired and from which they borrowed elements in the fields of both mythology and proto-scientific thinking. The article dismisses the stage theories of consciousness that stress that logical thinking developed itself at the cost of mythological thinking and that the ancient Egyptians were primitive minded people incapable of producing rational thoughts. The author demonstrates that the multiple soul theory of the ancient Egyptians was shared by the pre-Socratics and that this theory hindered neither of them in thinking logically. The self-reflexivity of the soul, which enables both mythological and logical thinking, is hardly related to singularity or multiplicity of soul-belief in antiquity. In fact, life force does not suppress the death force of one or multiple souls. Instead, the intuitive knowledge that is felt to be channelled to a person by the gods, regardless of his or her number of souls, links that person to a collective culture consisting of stories, myths and symbols. This process enables him or her to harmonise the opposing forces, the life and death force, within oneself. By doing so, it also makes the person a better thinker freed from the displacements of the unconsciousness by the rational soul. The history of the ancient Greeks and Egyptians demonstrates that (their) creations and discoveries are the results of the harmonious interactions between the right hemisphere and the left hemisphere of the brain. This is an important conclusion since our contemporary world is dominated by computers and other technology that enforces cognitive thinking, which can lead to feelings of depression and alienation. The scientific history of mankind is often described as a detachment process of human beings of their gods. The discovery of the origin and working of human nature enabled them to free themselves in the belief of gods and to celebrate rationality. Creativity was thereby most often depicted as something unrelated to human discoveries, imagination was perceived as a waste product of the mind. However, it is the imagination, the ability to think about the future, to create stories, myths and to understand symbols that keep our mind healthy, connected to others and socially engaged. The twenty-first century is heralded as the age of creativity, an era in which we will increasingly have to compete against computers by using our imagination and our creative mind. Of all the things that I learnt from Professor Wim van Binsbergen the most valuable one is to use my creativity in my pursuit to wisdom. Van Binsbergen's specific gift, Norman O. Brown's *Life against Death,* has certainly helped me to balance the opposing souls inside myself. This chapter, among others, is one of those creative results.

References cited

Adogame, A., 2012, 'Dealing with Local Satanic Technology: Deliverance Rhetoric in the Mountain of Fire and Miracles Ministries,' *Journal of World Christianity* 5 (1): 75-101.

Allen, J. P., 2003, 'The Egyptian Concept of the World', *Mysterious lands*, D. O'Connor and S. Quirke. New York: Routledge: 23-30.

Basham, A. L., 1976, 'The Practice of Medicine in Ancient and Medieval India, Asian Medical

Systems'. C. Leslie. Berkeley: University of California Press, 18-43.

Bremmer, J., 1987, *The Early Greek Concept of the Soul*, Princeton: Princeton University Press.

Brock, J. F., 2004, 'Pyramids to Pythagoras: Surveying from Egypt to Greece – 3000 B.C. to 100 A.D.', Workshop – History of Surveying and Measurement, from https:// www.researchgate.net/file.PostFileLoader.html?id=56f96c9293553b9af6019395&assetKey=A S%3A344668710686721%401459186832121.

Brown, N. O., 1959, *Life against Death: The Psychoanalytical Meaning of History*, Middletown / Connecticut: Wesleyan University Press.

Carpenter, T. H., 1990, *Dionysian Imagery in Archaic Greek Art*, Oxford: Clarendon Press.

Cassirer, E., 1944, *An Essay on Man: An Introduction to a Philosophy of Human Culture*, New Haven, Yale University Press.

Cassirer, E., 1957, [first published in German 1929]). *Phenomenology of Knowledge*, New Haven: Yale University Press.

Comba, A., 2001, Carakasamhita, Sarirasthana, I and Vaisesika, *Philosophy: Studies on Indian Medical History*, G. J. Meulenbeld and D. Wujastyk, Delhi: Motilal Banarsidass: 39-55.

Csapso, E., 2016, 'The 'Theology' of the Dionysia and Old Comedy'. *Theologies of Ancient Greek Religion*. E. Eidinow, J. Kindt and R. Osborne, Cambridge: Cambridge University Press, 117-152.

Damasio, A. R., 2000, *The Feeling of What Happens: Body, Emotion and the Making of Consciousness*, Vintage Publishing.

Faulkner, L.A., 2015, *Ancient Medicine: Sickness and Health in Greece and Rome*, Ichabod Press.

Finger, S., 2005, *Minds behind the Brain: A History of the Pioneers and Their Discoveries*, Oxford: Oxford University Press Inc.

Geru, M. A., 2013, *Het Egyptische dodenboek* [The Egyptian Book of the Death], Deventer: Ankh-Hermes.

Gwyn, G., 1966, 'Hecataeus and Herodotus on 'A Gift of the River', *Journal of Near Eastern Studies* 25(1): 57-61.

Hart, G., 1990, *Egyptian Myths*, London: British Museum Press

Herodotus, 2015, [first edition 440 BCE]), *The History of Herodotus*, New York: Palatine Press.

Hippocrates, ([460BCE-380BCE]), *On Regimen in Acute Diseases*, Alexandria: Library of Alexandria.

Homer., 1998, [original 800 BCE]), *The Iliad*. New York: Penguin Classics, tr. Fagles.

James, G. G. M.,1954, *Stolen Legacy*, New York: Philosophical Library.

Jones, W. H. S., 1868, 'Hippocrates Collected Works I' from: http: // daedalus.umkc.edu/hippocrates / HippocratesLoeb1 / index.html.

Joseph, R., 2012, 'Right Hemisphere, Left Hemisphere, Consciousness & the Unconscious, Brain and Mind', E-book: University Press.

Karenga, M., 2004, *Ma'ät, The Moral Ideal in Ancient Egypt: A Study in Classical African Ethics*, Taylor and Francis Ltd.

Lamy, L., 1989, *Egyptian Mysteries: New Light on Ancient Knowledge*, London: Thames and Hudson.

Littleton, C. S. & F. Fleming, 2004, *Mythologie; Een Geïllustreerde Geschiedenis van Mythen en*

Verhalen uit de Hele Wereld [Mythology: An Illustrated History of Myths and Stories all over the World], Kerkdriel: Librero.

Lorenz, H., 2003, 'Ancient Theories of Soul' from: https://plato.stanford.edu/entries/ancient-soul/.

Mann, W. N. & G.E.R. Lloyd, 1983, *Hippocratic Writings*, Harmondsworth: Penguin.

Massey, G., 2008, *Sign Language and Mythology as Primitive Modes of Representation*, New York: Cosimo Classics.

McCullough, J., 2013, *Dragonslayers: From Beowulf to St. George*, London: Bloomsburry.

Meeks, D. & C. Favard-Meeks, 1993, *Daily Life of the Egyptian Gods*, Ithaca: Cornell University Press.

Merkel, L.M. & R.A. Joyce, 2003, *Embodied Lives; Figuring Ancient Maya and Egyptian Experience*, Taylor and Francis Ltd.

Mittermaier, A., 2011, *Dreams that Matter: Egyptian Landscapes of the Imagination*, London: University of California Press.

Nietzsche, F., 1995, [first published 1871]). *Die Geburt der Tragödie aus dem Geiste der Musik* [The Birth of Tragedy from the Spirit of Music], New York: Dover Publications.

Olela, H., 1997, 'African Foundations of Greek Philosophy', *African Philosophy; An Anthology*, E. C. Eze, John Wiley And Sons Ltd: 43-50.

Olshewsky, T. M., 1976, 'On the Relations of Soul to Body in Plato and Aristotle,' *Journal of the History of Philosophy* 14 (4): 391-404.

Parkinson, R. B., 2002, *Poetry and Culture in Middle Kingdom Egypt: A Dark Side to Perfection*, London: Continuum.

Quirke, S., 2014, *Exploring Religion in Ancient Egypt*, John Wiley And Sons Ltd.

Remler, P., 2010, *Egyptian Mythology: A to Z*, New York: Chelsea House.

Rossiter, E., 1974, *Het Egyptische Dodenboek: Beroemde Egyptische Papyri* [The Egyptian Book of the Death: the Famous Egyptian Papyri], Alphen aan den Rijn: Atrium.

Schwaller de Lubicz, I. & L. Lamy, 1978, *Her-Bak: Egyptian Initiate*, New York: Inner Traditions International.

Shafer, B. E., J. Baines, L. H. Lesko & D. P. Silverman, 1991, *Religion in Ancient Egypt: Gods, Myths and Personal Practice*, Ithaca: Cornell University Press.

Traunecker, C., 2001, *The Gods of Egypt*, Ithaca: Cornell University Press.

van Binsbergen, Wim M.J., 2012a, 'The Relevance of Buddhism and Hinduism for the Study of Asian-African Transcontinental Continuities', paper read at the International Conference 'Rethinking Africa's transcontinental continuities in pre-and protohistory', Leiden: African Studies Centre; now published as: van Binsbergen, Wim M.J., 2017, 'The relevance of Taoism, Buddhism, and Hinduism, for the study of African-Asian transcontinental continuities', in: *idem, Religion as a social construct: African, Asian, comparative and theoretical excursions in the social science of religion*, Papers in Intercultural Philosophy / Transcontinental Comparative Studies, Haarlem: Shikanda, pp. 361-412.

van Binsbergen, Wim M.J., 2012b, *Before the Presocratics: Cyclicity, transformation, and element cosmology: The case of transcontinental pre- or protohistoric cosmological substrates linking Africa, Eurasia and North America*, special issue, *QUEST: An African Journal of Philosophy/Revue Africaine de Philosophie*, Vol. XXIII-XXIV, No. 1-2, 2009-2010, pp. 1-398, book version: Haarlem: Shikanda.

Van den Berk, T., 2015, *Het Oude Egypte: Bakermat van het Jonge Christendom*, Zoetermeer/ Meinema: Pelckmans.

Van Lommel, P., 2011, *Consciousness Beyond Life, the Science of the Near-Death Experience*, London: Harper Collins.

Watterson, B., 2013, *Gods of Ancient Egypt*, Stroud: The History Press.

Whitmarsh, T., 2015, *Battling the Gods: Atheism in the Ancient World*, New York: Penguin Random House.

Wilkinson, R. H., 2008, *Egyptian Scarabs*, London: Bloomsbury.

Wittendorff, A., 1994, *Tyge Brahe*, Copenhagen: G.E.C.Gad Fund.

An African itinerary towards intercultural philosophy

A conversation with Wim van Binsbergen

by Pius M. Mosima

Abstract: In this chapter, I explore the itinerary of Wim van Binsbergen from anthropology through intercultural philosophy to comparative mythology and examine his arguments for an intercultural philosophy and the need to take African knowledge systems seriously. I focus on two of his books: *Intercultural Encounters: African and Anthropological Lessons towards a Philosophy of Interculturality* (2003) and *Vicarious Reflections: African explorations in empirically- grounded Intercultural Philosophy* (2016). I argue that van Binsbergen has been very passionate and consistent for the past two decades in his output on the epistemological challenges of interculturality; notably in his explicit rendering of, 1) not only of the pitfalls of North Atlantic, potentially hegemonic (and occasionally racist) knowledge formation about past and present African social and cultural realities; but also, 2) articulating on the deeply emotional and political question as to the validity, global relevance and global applicability of African knowledge systems, and finally, 3) contemplating on the possibility, beyond specific Northern or Southern concerns of getting to intercultural forms of knowledge, beyond local cultural boundaries, in a bid to celebrate our common humanity.

Keywords: Anthropology, Intercultural philosophy, African knowledge systems, *Sangoma*, Afrocentricity, Hegemony, Intercultural encounters, Vicarious reflections

Introduction

In this chapter, I examine the contributions and the path taken by the Dutch Africanist anthropologist / intercultural philosopher, Wim van Binsbergen in his conception of intercultural philosophy in particular and African knowledge systems in general. Is there a specific African knowledge system and how does van Binsbergen negotiate the tension between the local practice and global relevance of African knowledge systems? What specific themes may be discerned from van Binsbergen's treatment of African knowledge systems? To what extent is van Binsbergen's conception of intercultural philosophy a process of boundary production and boundary crossing at the same time? In an attempt to answer these questions and examine his contributions, I explore two very huge (each book is over 600 pages), different but related publications on the subject: *Intercultural Encounters: African and Anthropological Lessons towards a Philosophy of Interculturality* (2003) and *Vicarious Reflections: African explorations in empirically- grounded Intercultural Philosophy* (2016). I argue that van Binsbergen has been very passionate and consistent for the past two decades in his output on the epistemological challenges of interculturality; notably in his explicit rendering of, 1) not only of the pitfalls of North Atlantic, potentially hegemonic (and occasionally racist) knowledge formation about past and present African social and cultural realities; but also, 2) articulating on the deeply emotional and political question as to the validity, global relevance and global applicability of African knowledge systems, and finally, 3) contemplating on the possibility, beyond specific Northern or Southern concerns of getting to intercultural forms of knowledge, beyond local cultural boundaries, in a bid to celebrate our common humanity. Wim van Binsbergen began his career as an anthropologist of religion but is ending as an empirically grounded intercultural philosopher cum comparative mythologist. Throughout his career he has consistently dialogued with several North Atlantic and Asian philosophers and anthropologists on the possibility and requirements of valid transcontinental intercultural knowledge. He argues convincingly that participant observation, humbly, receptively and patiently living the life of the host community, learning its language and culture, remains the most effective and convincing method for such intercultural knowledge construction. This explains his passion throughout his scholarly endeavours to bring Africa, his continent of expertise in addition to the European one in which he was born, in dialogue with other continents in a bid to forge the emergence of a truly comprehensive, interdisciplinary, historicising, global and counter-hegemonic vision of the world.

The limitations of anthropological fieldwork as a mode of constructing intercultural knowledge

In this section, I discuss van Binsbergen's reflections on the potentials and pit-falls of cultural anthropology as a form of intercultural knowledge production. His training and wide-ranging anthropological explorations for close to half a decade have led him through at least seven different African cultural complexes: rural Tunisia, rural Zambia, urban Zambia, rural Guinea-Bissau, urban Botswana, the Bamileke of Cameroonian Grassfields and the Bakweri, especially Bimbia, of the South West Region of Cameroon; Asia, in addition to the European complex into which he was born. Nevertheless, he has been very critical of the project of anthropology in general and Africanist anthropology in particular as he pinpoints the assumptions of inequality on which that model is based.[1] During his first major religious fieldwork in Khumiri society in Tunisia he noticed that he was a long way off from:

> '... problems of power, social change, the interplay between heterogeneous semantic, social and economic systems within one field of interaction, corporeality, self-reflection and interculturality– later to become the predominant themes of my scholarly work – but I nevertheless was starting to feel like an anthropologist (van Binsbergen 2003: 64).

He takes anthropology so seriously that he wishes to go beyond its historical and knowledge / political (Foucault) built–in limitations, as well as the defective general and intellectual education (unusually very unscholarly and rather blinkered) of most anthropologists. He opines that anthropological fieldwork can be a very problematic mode of intercultural knowledge production as it is done by specialists who mostly identify with the North Atlantic region. This is because anthropology depends on manipulated face-to-face relation, personal history, transference, and North-South hegemonic power (van Binsbergen 2003). He asserts that 'anthropology is more than just a sublimated form of sleuthing or espionage' (van Binsbergn 2003: 73), with an ideological nature in the production of intercultural knowledge. He maintains anthropology's complicity with other forms of socio-political oppression such as imperialism and colonization in these words: 'When at home and when out doing fieldwork, North Atlantic anthropologists implicitly share in the privileges and the power

---

[1] For more criticisms of anthropology, see van Binsberen 2003, especially chapters 2 and 4.

of the Northern part of the world, as against the South' (van Binsbergen 2003: 131). In this way, anthropology only perpetrates and perpetuates discourses and ideologies of imperialism as it 'represents a form of intellectual appropriation and humiliation against which Africans in the nationalist era rightly protested' (van Binsbergen 2003: 131). He discusses the works of one of the leading Marxist-inspired anthropologists of the 1970s and 1980s, Pierre-Philippe Rey. Rey's works make special theoretical contributions on the conditions and mechanisms of which an encroaching capitalist mode of production manages or fails to impose itself upon the non-capitalist societies of the Third World like Africa. In van Binsbergen's assessment, Rey's works demonstrate the impact of colonialism and capitalism and he (Rey) does not turn a blind eye on both local and imported forms of exploitation, and how they are interrelated. In spite of his admiration for Rey's stance on anthropology, van Binsbergen finds Rey's approach lacking in several ways. Rey is not self-critical enough and he does not explicitly criticize the ideologies of North Atlantic capitalist modes of production. He reminds us that all contemporary anthropology, even with Marxist or revolutionary versions as Rey articulates, is being produced by intellectual producers whose class position in the world system is based on dependence from capital. This dependence is mediated by the modern state or by large funding agencies, given that individuals cannot get the mammoth funding necessary for academic production today. It would therefore, be sociologically impossible for this capitalist context of intellectual production not to totally determine the nature and contents of the intellectual products of the anthropologists. These limitations of the anthropology of ideology could be redeemed with a liberating ideology, which is primarily a reflexive and up-to-date Marxist anthropology of ideology (van Binsbergen 2003: 91). Earlier on van Binsbergen had published in his first major scholarly book *Religious change in Zambia* (1979), following the Marxist idiom of the time, articulates a theory of the peripheral class struggle, one which exposed the hegemonic assumptions, the subjugating and exploitative relations of production that surrounded the practice of anthropology, which at the time was the dominant form of intercultural knowledge production. For example, when he gets totally immersed in the Nkoya culture during the girls' initiation rite, he discovers the beauty of this ritual and is caught between the dialectics between a Western adult male researcher and Nkoya sexuality. These experiences, he claims, cannot be properly represented by an anthropologist. In *Vicarious Reflections*, especially in chapters 2 and 3, van

Binsbergen continues to discuss the ethical aspects of fieldwork as a stepping stone towards more profound, self-critical and anti-hegemonic approaches to anthropological knowledge constructions. In chapter 3, for example, he demonstrates how the shortcomings of fieldwork experiences urge his denouncing of anthropology to intercultural philosophy. When he becomes a *sangoma* van Binsbergen is able to cross not just disciplinary but also cultural boundaries (van Binsbergen 2003: 155-297). The epistemological, moral and methodological limitations of anthropology in the construction of intercultural knowledge urge van Binsbergen to transcend ethnography to take the trajectory into intercultural philosophy. His conception of intercultural philosophy is the focus of the next section.

Van Binsbergen's conception of intercultural philosophy

Wim van Binsbergen's conception of intercultural philosophy comes from his explorations as an anthropologist and oral historian. It operates at the borderline between anthropology and philosophy; even though occasionally it spills over into *belles lettres*, ancient history, and comparative cultural and religious studies. Intercultural philosophy investigates as its central theme interculturality.[2] It is that branch of philosophy that was explicitly established in order to address the globalisation of difference. This explains the theory of interculturality; using philosophy to critique anthropology and anthropology to critique philosophy. Intercultural philosophy seeks to develop such a discourse that will allow for a discussion of all philosophical problems from an intercultural perspective. It does this by a theoretical reflection on concepts like culture, cultural diversity, cultural relativism, multiculturality, power, hermeneutics and dialogue. Intercultural philosophy, with the use of such concepts listed above, critically explores the conditions under which we could talk of interculturality. Intercultural philosophy seeks to prevent any philosophical position from assuming an absolute position. The central idea that runs across the views of

---

[2] *Cf.* Mall 1995; Mall & Lohmar 1993; Kimmerle & Wimmer 1997; van Binsbergen 2003. I have benefitted from discussions and translations of the works of Kimmerle and Mall from van Binsbergen 1999, 2003, 2016.

authors like Kimmerle, Mall, Lohmar and Wimmer, as far as intercultural philosophy is concerned, is that there are many philosophical traditions of significance in all regions of the world, rather than just a few or one. They posit that the meeting of different cultural orientations and philosophical tendencies calls for an intensive and qualified discourse on the part of all concerned.

Hence, in its general sense, such 'intercultural philosophy' as conceived by Mall, of an earlier vintage than van Binsbergen, investigates the conditions under which an exchange can take between two or more different 'cultures', especially an exchange under such aspects as knowledge production of one culture about another; tolerance or intolerance; conflict or co-operation in the economic, social and political domain.[3] Mall's approach is similar to a comparative approach philosophy, (which is tantamount to studying the similarities and differences between different 'cultures'), and this could foster relativism in philosophy. In this case, there would be no mutual exchange and enrichment, as it could not really help the self-understanding and practice of philosophy itself. I submit that intercultural philosophy is not specifically devoted to the comparison of the world's major philosophical traditions (African, Chinese, Indian, European, Jewish, and Islamic). Instead, it envisages an abstract and formal, rather than substantive, investigation for interculturality. Doing intercultural philosophy would entail putting into contrast, rather than merely comparing, different philosophical traditions. The Chinese intercultural philosopher, Vincent Shen (2010) understands 'contrast' as the rhythmic interplay between difference and complementarity, discontinuity and continuity, which pave the way for real mutual enrichment between the different traditions of philosophy.

We must admit that the term intercultural philosophy existed before van Binsbergen succeeded to Heinz Kimmerle's chair of that designation; however, he re-defined the concept. In his attempt to re-define intercultural philosophy in a more specific form, van Binsbergen does not endorse mere philosophical pluralism. He goes further and investigates, specifically, how the philosophical traditions relate with one another, how it is possible (or impossible) for them to create valid knowledge about one another, and about the life worlds that each

---

[3] *Cf.* Mall 1995, ch.1; Mall 1993.

of these philosophical traditions builds for their adherents.

Van Binsbergen asserts that:

> 'In a more specific form of the above we would conceive of intercultural philosophy as the search for a philosophical intermediate position where specialist philosophical thought seeks to escape from its presumed determination by any specific distinct 'culture'...rendering explicit the traditions of thought peculiar to a number of cultures and subsequently exploring the possibilities of cross-fertilization between these traditions of thought' (van Binsbergen 2003: 468-469).

Summarily, van Binsbergen proposes a radically dialogical version of intercultural philosophy which differs from the more static approach of comparative philosophy. He argues that 'cultures' (plural) do not exist as he takes into account the performative, interconnectedness and oneness of humanity.[4] Performativity implies that culture is not fixed, but flexible and based on changing experiences or contextual considerations. In times of global interaction, van Binsbergen prefers to speak of 'cultural orientations' and not 'cultures'. He prefers the former because it incorporates the overlapping and the dynamics of space and time in these patterns of collective programming. Thus, I think if we follow this proposal, philosophy will go beyond many boundaries simultaneously: first, it will leave behind the Western normative idea that 'real' philosophy consists of abstract thought and should be practiced only by professional philosophers; second, it will move beyond the idea that local wisdom is contained within fixed cultures (but rather is all the time anew performed, while cultures develop and interact with their context); and third it will move towards the most uncommon idea that philosophy cannot just be detected or unearthed in human practices (*e.g.* of justice, of mythological storytelling, or of healing) – but that these practices themselves *are* philosophical. Philosophy cannot be identified with reason, but is love of wisdom, be it present in abstract thought, in healing practices, or in therapeutic storytelling. Interestingly enough, all this is motivated by a commitment which reminds one of the pragmatism of William James, understanding philosophy as a way of dealing with shared human challenges of survival, and inviting into it therefore practical wisdom from all kinds of venues.[5]

---

[4] See van Binsbergen 2003, especially chapter 15.

[5] I am very grateful to Angela Roothaan for this argument.

Another major theme van Binsbergen addresses is the nature of intercultural knowledge and under what conditions reliable, valid and relevant intercultural knowledge be produced across (what is commonly taken to be) cultural boundaries. His discussion of the theoretical problems raised by globalisation and the possibility of an intercultural hermeneutics as proposed by the leading intercultural philosopher in Germany, Ram Adhar Mall proves very inspiring. Van Binsbergen identifies globalisation as a central problem for intercultural philosophy and conceptualises it under the following aspects: proto-globalization, the panic of space, the panic of time, the panic of language, rebellion against older inequalities, the virtualization of experience, the new inequality, and the new body (van Binsbergen (2003: 377-381). In chapter I of *Vicarious Reflections*, van Binsbergen focuses on virtuality as a key concept in the study of globalisation. He explores the concept of virtuality, by entering into a critical discussion with some of his closest friends and colleagues like Peter Geschiere, Rene Devisch and supervisor Matthew Schoffeleers (1928-2011), and applies it in different African situations, from village to town and from female puberty rites to witchcraft, healing and ethnic festivals (van Binsbergen 2016: 85-168). Mall (1995) conceives of distinct 'cultures' existing side-by-side, rejects the idea of one universal world philosophy and also insists on the need for any comparative philosophy to be impartial. In this way he argues that in post-modern hermeneutics, no tradition, place or language should be privileged as that could trigger the dangers of absolutization. It is an intercultural philosophy, *a la* Mall, that is 'placelessly localizing and localizingly placeless', an open hermeneutics where one tolerantly acknowledges that the Other differs from that which one considers one's own (Mall 1995: 99). However, for van Binsbegen, the 'placelessly localizing' character of intercultural philosophy defended by Mall tends to conceal the fact that localization undeniably takes place in this hermeneutic process. Localization does not necessarily take the form of any geographical domain the size of a language region or a nation-state; the philosophical interpreter, with the use of specialist language, explicitly constructs this kind of localization. Moreover, such intercultural hermeneutics would be language-based and this has a major shortcoming in that it gives the philosophers a privileged position in terms of intercultural hermeneutics and communication. Even though language has a clear structuring potential, it does not finally and completely establish the cultural domain, nor the entire limits of human cognition. In fact, van Binsbergen points to the violent, divisive role language plays as a tool when used as a tool for thinking interculturality and proposes encounters where we meet the genuine fusion of publicly constructed identities on a global scale. Here we are thinking of celebrating the human body which is a transcultural common given, bodily contact, songs, dances, rituals, and rhythms and other forms of intercultural modesty such as silence, love, empathy, and introspection (van Binsbergen 2008).

This leads us to another form of intercultural communication which is the use of the Information and Communication Technologies (ICT).[6] With a close look at Africa, he seeks to examine the place of ICT in Africa in a world increasingly dominated by ICT. Van Binsbegen asks some interesting questions about ICT in Africa: Does electronic ICT in Africa lead to creative and liberating cultural appropriation by Africans? Does it lead to the annihilation of the African cultural heritage? Or do both propositions apply somehow? Is the computer in Africa to be taken for granted or does it remain an alien element? He confronts African thinkers such as Kwame Gyekye (1997) and Ali Mazrui (1977, 1978) who argued that the ICT and African culture are incompatible. He thinks that ICT is just as much and as little owned by Africans as by any other collectivity in the contemporary world. He shows, with empirical examples, how ICT is a metalocal global culture in an African context and how African enculturation of ICT is taking place. This enculturation could be quite beneficial in the liberation of Africa.

Intercultural philosophy permits him to go beyond the North Atlantic borders and the particularising tendencies in non-Western cultural orientations. It rather unmasks concepts like 'Africa', 'Europe' and shows how such concepts engender a certain geopolitical hegemonic agenda that often leads to exclusion, hegemony and exploitation. It tries to investigate the conditions of possibility of a proper form of interaction between different cultural orientations. This is what van Binsbergen tries to do when he discards fieldwork in southern Africa and is personally transformed into a *sangoma*. In Southern Africa, *sangoma* is a term for a diviner priest in the tradition of the Nguni-speaking peoples (Zulu, Xhosa, Ndebele, and Swazi). Characteristically, they are specialists in the dynamics of collective healing at the level of the kin group and local community

---

[6] See van Binsbergen 2003, chapter 13.

in Southern Africa.[7]*Sangomas,* according to van Binsbergen:

> "...are people who consider themselves, and who are considered by their extended so-
> cial environment, as effective healers: as mediators between living people, on the one
> hand, and the ancestors, spirits and God (Mwali) on the other – in a general context
> where most bodily afflictions and other misfortunes of a psychological, social and eco-
> nomic nature, are interpreted in religious terms" (van Binsbergen 2003: 202).

Becoming a *sangoma*: An experiment in intercultural philosophy

In his section I discuss the implications of *sangomahood* on the personal and professional life of van Binsbergen. He and the wife Patricia had been sick and had gone through numerous trials and tribulations, and so he joined them in search of emotional and psychological therapy (van Binsbergen 2003: 2013). He asserts that

> 'From an ancestor-less piece of flotsam of human history, I became a priest in an ances-
> tral cult, in a decisive step not only of professional independence and Africanist explo-
> ration, but also of self- construction" (van Binsbergen 2003: 193).

He acknowledges that becoming a *sangoma,*

> "amounts to the most drastic rebuilding of my personality" (van Binsbergen 2003: 192).

This process goes beyond just getting ethnographic data as some anthropologists of religion had done and acknowledged but also one of crossing disciplinary and cultural boundaries.[8] This has intercultural implications in his personal and professional life

> "I was seeking existential transformation, fulfilment and redress, much more than an-
> thropological data, across cultural, geographical and boundaries" (van Binsbergen
> 2003: 171).

---

[7] According to van Binsbergen (1991, 2003, 2007), *sangomahood* is a source of valid knowledge and is translatable to a global format.

[8] Yet, some other anthropologists of religion have considered such a move simply as a form of adaptation to specific, hard field conditions in a bid to get valid information. Even when they got initiated into a local ritual status, their experiences have been presented, even by them-selves, as simple strategies of adaptation in the field. We may cite the cases of Devisch, Fidaali, Jaulin and Schoffeleers.

As a *sangoma* diviner, van Binsbergen intimates that *sangoma* divination is meant to release victims from psychic and existential schizophrenia. He opines that:

> The aim of sangoma divination is primarily therapeutic: to reinsert the client in what may be argued to be her or his proper place in the universe, so that the life force in principle available for that person but temporarily blocked by their drifting away from the proper place, can flow once more(van Binsbergen 2003: 256).

What is interesting is that van Binsbergen is able to translate *sangomahood* and its divinatory practices beyond the African context. As a *sangoma*, he is able to extend an African idiom of spiritual diagnosis, signification, and therapy to clients both in Africa and worldwide. He argues that it produces valid knowledge, comparable to North Atlantic science, beyond the ethnographic premises of the Southern African context. Moreover, he has been able to translate *sangoma* divination into a modern setting by making use of the information and communication technologies, rather than just throwing the oracular tablets on the floor. The divinatory sessions take a global format as he began to be approached by e-mail. He says:

> 'I began to be approached via e-mail by distant prospective clients, whom I could refer to my web pages for background information. Thus gradually a global practice emerged, where I would no longer meet my patients in person, but they – invariably total strangers to me – would contact me via an electronic intake form on my website, and they would subsequently receive via e-mail the outcome of the session I would conduct in their absence. I came to prefer this globalised format, especially because its communication is exclusively taking place via e-mail. In this way I could fit the sessions and the correspondence much more easily into my tight timetable, and I was relieved from nearly unbearable pressure' (van Binsbergen 2003: 236).

Another instructive and interesting turn in van Binsbergen's career is his knowledge of many other forms of divinatory systems (van Binsbergen 2003, 2016).[9] His capacity in dealing with diverse forms of communication, orders of knowledge and healing in varying intercultural contexts, portray that he is most likely to be a better *sangoma* than many of his co-practitioners in Africa and beyond. Moreover, he is able to draw examples from many different local linguistic and cultural settings in Africa, *i.e.* some of the fundamental techniques

---

[9] His knowledge, as Sanya Osha interprets, of *I Ching*, the New Age intellectual movement, Arabic geomantic divination, runic, divination, tarot, the Zulu bones oracle, *Ifa* divination, astrology and Native American varieties are very likely to enrich his practice of *sangomahood* (Osha 2005) and also to dilute it perhaps, and make it less authentically Southern African.

of sociability and anti-sociability around which African community life is organised namely; reconciliation and witchcraft, and tries to compare it with North Atlantic ones.[10] In this way it is both an attempt and at *emic* and *etic* rendering of such concepts as the African village, cosmology, community, virtuality, globalisation, kinship, class and class conflict.[11]

However, van Binsbergen exhorts us to be prudent when taking philosophical concepts from the ivory tower and applying them to the rich empirical reality of African societies as this could bring in several misinterpretations. He does this in his re-reading of the contents, format and societal locus of *ubuntu / hunhu* philosophy as conceived by southern African academic philosophers, managers and politicians.[12] Ramose, for example, in his articulation of *ubuntu* philosophy of reconciliation and social healing, thinks that it is an unadulterated form of African social life before European conquest. This implies that globalisation, is an outside phenomenon which is driven by North Atlantic conquest, and has resulted in the destruction of *ubuntu*-based communities. Hence, *ubuntu* could be seen as a way to revive African social life and remedy the corrosive effects of colonialism. Even though van Binsbergen sees the potentials of *ubuntu*, he also warns that both contemporary southern African life and *ubuntu* itself are products of globalisation. He sees *ubuntu* as a contemporary academic construct, called forth by the same forces of physical oppression, economic oppression, and cultural alienation that have shaped southern African society over the past two centuries. In this way *ubuntu* can mask real conflict, perpetuate resentment and hide the fact that someone is using *ubuntu* for the pursuit of individual

---

[10] However, see his reflections on the production of knowledge and the circulation of ignorance under the specific politics of intercultural knowledge production obtaining in the same part of Africa in the context of international development intervention (van Binsbergen 2003: 333-347).

[11] We may cite van Binsbergen's inspiration from Kant's theory of *sensus communis* in his discussion of symbolic production as affirmation of either universality or particularist conceptions of community, based on ethnographic research among the Nkoya (van Binsbergen 2003: 317-332). In the present volume, his Indonesian PhD student Dr Stephanus Djunatan takes up his former supervisors's argument and brings it to a new and illuminating application.

[12] *Ubuntu / hunhu means* 'being human, humanity, the act of being human'. For more on *ubuntu* philosophy, see for example, Samkange & Samkange 1980; see also: Bhengu 1996; Khanyile 1990; Khoza 1994; Makhudu 1993; Mbigi 1992, 1996; Mbigi & Maree 1995; Prinsloo 1996, 1998; Ramose 1999, van Binsbergen 2003, especially chapter 14. However, van Binsbergen's judgment of *ubuntu* as a major philosophical achievement has generated some controversy. For more on this debate, see, for example, Bewaji & Ramose 2003.

gain (van Binsbergen 2003: 450-454).

Moreover, *sangomahood* permits van Binsbergen to make some interesting counter-hegemonic claims, as he questions the superiority of North Atlantic science and philosophy vis-à-vis- other traditions of scholarship. This burning zeal could be noticed when he took over as Editor of *Quest: An African Journal of Philosophy / Revue Africaine de Philosophie* in 2002 when he states, concerning the various contributions in the *Journal*, inter alia,

> '...my aim is to bring out their potential to contribute to what, through, major debates featuring some of the great names in African philosophy, have been the leading themes in Quest over the years:
>
>> *the reflection of the philosophical canon, both in the North Atlantic and in Africa (with possible extensions towards the world's other philosophical traditions, in Islam, Judaism, India, China, the New World, Oceania, etc);*
>>
>> *the conceptual and theoretical effort to develop African philosophy into a tool that illuminates, by comparison and contrast, current socio-political developments on the African continent;*
>>
>> *the critical reflection on the North-Atlantic-dominated, hegemonic context in which African knowledge production takes place today, and the formulation of radical anti-hegemonic alternatives; and finally*
>>
>> *the exploration of the possibilities for an intercultural production of knowledge, that while affirming its specific (e.g. African) roots in space and time, yet situates itself in a field of tension between the universal and the particular.*[13]

In *Intercultural Encounters*, van Binsbergen indicates some of the directions, beyond hegemony, where dynamic alternative logics may be found. In *Vicarious Reflections,* he is much more explicit in exploring the extent to which an intercultural philosophical perspective may enrich the existing social-science approaches to religion. He argues that such an approach could act as a way out from the usually implicit hegemonism that assumes that one's own perspective is, self-evidently, the most central, obvious and truth-producing one (van Binsbergen 2016: 56). The focus of Part II of the book is on *Religious hegemony and*

---

[13] See van Binsbergen 2002: 238-271, in the special issue: van Binsbergen, W.M.J., 2004, 'Postscript: Aristotle in Africa-Towards a comparative Africanist reading of the South African Truth and Reconciliation Commission', in: Philippe- Joseph Salazar, Sanya Osha, & Wim van Binsbergen, 2002, eds., *Truth in Politics: Rhetorical Approaches to Democratic Deliberation in Africa and beyond*, special issue of: *Quest: An African Journal of Philosophy / Revue Africaine de Philosophie*, XVI, pp.238-272; reprinted in van Binsbergen 2016: 289-320.

some of its remedies; and he explores the transcultural study of evil (chapter 4), the interpretation of violent ideologies in the context of today's militant Islam and the commensurately violent reactions it has met from the West (chapter 5); the promise of a viable intercultural theoretical approach to religion to be derived from the work of the greatest French post-structuralist philosopher, the late lamented Jacques Derrida (chapter 6); and an application of the Derridean approach in search of an African spirituality (chapter 7) and an affirmation of the politics of sociability as a constitutive element of African spirituality today (chapter 8).

In Part III, van Binsbergen continues to formulate his radical criticism following the hegemonic context in which knowledge production takes place with the very evocative title; 'Names-Dropping: How not to crush Africa under North Atlantic Thought?'. He argues in chapter 9 that the use of Aristotle's 'rhetorical' approach to South African Truth and Reconciliation Commission as an attempt through which Post-Apartheid South Africa sought to come to terms with its conflictuous recent history was an attempt to crush Africa and African thought. Or, better still, his discussion of the late French philosopher Felix Guattari eclectic scientism (chapter 10). This is because Africa has its own indigenous models of reconciliation (van Binsbergen 2003: 349-376). Yet, he comes to terms with the late Cameroonian philosopher (and his critic Jean Bertrand Amougou), Meinrad Hebga, who does not fall prey to North Atlantic hegemony. Van Binsbergen argues that it would be wrong to project alien subordinating perspectives onto Africa as if there are in that continent no local African traditions of worldview and philosophy available to make sense of African situations in the first place (chapter 11).

Part IV has a rather pathetic title: *Beyond Africa: The price of universalism*, with one very important chapter (chapter 12), in which van Binsbergen discusses the work of his friend, the great African classicist, Romance languages specialist and philosopher, Valentin Yves Mudimbe from former Zaire, today the Democratic Republic of Congo. In this chapter, van Binsbergen discusses issues of hegemony and the liberation of African difference, the 'Colonial Library' (Mudimbe's aggregate term for the accumulated textual records of 'North' colonial knowledges – inevitably warped- concerning 'the South'), and Mudimbe's rejection of African religion and Afrocentricity as potential sources of empowerment and pride, even though these two achievements hcan be seen to be vocally articulated in the self construction of African pride, not to say (in Mudimbe's own terms): of African difference. He judges that the path Mudumbe has taken is to exile himself away from Africa, not just physically, but also existentially. In this way, Mudimbe has tended to radically reject the specificities of the African historical, political, cultural and spiritual heritage, in favour of a placeless, homeless universalism that cherishes its (largely North Atlantic and ancient , and only sporadically African) classics and their epiphanies in the form of texts. The price to be paid for such universalism, van Binsbergen asserts, can be cultural and spiritual self denial, not as a time-honoured spiritual virtue, but as a form of self-destruction. However, do we as Africans have a *choice* in the face of such hegemony, such individual death and of the reasonable doubt cast on the independent ontological existence, life, or gods, ancestors and other spirits? This answer to this question, according to van Binsbergen, lies in making a choice which he calls the *collective construction of enduring, self-reflexive culture* (van Binsbergen 2016: 57ff). For van Binsbergen, Mudimbe's inability to affirm his Africanity seems to come from his education within the folds of Christianity, and this scholarship has induced him to deny the intrinsic value, except as metaphors and poetry, of African knowledge systems.[14] He (van Binsbergen) proposes the rehabilitation of African knowlwdge systems, not as the untutored and incoherent stammering of 'savages' waiting to be finally rescued, that is enlightened by initiatives from other continents, but as regional provinces of meaning and truth that have never been totally isolated from the flows of knowledge and truths in other continents, but that yet, more importantly, constitute their own indispensable contributions to the sum total of human knowledge- irreplaceable, immensely valuable, and in principle capable of global circulation and global relevance.[15]

But what is inside African knowledge systems? Are they worth studying given the onslaught of globalisation and the influence of North Atlantic hegemony?

---

[14] Elsewhere (Mosima 2016), I have illustrated this homelessness even in Mudimbe's spiritual life. Even in an attempt to affirm the existence of an African philosophy, Mudimbe together with another Cosmopolitan African philosopher Kwame Appiah, are rather hesitant in such an affirmation.

[15] See van Binsbergen's discussion of *sangomahood*, the validity of knowledge it produces and its transcultural relevance (van Binsbergen 2003, especially chapters 5, 6, and 7). See also van Binsbergen 2016, especially chapters 13, 14, 15 and 16. He argues that these unique, local African knowledge systems deserve global circulation.

These are the questions van Binsbergen attempts to answer in Part V with the triumphant title: *Inside African Knowledge Systems.*[16] He argues, in his discussion of the path-breaking feminist philosopher of science, Sandra Harding, for what underpins scientific knowledge systems, including the North Atlantic ones with their claims to universality, objectivity and rationality. Is it epistemological superiority, as North Atlantic chauvinists would claim or sheer hegemonic power, privileging one set of knowledge over others? Sandra Harding (1994, 1997), for example, argues that modern sciences are all multi cultural, ethnosciences, which need to complement other knowledge systems around the globe. This permits van Binsbergen (chapter 13) to ask one pertinent question: Is it hegemonic power that keeps the airplane in the air above the North Atlantic region, but makes it crash when it begins to fly over lands where the West holds not effective sway? The answer to this rhetorical question lies in the fact that it cannot just be hegemonic power and nothing more, for the plane does not crash at all, but simply flies on beyond the geographical frontier of its ideological and military support. Within this epistemological context, van Binsbergen (chapters 14 and 15), using long-range historical perspective rather a recognisable intercultural-philosophical one, looks at divination and board-games as formal systems reflecting upon space and time. He attempts to ask if there is any truth in African divination, in other words, Does African divination 'work', and if so, how is it possible?[17] Finally, van Binsbergen focuses on a concept that is much relevant in the discussion of interculturality and of African knowledge systems; wisdom (chapter 16). He argues that wisdom should not be conceived as a static receptacle of seemingly lasting and stable unequivocal truths of the elderly, but as a dynamic method to articulate, reconcile, negotiate and salvage contradictory truths such as are at the heart of any society, and such as, a fortiori, inevitably arise in any intercultural encounter. He asserts that truth only exists and is constructed in a context (especially ritual and symbolic) and his forms the practices of that community. Consequently, a community's truth, cannot subjectively, be beyond that community, given that it is seen as justified and true, based on the internal standards defined within the

---

[16] This title reminds me of the various inspirational contributions in Devisch & Nyamnjoh (2011) to which van Binsbergen contributed.

[17] Much of the affirmative answer to this question will be found in his forthcoming *Sangoma Science*.

local and contemporary social horizon. Hence, van Binsbergen opines that if North Atlantic knowledge systems find those of Africa fundamentally untrue and incredible, this is not in itself an objective sign of the intrinsic invalidity of African knowledge systems, but merely the result of social truth construction within communal boundaries. This implies that only what the West constructs as true, can be true in the West. However, it is wisdom's boundary crossing, in the context of globalisation, which could solve the problem of intolerance to alternative truth claims. Here, interculturality permits us not to apply rigid, strict, procedural logic but depend on other forms of such as bodily contacts, dances, rituals, songs, and other forms of epistemological modesty such as silence, love, introspection and love.

Against Euro/(Afro)/centrism: Long-range correspondences in space and time

In his conception of intercultural philosophy van Binsbergen has been concerned with the fundamental unity of humankind. Even though he is a declared and recognised Afrocentrist,[18] his conception of intercultural philosophy is one that allows for a non-essentialist perception of intercultural relations. This permits him to deconstruct the absolute and essentialist differentiations that are brought to the global reality with concepts like 'Africa', 'Orient', and 'Europe' and rebut the racialist variant of Afrocentrism. He thinks it is weird to talk of quintessential African values that separate them from the rest of humanity. In his contribution to the *Black Athena* debate initiated by Martin Bernal, van Binsbergen argues that Bernal does not take into consideration the mutual interpenetration of Ancient Egyptian and sub-Sahara-Africa, in the way of concepts and structures of thought, myths, symbolism, productive practices and state formation (van Binsbergen 1997, 2011, 2012). Moreover, by the end of the Fourth Millennium before the Common Era, Ancient Egypt owed its emergence as a civilization (contrary to what Bernal thinks to be the case), to the interaction between Black African and Eastern Mediterranean / West Asian cultural orientations. Van Binsbergen, for example, rather draws on heterogeneous and

---

[18] See Amselle 2001: 109; Obenga 2001: 23; van Binsbergen 2000a, 2000b.

fragmented 'Pelasgian' continuities that could hardly be relegated to a primal and exclusive African origin. He contends that the Aegean region looks similar to Ancient Egypt, not primarily because of diffusion from Egypt in the Late Bronze Age, but primarily because both were the recipients of the 'Pelasgian' demic, linguistic and cultural movement from West (ultimately Central) Asia. Subsequently, this movement also extended to sub-Saharan Africa, producing the same similarities there (van Binsbergen 2011: 327ff.). Moreover, his trans-continental study and practice of divinatory systems point to the undeniable empirical reality of massive cultural continuities in space and time and the fundamental unity of humankind. They also invite us to stress the transconti-nental complementarity of the intellectual achievements of us, Anatomically Modern Humans, in the course of millennia. This implies that cultural orienta-tions have been overlapping and so we cannot afford to overlook the history of cross-cultural influences and mutual exchanges between sub-Saharan African and East Mediterranean / West Asian cultural orientations. This invalidates the use of homogenous categories like Europe, Africa, and Asia and gives us a fluid conception of culture; one that is not bounded, not tied to a place, not unique but multiple, and easy to combine, blend and transcend.

Conclusion

Throughout this chapter, I have tried to explore the itinerary and contributions of Wim van Binsbergen in his conception of intercultural philosophy in particu-lar and African knowledge systems in general. I think he has been able to go against the grain by challenging North Atlantic universalising rationality, the hegemonic context in which knowledge production takes place and bring to the fore Africa's contributions in the domain of intercultural philosophical studies. His efforts at patiently studying African knowledge systems, showing their global relevance and applicability in a bid to cross-fertilise them with other global available philosophical traditions point to his transcontinental career. He has persistently demonstrated the hallmarks of a courageous, creative and interdisciplinary scholar who has explored and grappled with data of over-whelming complexity and heterogeneity. His transformation from an anthro-pologist to an intercultural philosopher and subsequently as comparative mythologist has enriched his knowledge of the life –world of his hosts, reshaped his initial Eurocentric mindset and given him a more modest argument for thinking interculturality. According to van Binsbergen (2003: 37), if we have to make any progress in intercultural philosophy, we must discard the following illusory assumptions:

- that 'cultures' (plural) do exist;

- that identities proclaimed within the public arenas of the contemporary multicultural society are authentic and free from performativity;

- that everyone has one and only one culture, and is inevitably tied to that one;

- that it is meaningful to speak of intercultural philosophy in the sense of comparing the ways various philosophical traditions of the world have dealt with perennial themes (such as the nature of world, the person, morality, time, force, life) without first investigating the conditions and the distortions of *interculturality* as such;

- that philosophy can afford to ignore empirical evidence as produced by the social and natural sciences;

- that Africa is a patchwork quilt of discrete, localised, bounded cultures;

- that ethnography is ipso faco a valid form of intercultural knowledge production;

- that intercultural philosophy as it has been pioneered over the past decade has already substantiated its claims of constituting a valid form of intercultural knowledge;

- that in cultural analysis we can afford to ignore comprehensive long-range correspondences in space and time, across millennia and across thousands of miles;

- that Greece is the cradle of the North Atlantic, and subsequently global, civilisation;

- and that cultural fragmentation is the original condition of humankind instead of a secondary product of historical group interaction.

Yet, on the positive side, van Binsbergen (2003: 39-40) advocates for the need for particular kinds of dynamic logic which acknowledge difference without being entrenched in difference. Such dynamic logics may be found:

- in Derridean deconstruction and *différance*;

- in Lévistraussian *savage thought*, which (whether we intellectuals like it or not) is the inconsistent standard mode of thought of most human beings in most situations world-wide, and also of ourselves unless we are in a specifically marked technical academic mode;

- in dialogue;

- in the mythomaniacal logic of association and projection that a century of psychoanalysis has taught us to recognise and that we should first of all use, not condescendingly and gleefully to detect the transference in *other's* behaviour, but self-critically, to identify such transference in our own construction of knowledge;

- in African techniques of reconciliation through creative and selective hermeneutics;

- in intercultural hermeneutics for which both cultural anthropology and intercultural philosophy, despite all their shortcomings as highlighted in this book, have proposed promising models and methods in the course of the twentieth century CE.

References cited

Amselle, J.-L., 2001, Branchements: Anthropologie de l'universalité des cultures, Paris: Flammarion.

Bewaji, J.A.I., & Ramose, M.B., 2003, 'The Bewaji, van Binsbergen and Ramose Debate on Ubuntu', South African Journal of Philosophy 22,4:378-415.

Bhengu, M.J., 1996, Ubuntu: The essence of democracy, Cape Town: Novalis.

Gyekye, K., 1997, 'Philosophy, culture, and technology in the postcolonial', in: Eze, E.C., ed., Postcolonial African philosophy: A critical reader, Oxford: Blackwell, pp. 25-44.

Devisch, R., & Nyamnjoh, F.B., 2011, eds., The Postcolonial turn: Re-imagining anthropology and Africa, Bamenda (Cameroon) / Leiden (the Netherlands): Langaa / African Studies Centre.

Harding, S., 1986, The science question in feminism, Ithaca NY: Cornell University Press.

Harding, S., 1991, Whose science, whose knowledge?, Ithaca NY: Cornell University Press.

Khanyile, E., 1990, Education, culture and the role of ubuntu (non vidi).

Khoza, R., 1994, African humanism, Ekhaya Promotions: Diepkloof Extension SA.

Kimmerle, H., & Wimmer, F.M., (eds.), 1997, Philosophy and democracy in intercultural per-spective, Amsterdam / Atlanta: Rodopi.

Makhudu, N., 1993, 'Cultivating a climate of co-operation through ubuntu', Enterprise, 68: 40-41.

Mall, R.A., 1993'Begriff, Inhalt, Methode und Hermeneutik der interkulturellen Philosophie', in: Mall, R. A., & Lohmar, D., eds., Philosophische Grundlagen der Interkulturalitat, Amster-dam/Atlanta: Rodopi, pp.1-27.

Mall, R. A., 1995, Philosophie im Vergleich der Kulturen: InterkulturellePhilosophie, eine neue Orientierung, Darmstadt: Wissenschaftliche Buchgesellschaft.

Mall, R.A., 2000, Intercultural Philosophy, Lanham: Rowan & Littlefield.

Mall, R.A., & Lohmar, D., 1993, eds., Philosophische Grundlagen der Interkulturalität, Amster-dam/Atlanta: Rodopi.

Mazrui, A.A., 1977, 'Development equals modernization minus dependency: A computer equa-tion,' in: Taylor, D.R.F., & Obudho, R.A., eds., The computer and Africa: Applications, prob-lems, and potential, New York: Praeger, pp. 279-304.

Mazrui, A.A., 1978, 'The African computer as an international agent', in: Mazrui, A.A., Political values and the educated class in Africa, London: Heinemann, pp. 320-342.

Mbigi, L., 1992, 'Unhu or Ubuntu: The basis for effective HR management', Peoples Dynamics, October, pp. 20-26.

Mbigi, L., 1996, Ubuntu: The African dream in management, second impression, Randburg South Africa: Knowledge Resources.

Mbigi, L., & Maree, J., 1995, Ubuntu: The spirit of African transformation management, Rand-burg South Africa: Knowledge Resources.

Mosima, Pius Maija, 2016, 'Philosophic sagacity and intercultural philosophy: Beyond Henry Odera Oruka', PhD thesis, Tilburg University.

Obenga, T., 2001, Le sens de la lutte contre l'africanisme eurocentriste, Paris: L'Harmattan/Gif-sur-Yvette: Khepera.

Osha, S., 2005, Review article: 'The Frontier of Interculturality. A review of Wim van Binsber-gen's Intercultural Encounters: African and Anthropological Lessons towards a Philosophy of Interculturality' (2003), Africa Development, 30, 1-2, 2005: 239-250.

Prinsloo, E.D., 1996, 'The Ubuntu style of participatory management', in: Malherbe, J.G., ed., Decolonizing the mind: Proceedings of the 2nd Colloquium on African philosophy held at the University of South Africa, October 1995, Pretoria: Unisa, pp. 112-122.

Ramose, M.B., 1999, African philosophy through ubuntu, Avondele Harare: Mond.

Samkange, S., & Samkange, T.M., 1980, Hunhuism or ubuntuism: A Zimbabwe indigenous political philosophy, Salisbury [Harare]: Graham.

Shen, V., 2010, Intercultural Philosophy, Confucianism and Taoism, Haarlem: Papers in Inter-cultural Philosophy-Transcontinental Comparative Studies, No.1.

van Binsbergen, W.M.J., 1981, Religious Change in Zambia: Exploratory Studies, London:

Routledge &Kegan Paul.

van Binsbergen, W.M.J., 1991, 'Becoming a sangoma: Religious anthropological fieldwork in Francistown, Botswana', Journal of Religion in Africa, 21, 4:309-344.

van Binsbergen, W.M.J., 1997, ed., Black Athena: Ten Years After, Hoofddorp: Dutch Archaeological and Historical Society, special issue, Talanta: Proceedingsof the Dutch Archaeological Historical Society, vols 28-29, 1996-1997.

van Binsbergen, W.M.J., 1999, 'Culturen bestaan niet': Het onderzoek van interculturaliteit als een openbrekenvan vanzelfsprekendheden, Erasmus University Rotterdam, Rotterdam: Rotterdamse Filosofische Studies, inaugural address; English version in: van Binsbergen, Wim M.J.,1999b, 'Cultures do not exist', Exploding self-evidences in the investigation of interculturality', Quest, An AfricanJournal of Philosophy, 13, 1-2, Special Issue: Language & Culture, pp. 37-114, also at: http://www.quest-journal.net/Quest 1999 PDF articles/Quest_13_vanbinsbergen.pdf ; also incorporated as ch. 15 in: van Binsbergen, Wim M.J., 2003, Intercultural encounters: African and anthropological towards a philosophy of interculturality, Berlin / Boston / Muenster: LIT, pp. 459-522.

van Binsbergen, W.M.J., 2000a, 'Dans le troisième millénaire avec Black Athena?', in: Fauvelle-Aymar, F.-X., Chrétien, J.-P., & Perrot, C.-H., Afrocentrismes: L'histoire des Africains entre Égypte et Amérique, Paris: Karthala, pp. 127-150.

van Binsbergen, W.M.J., 2000b, 'Le point de vue de Wim van Binsbergen', in: 'Autour d'un livre. Afrocentrisme, de Stephen Howe, et Afrocentrismes: L'histoire des Africains entre Egypte et Amérique, de Jean-Pierre chrétien [sic], François-Xavier Fauvelle-Aymar et Claude-Hélène Perrot (dir.), par Mohamed Mbodj, Jean Copans et Wim van Binsbergen', Politique africaine, no. 79, Octobre 2000, pp. 175-180.

van Binsbergen, W.M.J., 2007, 'The underpinning of scientific knowledge systems: Epistemology or hegemonic power? The implications of Sandra Harding's critique of North Atlantic science for the appreciation of African knowledge systems' in : Hountondji, Paulin J., ed.,La rationalité, une ou plurielle, Dakar: CODESRIA [Conseil pour le développement de la recherche en sciences sociales en Afrique]/ UNESCO [Organisation des Nations Unies pour l'éducation, la science et la culture], pp. 294-327.

van Binsbergen, W.M.J., 2008, 'Traditional wisdom-its expressions and representations in Africa and beyond: Exploring intercultural epistemology', in: Quest: An African Journal of Philosophy/Revue Africaine de Philosophie, Vol. XXII, No. 1-2, pp.49-120.

van Binsbergen, W.M.J., 2011, ed., Black Athena comes of age: Towards a constructive reassessment, Berlin-Muenster-Wien-Zurich-London: Lit.

van Binsbergen, W.M.J., 2012,Before the Presocratics: Cyclicity, transformation, and element cosmology: The case of transcontinental pre- or protohistoric cosmological substrates linking Africa, Eurasia and North America, special issue, Quest: An African Journal of Philosophy/Revue Africaine de Philosophie, Vol. XXIII-XXIV, No. 1-2, 2009-2010, pp.1-398 (effective date of publication April 2013), book version: Haarlem: Shikanda.

van Binsbergen, W.M.J., 2003, Intercultural encounters: African and anthropological lessons towards a philosophy of interculturality, Berlin/Boston/Muenster: LIT.

Notes on contributors

Dr. Pieter BOELE VAN HENSBROEK teaches Philosophy at the University of Groningen, the Netherlands. He is one of the co-founders in 1987, with the late Roni Khul Bwalya at the University of Zambia, of *Quest: An African Journal of Philosophy / Revue Africaine de Philosophie*. He was founder-editor of that Journal till 2001. (p.boele@rug.nl)

Dr. Stephanus DJUNATAN teaches Philosophy at the Parahyangan Catholic University in Bandung, Java, Indonesia. He defended his dissertation titled *The principle of Affirmation, an ontological and epistemological ground of Interculturality* in 2011 under the co-supervision of Professor Wim van Binsbergen and Professor Bambang Sugiharto. He has jsut completed a postdoctoral research on *Epistemological Roots of Violence, an elaboration on violence against Chinese Indonesians* in 2015. Now he teaches Eastern Philosophy which includes Indian and East Asian philosophies; basic Logics, Education of Five Moral Principles (Pancasila), Ethics, Language and Logics. His interest mainly covers philosophy of culture, epistemology and Ethics. Nowadays he focuses his elaboration on philosophical thought of Sundanese people, an Indonesian ethnic group in West Java (.djunatan@unpar.ac.id)

Dr. Frans DOKMAN is senior researcher and director of *Basileia Business & Research*, Portugal. He specializes in 'International Management and Religions'. Prof. Wim van Binsbergen and Dr. Frans Dokman collaborated particularly in organizing, together with Dr. Pius Mosima, a symposium about African philosophy at Nijmegen University, The Netherlands, February 2016 (f.dokman@basileiabusiness.com)

Dr Bongasu Tanla Kishani is emeritus Professor of Philosophy at the Department of Philosophy, Ecole Normale Superieur, University of Yaounde I, Cameroon (bongasutk@yahoo.com).

Dr Dierk Lange is emeritus Professor of African History at the University of Bayreuth, Germany. Having completed his university edutation in France, having lived in Egypt and West-Africa, and blessed with an African wife and African-born children, his has been is a transcontinental career if ever there was one. Equipped by education with far more transcontinental philological expertise than common among Africanists, the main fruit of his mature years has been the pioneering of West Asian strands in West African history. Of this effort, his contribution to the present book is another installment. (info@dierklange.com)

Dr. Emily Lyle is an Honorary Fellow in the Department of Celtic and Scottish Studies, School of Literatures, Languages and Cultures, at the University of Edinburgh. She has had a long-term interest in mythology, and has explored the possibilities for restoring an understanding of the cosmological structures of the Indo-European and related cultures. She has enjoyed many international exchanges through the Traditional Cosmology Society and its journal *Cosmos* and through the International Association for Comparative Mythology, of which she has been one of the founding members and long-standing directors. Her most recent book is *Ten Gods: A New Approach to Defining the Mythological Structures of the Indo-Europeans* (2012)(e.lyle@ed.ac.uk)

Dr. Kazuo Matsumura, born 1953, is professor at the Department of Transcultural Studies, Wako University, Machida, Tokyo, Japan. His research interests include comparative mythology, history of religions, and Japanese mythology. Recent work includes *Mythical Thinkings: What can we learn from Comparative Mythology?*,Countershock Press, 2014; 'Aspects of Heroic Mythology' in Klaus Antoni & David Weiss eds. *Sources of Mythology*, LIT, 2014; and 'Heroic Sword God', *Cosmos* 31(2015) (kmat@wako.ac.jp)

Dr. Pius M. Mosima teaches Philosophy at the University of Bamenda Cameroon and is a Research Fellow at the African Studies Centre Leiden; the Netherlands. His research interests include: African / intercultural Philosophy; Globalization and Culture, Moral and Political Philosophy, Political Sociology and Anthropology, Citizenship and Development Studies, Gender Studies and

Bioethics. He has supervised research, presented papers in many conferences, facilitated in many workshops and given guest lectures in Summer Universities on these topics in many African and European universities. His main publication is *Philosophic Sagacity and Intercultural Philosophy: Beyond Henry Odera Oruka* (2016). (piusmosima@yahoo.com)

Dr. Louise F. MULLER is Lecturer in African Literature at the Department of African Languages and Cultures of Leiden University, the Netherlands. She studied Philosophy, World History and African Studies. Her main publication is *Religion and Chieftaincy in Ghana* (LIT). Keywords: African philosophy, film, history, literature and religion. (l.f.muller@umail.leidenuniv.nl)

Dr. Julie NDAYA TSITEKU est Professeure au Département d'Anthropologie à l'Université de Kinshasa(j.ndaya@gmail.com)

Dr. Sanya OSHA is a research fellow at the DST-NRF CoE in Scientometrics and STI Policy in the Institute for Economic Research in Innovation at Tshwane University ofTechnology, South Africa and fellow of Africa Studies Centre, Leiden, the Netherlands. His previous work has appeared in *Transition, Socialism and Democracy, Research in African Literatures, QUEST: An African Journal of Philosophy, Africa Review of Books* and *The Blackwell Encyclopedia of Twentieth Century Fiction.* He is the author of *Kwasi Wiredu and Beyond: The Text, Writing and Thought in Africa* (2005), *Ken Saro-Wiwa's Shadow: Politics, Nationalism and the Ogoni Protest Movement* (2007), *Postethnophilosophy* (2011) and *African Postcolonial Modernity: Informal Subjectivities and the Democratic Consensus* (2014). He is also a co-editor of *The Africana World: Fragmentation to Unity and Renaissance* (2012) and editor of *The Social Contract in Africa* (2014). (babaosha@yahoo.com)

Dr. Thera RASING is a Dutch anthropologist specialised in religion and gender studies. She was affiliated as researcher to the African Studies Centre (Leiden, The Netherlands) and to the Leiden University Medical Centre (LUMC), Department of Culture, Health and Illness, and was lecturer at the University of Amsterdam. Thereafter, she was appointed as Senior Lecturer and Head of Gender Studies Department at the University of Zambia (UNZA) where she worked for seven years. She was Senior Lecturer at the Catholic University Malawi (Cunima), and part time Lecturer at the University of Lusaka (Zambia), the University of Africa (Zambia) and Justo Mwale University (Zambia). She worked for the Zambian Ministry of Community Development Mother and Child Health (MCDMCH). Currently, she is consultant for several international organisations. Her main publications include: *Passing on the rites of passage: Girls initiation rites in the context of an urban Roman Catholic community on the Zambian Copperbelt* (Avebury 1995); *The Bush Burnt, the Stones Remain: Female initiation rites in urban Zambia* (LIT 2001), and (as co-editor) *Religion and the Challenges of Aids treatment in Africa: Saving Souls, Prolonging Lives.* (Ashgate 2014). (trasingster@gmail.com)

Dr. Sjaak VAN DER GEEST is emeritus Professor of Medical Anthropology at the University of Amsterdam, the Netherlands. He conducted fieldwork in Ghana and Cameroon and published books and articles on marriage and kinship, perceptions and practices concerning birth control, witchcraft beliefs, anthropological field research, Ghanaian Highlife songs, missionaries and anthropologists, anthropology of the night, and various topics in medical anthropology, in particular the cultural context of pharmaceuticals in non-Western communities, hospital ethnography, perceptions of sanitation and waste management, and social and cultural meanings of care and old age in Ghana (s.vandergeest@uva.nl ; also see www.sjaakvandergeest.nl

DR. FRED C. WOUDHUIZEN is well-versed in the interdisciplinary field of Mediterranean protohistory. He wrote his dissertation about the *Ethnicity of the Sea Peoples* (2006), and specialized as a Luwologist, of which fact numerous books and articles in the field of Luwology bear testimony (fredwoudhuizen@gmail.com).

BOOKS / INDEPENDENT PUBLICATIONS BY WIM VAN BINSBERGEN

(literary work not included, but see: http://www.quest-journal.net/shikanda/literary/index.htm)
'with' = equal contribution by co-authors / co-editors; 'with the assistance of' = co-author / co-editor in junior position

AUTHORED BOOKS

1981   *Religious Change in Zambia: Exploratory studies*, Londen / Boston: Kegan Paul International; also as Google Book.

1982   *Dutch anthropology of sub-Saharan Africa in the 1970s*, Research Report no. 16, Leiden: African Studies Centre, 1; also at: http://www.quest-journal.net/shikanda/publications/01PUB0000000592.pdf

1988   *J. Shimunika's Likota lya Bankoya: Nkoya version*, Research report No. 31B, Leiden: African Studies Centre; also at: http://www.quest-journal.net/shikanda/publications/ASC_1239806_002.pdf

1992   *Kazanga: Etniciteit in Afrika tussen staat en traditie*, inaugural lecture, Amsterdam: Vrije Universiteit; also at: http://www.quest-journal.net/shikanda/publications/ASC-1239806-014.pdf

1992   *Tears of Rain: Ethnicity and history in central western Zambia*, London / Boston: Kegan Paul International; also at: http://www.quest-journal.net/shikanda/ethnicity/Tearsweb/pdftears.htm and as Google Book.

1997   *Virtuality as a key concept in the study of globalisation: Aspects of the symbolic transformation of contemporary Africa*, Den Haag: WOTRO [Netherlands Foundation for Scientific Research in the Tropics, a division of NWO [Netherlands Science Foundation], Working papers on Globalisation and the construction of communal identity, 3; also at: http://www.quest-journal.net/shikanda/publications/ASC-1239806-018.pdf

1999   *'Culturen bestaan niet': Het onderzoek van interculturaliteit als een openbreken van vanzelfsprekendheden*, inaugural lecture, Rotterdam: Faculteit der Wijsbegeerte Erasmus Universiteit Rotterdam, Rotterdamse Filosofische Studies XXIV, also at: http://www.quest-journal.net/shikanda/publications/oratie_EUR_na_presentatie.pdf.

2003   *Intercultural encounters: African, anthropological and historical lessons towards a philosophy of interculturality*, Berlin / Hamburg / London: LIT; also at: http://www.quest-journal.net/shikanda/intercultural_encounters/index.htm

2009   *Expressions of traditional wisdom from Africa and beyond: An exploration in intercultural epistemology*, Brussels: Royal Academy of Overseas Sciences / Academie Royale des Sciences d'Outre-mer, Classes des Sciences morales et politiques, Mémoire in-8º, Nouvelle Série, Tome 53, fasc. 4. http://www.quest-journal.net/shikanda/topicalities/wisdom%20as%20published%20ARSOM_BETTER.pdf. .

2011   *Ethnicity in Mediterranean Protohistory*, British Archaeological Reports (BAR) International Series No. 2256, Oxford: Archaeopress; also at: http://www.quest-journal.net/shikanda/topicalities/Ethnicity_MeditProto_ENDVERSION%20def%20LOW%20DPI.pdf (with Woudhuizen, Fred. C.)

2012   *Before the Presocratics: Cyclicity, transformation, and element cosmology: The case of transcontinental pre- or protohistoric cosmological substrates linking Africa, Eurasia and North America*, special issue, *QUEST: An African Journal of Philosophy/Revue Africaine de Philosophie*, Vol. XXIII-XXIV, No. 1-2, 2009-2010; book version: Haarlem: Papers in Intercultural Studies and Transcontinental Comparative Studies; also at: http://www.quest-journal.net/2009-2010.htm

2012   *Spiritualiteit, heelmaking en transcendentie: Een intercultureel-filosofisch onderzoek bij Plato, in Afrika, en in het Noordatlantisch gebied, vertrekkend vanuit Otto Duintjers Onuitputtelijk is de Waarheid,* Haarlem: Papers in Intercultural Studies and Transcontinental Comparative Studies; also at: http://www.quest-journal.net/PIP/spiritualiteit.pdf .

2014   *Het dorp Mabombola: Vestiging, verwantschap en huwelijk in de sociale organisatie van de Zambiaanse Nkoya,* Haarlem: Papers in Intercultural Philosophy and Transcontinental Comparative Studies; also at: www.quest-journal.net/PIP/Mabombola%20TEXT%20lulu3%20%20ALLERBEST.pdf .

2015   *Vicarious reflections: African explorations in empirically-grounded intercultural philosophy,* Haarlem: Papers in Intercultural Philosophy and Transcontinental Comparative Studies; also at: http://www.quest-journal.net/shikanda/topicalities/vicarious/vicariou.htm.

2017   *Religion as a social construct: African, Asian, comparative and theoretical excursions in the social science of religion,* Haarlem: Papers in Intercultural Philosophy and Transcontinental Comparative Studies; also at: http://www.quest-journal.net/shikanda/topicalities/rel%20bk%20for%20web/webpage%20relbk.htm

2017   *Researching power and identity in African state formation,* Pretoria: University of South Africa Press (with Doornbos, Martin R.); also at: http://www.quest-journal.net/shikanda/topicalities/doornbos_&_van_binsbergen_proofs.pdf

2018   *Confronting the Sacred: Durkheim vindicated through philosophical analysis, ethnography, archaeology, long-range linguistics, and comparative mythology,* Hoofddorp: Shikanda; also at: http://www.quest-journal.net/shikanda/topicalities/naar%20website%208-2018/Table_of_contents.htm

EDITED COLLECTIONS

1976   *Religious Innovation in Modern Africa Society,* special issue, *African Perspectives,* 1976/2, Leiden: African Studies Centre, (with Buijtenhuijs, Robert)

1978   *Migration and the Transformation of Modern African Society,* special issue, *African Perspectives* 1978/1, Leiden: African Studies Centre; also at: http://www.quest-journal.net/shikanda/ethnicity/migratio.htm (with Meilink, Henk A.)

1984   *Aspecten van staat en maatschappij in Africa: Recent Dutch and Belgian research on the African state,* Leiden; African Studies Centre; also at: http://www.quest-journal.net/shikanda/ethnicity/asp_staat_mij/tableof.htm (with Hesseling, Gerti S.C.M.)

1985   *Old modes of production and capitalist encroachment,* London/Boston: Kegan Paul International, also at Google Books; selectively based on the original Dutch version: *Oude produktiewijzen en binnendringend kapitalisme: Antropologische verkenningen in Afrika,* Amsterdam: Free University, 1982 (with Geschiere, Peter L.)

1985   *Theoretical explorations in African religion,* London / Boston: Kegan Paul International, 389 pp; also as Google Book (with Schoffeleers, J. Matthijs)

1986   *State and local community in Africa,* Brussels: Cahiers du CEDAF / ASDOC Geschriften, 1986, 2-3-4; also at: http://www.quest-journal.net/shikanda/ethnicity/local_community_and_state_1986/local.htm (with Reijntjens, F., and Hesseling, Gerti S.C.M.)

1987   *Afrika in spiegelbeeld*, Haarlem: In de Knipscheer; also at: http://www.quest-journal.net/shikanda/literary/afrikain.htm (with Doornbos, Martin R.)

1993   *De maatschappelijke betekenis van Nederlands Afrika-onderzoek in deze tijd: Een symposium*, Leiden : Werkgemeenschap Afrika; also at: http://www.quest-journal.net/shikanda/publications/afrika-onderzoek%20final.pdf

1997   *Black Athena: Ten Years After*, Hoofddorp: Dutch Archaeological and Historical Society, special issue, *Talanta: Proceedings of the Dutch Archaeological and Historical Society*, vols 28-29, 1996-1997; also at: http://www.quest-journal.net/shikanda/afrocentrism/index.htm

1999   *Modernity on a shoestring: Dimensions of globalization, consumption and development in Africa and beyond: Based on an EIDOS conference held at The Hague 13-16 March 1997*, Leiden / London: EIDOS [European Interuniversity Development Opportunities Study group] ; also at: http://www.quest-journal.net/shikanda/general/gen3/expand_nov_99/shoestring_files/defshoe.htm (with Fardon, Richard, and van Dijk, Rijk)

2000   *Trajectoires de libération en Afrique contemporaine*, Paris: Karthala; also at: http://www.quest-journal.net/shikanda/ethnicity/trajecto.htm (with Konings, P., & Hesseling, Gerti S.C.M.)

2002   *Truth in Politics, Rhetorical Approaches to Democratic Deliberation in Africa and beyond*, special issue of *Quest: An African Journal of Philosophy*, 16, 1-2; also at: http://www.quest-journal.net/2002.htm (Salazar, Philippe-Joseph, Osha, Sanya, & van Binsbergen, Wim)

2003   *The dynamics of power and the rule of law: Essays on Africa and beyond: In honour of Emile Adriaan B. van Rouveroy van Nieuwaal*, Berlin / Münster: LIT, also at: http://www.quest-journal.net/shikanda/ethnicity/just.htm (with the assistance of Pelgrim, R.)

2004   *Situating globality: African agency in the appropriation of global culture*, Leiden: Brill, also at: http://quest-journal.net/shikanda/topicalities/situatin.htm (with van Dijk, R.)

2005   *Commodification: Things, agency and identities: The Social Life of Things revisited*, Berlin / Boston / Münster: LIT; also at: http://www.quest-journal.net/shikanda/ethnicity/commodif.htm.(with Geschiere, Peter L.)

2008   *African philosophy and the negotiation of practical dilemmas of individual and collective life*, vol. XXII-2008 of *Quest: An African Journal of Philosophy/ Revue Africaine de Philosophie*; full text at : http://www.quest-journal.net/2008.htm

2008   *African feminisms*, special issue of: *Quest: An African Journal of Philosophy/ Revue Africaine de Philosophie*, XX, 1-2, 2006; full text at: http://www.quest-journal.net/2006.htm (Osha, Sanya, with the assistance of van Binsbergen, Wim M.J.)

2008   *Lines and rhizomes: The transcontinental element in African philosophies*, special issue of *Quest: An African Journal of Philosophy/ Revue Africaine de Philosophie*, XXI -2007; full text at: http://www.quest-journal.net/2007.htm

2010   *New Perspectives on Myth: Proceedings of the Second Annual Conference of the International Association for Comparative Mythology, Ravenstein (the Netherlands), 19-21 August, 2008*, Haarlem: Papers in Intercultural Philosophy – Transcontinental Comparative Studies; also at: http://www.quest-journal.net/PIP/New_Perspectives_On_Myth_2010/toc_proceedings_IACM_2008_2010.htm (with Venbrux, Eric)

2011   *Black Athena comes of age*, Berlin / Boston / Munster: LIT; also at: http://www.quest-journal.net/shikanda/topicalities/20102011.htm under 'August 2011'

www.ingramcontent.com/pod-product-compliance
Lightning Source LLC
Chambersburg PA
CBHW060446240326
41599CB00062B/4260